QUANTUM INFORMATION AND QUANTUM OPTICS WITH SUPERCONDUCTING CIRCUITS

Superconducting quantum circuits are among the most promising solutions for the development of scalable quantum computers. Built with sizes that range from microns to tens of meters using superconducting fabrication techniques and microwave technology, superconducting circuits demonstrate distinctive quantum properties such as superposition and entanglement at cryogenic temperatures. This book provides a comprehensive and self-contained introduction to the world of superconducting quantum circuits and how they are used in current quantum technology. Beginning with a description of their basic superconducting properties, the author then explores their use in quantum systems, showing how they can emulate individual photons and atoms and ultimately behave as qubits within highly connected quantum systems. Particular attention is paid to cutting-edge applications of these superconducting circuits in quantum computing and quantum simulation. Written for graduate students and junior researchers, this accessible text includes numerous homework problems and worked examples.

JUAN JOSÉ GARCÍA RIPOLL is Senior Researcher at the Spanish Research Council (CSIC), and a leading researcher in the design of quantum hardware for quantum technologies. He has been at the forefront of superconducting quantum circuits research for several years and has published more than a hundred papers on quantum technologies.

QUANTUM INFORMATION AND QUANTUM OPTICS WITH SUPERCONDUCTING CIRCUITS

JUAN JOSÉ GARCÍA RIPOLL
Institute of Fundamental Physics (IFF), CSIC

CAMBRIDGE
UNIVERSITY PRESS

University Printing House, Cambridge CB2 8BS, United Kingdom

One Liberty Plaza, 20th Floor, New York, NY 10006, USA

477 Williamstown Road, Port Melbourne, VIC 3207, Australia

314–321, 3rd Floor, Plot 3, Splendor Forum, Jasola District Centre, New Delhi – 110025, India

103 Penang Road, #05–06/07, Visioncrest Commercial, Singapore 238467

Cambridge University Press is part of the University of Cambridge.

It furthers the University's mission by disseminating knowledge in the pursuit of education, learning, and research at the highest international levels of excellence.

www.cambridge.org
Information on this title: www.cambridge.org/9781107172913
DOI: 10.1017/9781316779460

First published 2022

A catalogue record for this publication is available from the British Library.

Library of Congress Cataloging-in-Publication Data
Names: García Ripoll, Juan José, author.
Title: Quantum information and quantum optics with superconducting circuits / Juan José García Ripoll.
Description: New York : Cambridge University Press, 2022. | Includes bibliographical references and index. |
Summary: "The dawn of the 20th century brought us the birth of quantum mechanics and a deeper understanding of the miscroscopic world. The new theory describing photons, atomic spectra and many other physical processes, postulates that the microscopic world is, in its truest essence, probabilistic. Particles such as electrons or photons move or "propagate" as probability waves to be detected at a given position, or in a given state. However, those waves or wavefunctions are very different from a mere representation of our ignorance about the world"– Provided by publisher.
Identifiers: LCCN 2021061918 (print) | LCCN 2021061919 (ebook) | ISBN 9781107172913 (hardback) | ISBN 9781316779460 (epub)
Subjects: LCSH: Quantum theory. | Quantum optics. | BISAC: SCIENCE / Physics / Quantum Theory
Classification: LCC QC174.12 .G359 2022 (print) | LCC QC174.12 (ebook) | DDC 535/.15–dc23/eng20220521
LC record available at https://lccn.loc.gov/2021061918
LC ebook record available at https://lccn.loc.gov/2021061919

ISBN 978-1-107-17291-3 Hardback

Contents

Figures

Tables

Notation

$\Delta\hat{O}$	Uncertainty of operator, $(\Delta\hat{O})^2 = \frac{1}{2}\langle\hat{O}\hat{O}^\dagger + \hat{O}^\dagger\hat{O}\rangle - \lvert\langle\hat{O}\rangle\rvert^2$. Variance of observable, $\hat{O}^\dagger = \hat{O}$, as $(\Delta\hat{O})^2 = \langle(\hat{O} - \langle\hat{O}\rangle)^2\rangle$.
C	Capacitance
E_J	Josephson energy
H.c.	Hermitian conjugate, as in $a + \text{H.c.} = a + a^\dagger$
L	Inductance
$\mathcal{L}$	Lagrangian or Lindblad operator
$\mathbf{x}$, $\mathbf{s}\ldots$	Vectors of numbers such as $\mathbf{x} = (x_1, x_2, \ldots, x_N)$
h	Planck constant, $6.626070040(81) \times 10^{-34}\ \text{J} \cdot \text{s}$
$\hbar = \frac{h}{2\pi}$	Reduced Planck constant
ϕ	Electric flux on a node or a branch of a circuit (Section 4.3)
$\Phi_0 = \frac{h}{2e}$	Magnetic flux quantum, $2.067833831(13) \times 10^{-15}$ Wb
$\varphi_0 = \frac{\Phi_0}{2\pi} = \frac{\hbar}{2e}$	Flux-to-phase conversion
$\sigma(H)$	Spectrum or collection of eigenvalues of an operator H
σ^x, σ^y, σ^z	Pauli matrices
$\boldsymbol{\sigma}$	Vector of Pauli matrices $\boldsymbol{\sigma} = (\sigma^x, \sigma^y, \sigma^z)$
—||—	Linear capacitor
-ᘐᘐᘐ-	Linear inductor
—✕—	Nonlinear inductance associated to a Josephson junction
—⊠—	Josephson junction: nonlinear inductor and capacitor in parallel
—|ı—	Constant voltage source

1
Introduction

The dawn of the twentieth century brought us the birth of quantum mechanics and a deeper understanding of the microscopic world. The new theory describing photons, atomic spectra, and many other physical processes postulates that the microscopic world is, in its truest essence, probabilistic. Particles such as electrons or photons move or "propagate" as probability waves to be detected at a given position, or in a given state. However, those waves or wavefunctions are very different from a mere representation of our ignorance about the world.

Quantum mechanics is the mathematical language that we use to describe the microscopic world. Quantum mechanical states – our *wavefunctions* – go beyond our classical understanding of the world, accepting the possibility of a system to coexist in a *quantum superposition* of two distinguishable configurations – i.e., an atom in two positions, a photon in two polarizations, or a neutron moving in two opposite directions. Only when we measure the state of the quantum mechanical object, it collapses to a well-defined configuration, which may be different on each realization of the experiment. Even more dramatically, multiple particles may be in a collective superposition or *entangled state,* allowing arbitrary measurements on these particle to produce random but perfectly correlated outcomes, irrespective of the space and time separation between those measurements. These and other predictions of quantum mechanics made some physicists such as Albert Einstein deem the theory as "incomplete" or even inconsistent.

The end of the twentieth century and the beginning of the twenty-first have witnessed an incredible collective effort to challenge the wildest predictions of quantum mechanics in the broadest variety of experimental systems possible, from photons, to atoms, all the way to macroscopic solid-state devices and circuits. Great experimentalists of our time have revealed quantum properties in objects that we can see with the naked eye, created complex quantum states of tens to thousands of particles, and entangled photons and spread them to different points of the planet, to test their quantum mechanical correlations. In the process, not only has quantum mechanics been validated, but we have become incredibly good at controlling quantum systems.

Trapped ions were one of the first systems to enter this controlled quantum regime (Leibfried et al., 2003). Using electromagnetic traps and laser cooling, experimentalists may isolate one or more charged particles, cooling some of their properties close to the absolute. This makes it possible to observe the discrete or *quantized* nature of the atomic excitations, not only at the level of the electronic states, but also with respect to quanta of energy stored in the motional states of the trapped atom.

If instead of charged particles we use neutral atoms, we can trap more particles and enter the regime of mesoscopic quantum systems. Experiments in 1995 showed how to to trap and evaporatively cool a cloud of 10^6 alkali atoms in an almost perfect vacuum down to few nanokelvins. At these temperatures, the ultracold atoms form a singular state of matter called Bose–Einstein condensate (BEC) (Townsend et al., 1997), in which all bosonic atoms are described by the same wave function, $\phi(x)$, whose dynamics and quantum properties can be engineered at will.

Starting from a perfectly controlled state of matter, such as the condensate, we can build extremely sophisticated quantum states and physical devices. We can split the atoms of a BEC into the pockets of an optical lattice, to recreate a crystal of bosons and simulate quantum magnetism and even Luttinger liquid physics. Bose–Einstein condensates can be used to cool down fermionic atoms and create Fermi seas, study Bardeen–Cooper–Schrieffer (BCS) superconductivity or even implement the Hubbard model for electrons in a solid. And using time-dependent controls, we can create large amounts of entanglement and squeezing that can be used for sensing or for foundational experiments in quantum information.

The beauty of atomic, molecular and optical (AMO) physics' bottom-up approach can hardly be exported to solid-state devices. Condensed-matter objects are simply too large. We cannot control all atoms in a piece of metal and convince them to adopt a perfect state. Moreover, solid devices rarely exist in perfectly isolated environments. They are always in contact with other elements – contacts, substrates, measurement apparatuses – at higher temperatures than the AMO setups we just discussed.

In a way, once we overcome the barriers associated to cooling and trapping, it is not surprising that we can reveal quantum mechanical phenomena in large atomic setups – it just confirms that the rules of quantum mechanics extend to very large composite systems. The million dollar question, however, was whether those rules extend all the way to solid-state and condensed matter systems!

This was indeed the great challenge for condensed matter systems around the 1990s. Can we prepare a perfect solid-state quantum system? Can we build a simple, atomic-like device, with a perfectly controlled state? The approach to this problem is subtly different from AMO. Starting from a macroscopic object – a gated or self-assembled quantum dot, impurities in diamond, nanoresonators, or superconducting circuits – we select one or two degrees of freedom in which we seek evidence of quantum phenomena. The candidate for a quantum degree of freedom may be the charge of a small superconducting island, the electrical current along a metallic loop, or the wavefunction of electrons in a dot or a color center, for example. Experimentalists focus on one property and work toward isolating it from their environments, cooling them down, and engineering better and better quantum states.

This challenge, the quest for an *artificial atom* – what we now call a *qubit*; see Chapter 6 – was still ongoing when I first learned about the field of superconducting circuits. The review article by Makhlin et al. (2001) showed some promising candidates in the forms of charge qubits (Section 6.2) or three-Josephson junction qubits (Section 4.8), but quantum superpositions were fragile and easily killed by the environment.

The surprise came in year 2004, when the Yale team (Wallraff et al., 2004) radically improved the lifetime of charge qubits by placing them inside microwave resonators. That work, and subsequent works at various groups in the University of California, Santa Barbara (UCSB), Saclay, Karlsruhe Institute of Technology (KIT), etc., sparked the beginning of a productive intersection between superconducting circuit technology, quantum optics, and quantum information, which we now call *circuit-QED*.

We live in the *second quantum revolution* (Dowling and Milburn, 2003), in which quantum mechanics becomes a tool for industrial applications and practical devices. These applications are generally known as *quantum technologies,* because they use the unique properties of quantum mechanics – quantum superpositions, entanglement, noncontextuality, etc. – to create actual technologies for *quantum communication*, *quantum cryptography*, *quantum sensing*, *quantum simulation*, and *quantum computing*.

Superconducting circuits are at the center the quantum revolution. Superconductors are the basis for the most accurate single-photon detectors, which are used in advanced setups for quantum communication. We rely on superconducting qubits for some of the most powerful and accurate quantum computers to date (see Chapter 8), and also to simulate large and complex problems from quantum magnetism (Chapter 9). Superconducting circuits have also found their way, as controls and interfaces for other microwave-based quantum technologies, from quantum dots to nanomechanical resonators. Understanding how these circuits work, how they are designed, and how they are operated is a natural path for both theoreticians and experimentalists alike. This book is a self-contained undergraduate manual designed to help you gain this understanding.

1.1 The Book

The book is structured in three parts. The first part introduces the platform, superconducting circuits, and a common language, formed by quantum mechanics and circuit quantization. The second part introduces the two fundamental objects we build using superconductors: these are photonic devices and artificial atoms. They are studied separately, and then combined to develop the theory of *circuit quantum electrodynamics*, or circuit-QED. The third part concerns the physics that can be explored and the technologies that can be built using superconducting circuits. Along separate chapters and sections, we learn about how to build quantum computers, quantum simulators, quantum optimizers, and other quantum technologies using circuits or hybrid setups.

The structure and the level of the book also allow different applications. This textbook was born and has been used as an introductory course in quantum optics. Using qubits and microwave photons, it is possible to review the basic concepts in light–matter interaction, the Rabi, Jaynes–Cummings and Dicke model, superradiance, and subradiance, and even understand sophisticated models of low-dimensional photonics. But the same book can be repurposed to focus on the design and construction of practical quantum computers or to learn about applications in the field of quantum simulation and many-body physics.

The book is mostly self-contained. It assumes familiarity with the formalism of quantum mechanics, including key concepts such as pure states, density matrices, observables, and the Schrödinger equation, but these concepts are reviewed as they are used. At different stages, we will also introduce minimal concepts in second quantization and field theory, such as the notion of modes, creation and annihilation operators, commutation relations, bosonic statistics, etc. We also review important concepts from quantum information theory, such as qubits, entanglement, quantum gates, etc. The most important ideas are reinforced through selected exercises that students are encouraged to do.

Building on these prerequisites, the reading order of this book would be the following one. In the first part of the book, Chapter 2 provides an overview of quantum mechanics, focusing on concepts that are used throughout the book. In Chapter 3, we introduce a minimal understanding of the theory of superconductivity, with a focus on London's mesoscopic models and the electrical properties of superconducting circuits. Chapter 4 introduces the theory of quantum circuits. This is a set of rules that allow us to find the mathematical model that describes a given superconducting circuit. We derive these rules using as practical examples all the elementary circuits that we will later study, such as qubits, microwave resonators, or superconducting quantum interference devices (SQUIDs).

In the second part of the book, Chapter 5 focuses on the linear models of superconducting circuits, such as microwave LC resonators and waveguides (the equivalent of coaxial cables), and showing how the quantization of the circuit gives birth to microwave photons that can be created, manipulated, and measured. Chapter 6 studies superconducting circuits in the highly nonlinear regime, in which the circuit acts as an artificial atom or qubit. We review the most popular qubit designs, such as the transmon or the flux qubit, and develop tools to prepare and characterize those qubits. Chapter 7 combines artificial atoms with microwave photons, developing the theory of light–matter interaction. This chapter is a primer on concepts from quantum optics – models for low-dimensional atom–light interaction, the physics of spontaneous emission, photon absorption, and open quantum systems. We then focus on the simpler setup of qubit–resonator interactions, introducing the Rabi and Jaynes–Cummings models, and showing how qubits can control the state of light and vice versa, cavities can be used to control and measure qubits.

The third part of the book focuses on the applications of superconducting circuits to various quantum technologies. We begin in Chapter 8 studying universal quantum computers in the circuit model. We offer a checklist of ingredients that are needed to build a practical quantum computer, and investigate how those elements are built and characterized in the superconducting platform. We also discuss the roadmap toward fault-tolerant, error-corrected, and fully scalable quantum computers, and what we can do with their small-scale, faulty versions in the near term. Finally, Chapter 9 builds on the previous formal developments to introduce a design for an adiabatic quantum computer, also known as *quantum annealer*. This chapter starts with a formal description of this computational model, where a physical system is adiabatically coerced into a configuration that represents the solution to a mathematical problem. It then moves on to the actual superconducting architecture for a quantum annealer using superconducting flux qubits, along the line of the D-Wave machine, but referencing other later designs.

If the book is to be used with a strong focus on quantum optics or quantum circuits, the student or teacher should at least cover Chapters 1–7. This should give students an overview of the most relevant experiments in the field from 2004 to this day, providing them with the language and tools to access more complex literature and explore their own ideas. Chapters 8 and 9 in the third part of the book are independent from each other, but require a good understanding of the first part of the book. A minimal set of exercises is provided within each chapter. Solutions and errata will be posted on the author's webpage.[1]

[1] http://juanjose.garciaripoll.com.

1.2 Acknowledgments

This book would not be possible without the help of many colleagues and friends who supported me both before and during the writing of the text. I was introduced to the exciting field of superconducting circuits by my friend and colleague Enrique Solano, who himself laid out the foundations to many ideas that are explored in the realm of circuit-QED and photon measurements.

I have learned a lot interacting with fellow experimentalists, including Frank Deppe, Achim Marx, Adrian Lupascu, Chris Wilson, Gerhard Kirchmair, Mathieu Juan, and Aleksei Sharafiev. I am indebted to Pol Forn-Díaz, whose curiosity and perseverance has strongly driven our interest in ultrastrong qubit-microwave interactions and quantum annealing.

This book would not be born without the spark and motivation to write it, which was ignited by my friend and outstanding physicist Oriol Romero-Isart. My visits to Innsbruck – always a place for inspiration at various inflection points in my career – and the lectures I gave there helped me move on and were the seed for this book. I obviously owe many thanks to the students who have suffered various iterations of this work, and also my former PhD student Borja Peropadre, whose thesis inspired the first iterations of this project.

Special thanks must go to Carlos Navarrete, Guillermo Romero, Manuel Pino, Paula García-Molina, Pol Forn Díaz, and Alp Sipahigil for carefully reading various iterations of this manuscript.

None of this would be possible without the support of a family that has always been there, even when I was not. I owe to them, not this book, but the privilege of a life as scientist.

And finally, this book exists because of you, the reader. I hope it will help you approach this field, learn about its beauty and its challenges, and inspire you to be a better “quantum mechanic.”

2
Quantum Mechanics

As mentioned in the introduction, this book assumes some basic understanding of quantum mechanics at the undergraduate level. This understanding is essential for many reasons: not only to be able to write down realistic models, compute things, and make predictions, but also because quantum physics is a world full of subtleties and unexpected phenomena which may cause major confusion and headaches when seen for the first time.

Consequently, here is a friendly warning: If you never took a course on quantum physics and the formalism of quantum mechanics, take it, or grab some books that can help you overcome this initial barrier. There are many good books to choose from, such as the two volumes of Cohen-Tannoudji et al. (1977) or the wonderful and complete book by Ballentine (1998). Simply take the one that resonates best with you.

This said, I felt the need to write down a chapter where I could summarize many of the concepts and tools from quantum mechanics that are repeatedly invoked in this book. Some of these are pretty basic, such as the relation between the Schrödinger equation and unitary operators, but others are a bit more subtle and extend beyond a typical course on quantum mechanics, such as the notion of density matrices and master equations. Please take this therefore as a "unifying" chapter that provides a common language and definitions, and which you may freely skip if you have a graduate or postgraduate level in the study of quantum physical systems.

2.1 Canonical Quantization

Quantum physics is an old field of research, whose birth we attribute to Planck's theory for the black-body radiation. A black body is an ideal object that can absorb energy at any frequency of the spectrum, which incidentally implies that it is the body that can emit energy most efficiently at any given frequency. In 1900, the physicist Max Planck showed that experiments measuring the radiation from black bodies could be explained by assuming that these objects – which at the time were just perfect cavities with a tiny hole – could only exchange energy in fixed amounts or *quanta*, determined by the frequency ν of the light emitted or absorbed by the black cavity.

In his treatment, Planck models the excitations of the black body as a collection of harmonic oscillators with frequencies that cover the measured spectrum – i.e., $\nu(\mathbf{k})$, labeled by wave vectors $\mathbf{k}$. The energy of those oscillators is quantized, which means that the oscillators equilibrate to the same temperature by exchanging discrete units of energy or *quanta* with the environment. If we could measure the state of the black body, its energy would be a sum of the *quanta* $n_{\mathbf{k}}$ that are stored in each electromagnetic mode $\mathbf{k}$:

$$E = \sum_{\mathbf{k}} h\nu(\mathbf{k}) \times n_{\mathbf{k}}, \quad n_{\mathbf{k}} \in \{0, 1, 2, \ldots\}. \tag{2.1}$$

In this model, each oscillator has associated a quantum of energy $h\nu(\mathbf{k})$ determined by the frequency[1] $\nu(\mathbf{k})$ and Planck's constant $h \simeq 6.62607004(81) \times 10^{-34}$ J/Hz. The collection of all integers $|n_{\mathbf{k}_1}, n_{\mathbf{k}_2}, \ldots\rangle$ is a unique configuration of the black body, which we call the *quantum state*, and the collection of all states is used to develop a statistical model of the black body's spectrum.

Despite his success, Planck was very wary of extending the idea of quanta to the actual electromagnetic field. It was Einstein who made the connection between Planck's quanta and the existence of a particle of light, the *photon*. With this particle, Einstein could explain in 1905 the photoelectric effect: Some materials may convert light into an electrical current, but when the intensity of light is lowered enough, this current becomes a series of discrete random bursts, which Einstein associated with the absorption of photons. This successful explanation was shortly followed by Bohr and Rutherford's model of the atom, based on quantized electronic orbits that explained the discrete spectra of light-emitting atoms. Barely a decade later, Schrödinger (1926) and Heisenberg (1925) replaced all ad hoc quantization ideas with two equivalent formulations of quantum mechanics based on wave and matrix equations. In both theories, the discretization of energies is a mathematical consequence of the discrete spectra of the operators that govern the evolution of light and matter. Put to work, the newly born theory provided quantitative explanations for the spectra of atoms, molecules, solids, and the electromagnetic field itself, consolidating nonrelativistic quantum mechanics as

[1] Ordinary frequencies are typically denoted by the letter ν and are measured in the S.I. unit of Hertz (Hz). Quite often we will also use angular frequencies $\omega = 2\pi\nu$, sometimes denoted in rad/s or s^{-1}. In the first case, the quantum of energy is given by $h\nu$, while in the second case it is given by Planck's reduced constant $\hbar\omega$ with $\hbar = h/2\pi$.

the tool for understanding chemistry, solid-state physics, nano-electronics, and photonic devices, to name a few examples.

Schödinger's and Heisenberg's formulations are powerful theories that explain the microscopic behavior of Nature in a bottom-up fashion: starting from elementary components – the electron, the proton, and the neutron – and fundamental interactions – for instance, the Coulomb attraction between protons and electrons mediated by the electromagnetic field – one builds a many-body equation whose solution accounts for all the physics we observe in the laboratory. Unfortunately, the bottom-up approach does not always scale well as we move on to larger systems. In this book, we are concerned with solid-state superconducting devices that include more than 10^{24} atoms, all collectively exhibiting quantum mechanical phenomena. It is unfathomable to even think of writing an equation for all those particles, and we are forced to seek effective descriptions that are consistent with the principles and rules of the underlying quantum mechanical theory.

Shortly after the publication of Schrödinger's and Heisenberg's work, Paul Dirac developed an alternative derivation of quantum mechanics, known as *canonical quantization*, that establishes a link between the quantum model for a given object (particle, field, etc.) and the dynamics that we would expect from it in a classical world. In Chapter 4, we will apply this procedure to the quantization of an electrical circuit, developing a quantum theory of superconducting circuits. This theory will be consistent with the microscopic description introduced in Chapter 3, and it will provide the appropriate limit of the circuit when temperatures are high enough that superconductivity is lost, or quantum phenomena are masked.

2.1.1 Hamiltonian Equations

For simplicity, we will describe how canonical quantization works for a simple object: a point-like particle[2] with position $\mathbf{x}$ and momentum $\mathbf{p} = m\dot{\mathbf{x}}$, moving in an external potential $V(\mathbf{x})$. The particle's trajectory is governed by a set ordinary differential equations – Newton's equations – which we write in terms of the particle's acceleration $\ddot{\mathbf{x}}$ and the force $\nabla V(\mathbf{x})$ experienced by the particle

$$\ddot{\mathbf{x}} = -\frac{1}{m}\nabla V(\mathbf{x}). \tag{2.2}$$

Newton's equation can be derived from a *stationary action* principle, as the trajectory that minimizes the action $S = \int_{t_1}^{t_2} \mathcal{L}(\dot{\mathbf{x}}, \mathbf{x})\mathrm{d}t$. The functional S maps orbits $x(t)$ to real numbers according to the *Lagrangian*

[2] As we will see in Chapter 4, this is not a futile exercise, because the harmonic potential $V(\mathbf{x}) = \frac{1}{2}m\omega^2\mathbf{x}^2$ is formally analogous to the simplest electrical circuit, an LC resonator, and describes how this circuit is actually quantized.

$$\mathcal{L}(\dot{\mathbf{x}}, \mathbf{x}) = \frac{1}{2} m \dot{\mathbf{x}}^2 - V(\mathbf{x}). \tag{2.3}$$

According to the stationary principle, a small perturbation of the particle's true trajectory $x_\varepsilon(t) = x(t) + \varepsilon(t)$ should leave the action unperturbed up to second-order corrections $S[x_\varepsilon] = S[x] + \mathcal{O}(\varepsilon^2)$. This stationary principle produces Lagrange's equations

$$\frac{\mathrm{d}}{\mathrm{d}t} \frac{\partial \mathcal{L}}{\partial \dot{x}_n} = \frac{\partial \mathcal{L}}{\partial x_n}, \tag{2.4}$$

which are equivalent to the original Newtonian equations (2.2).

We now introduce a *Hamiltonian formulation*, where the functional that generates the dynamical equations is a function of two canonically conjugate variables, $\mathbf{x}$ and $\mathbf{p}$, with the prescription

$$H(\mathbf{x}, \mathbf{p}) = \mathbf{p}\dot{\mathbf{x}} - \mathcal{L}(\dot{\mathbf{x}}, \mathbf{x}), \quad \text{with } \mathbf{p} = \frac{\partial \mathcal{L}}{\partial \dot{\mathbf{x}}}. \tag{2.5}$$

This *Legendre transform* establishes a link between the particle's velocity $\dot{\mathbf{x}}$ and its canonical momentum $\mathbf{p}$, which now replaces the former in all equations. The transform also produces an object, the *Hamiltonian* $H(\mathbf{x}, \mathbf{p})$, governing the orbits of the particle. More precisely, any observable $O(\mathbf{x}, \mathbf{p}, t)$ that we can construct as a function of the canonical variables and time evolves according to the Hamiltonian equation

$$\frac{\mathrm{d}}{\mathrm{d}t} O = \{O, H\} + \frac{\partial O}{\partial t}, \tag{2.6}$$

with the classical *Poisson brackets*

$$\{A, B\} = \sum_j \left(\frac{\partial A}{\partial x_i} \frac{\partial B}{\partial p_i} - \frac{\partial B}{\partial x_i} \frac{\partial A}{\partial p_i} \right). \tag{2.7}$$

In particular, since $\{x_i, p_j\} = \delta_{ij}$ this prescription trivially recovers Newton's equations, but now expressed as a set of first-order differential equations:

$$\frac{\mathrm{d}}{\mathrm{d}t} \mathbf{x} = \frac{\partial H}{\partial \mathbf{p}} = \frac{1}{m} \mathbf{p}, \qquad \frac{\mathrm{d}}{\mathrm{d}t} \mathbf{p} = -\frac{\partial H}{\partial \mathbf{x}} = -\nabla V(\mathbf{x}). \tag{2.8}$$

2.1.2 Quantum Observables

Dirac's canonical quantization describes the transition from a Hamiltonian theory of a classical particle to a quantum mechanical theory that preserves the particle's dynamical equations (2.6). First of all, following the axioms of quantum mechanics, it introduces a Hilbert space of *vector states* describing our system. Next, it replaces any measurable quantity, including $\mathbf{x}$, $\mathbf{p}$, and any other function thereof $O(\mathbf{x}, \mathbf{p})$, with linear Hermitian operators $\hat{\mathbf{x}}, \hat{\mathbf{p}}, \hat{O}$ acting on this Hilbert space. Each observable will have a spectrum of eigenvalues and eigenstates determining all possible measurement outcomes.

In our toy model, the quantum state of a particle that is at position $\mathbf{r} \in \mathbb{R}^N$ is associated a vector $|\mathbf{r}\rangle$ in the Hilbert space, with the property $\hat{x}_n |\mathbf{r}\rangle = r_n |\mathbf{r}\rangle$. Generic states are constructed as quantum superpositions of different measurement outcomes, such as the wavefunction $\psi(\mathbf{r})$:

$$|\psi\rangle = \int \psi(\mathbf{r}) |\mathbf{r}\rangle \, \mathrm{d}^N\mathbf{r}. \tag{2.9}$$

The weights of the wavefunction $\psi(\mathbf{r}) \in \mathbb{C}$ are complex numbers whose modulus gives the probability distribution $P(\mathbf{r}) = |\psi(\mathbf{r})|^2$ that the particle is found at the position $\mathbf{r}$, if the observable $\hat{\mathbf{x}}$ is ever measured. States are normalized, so that the total probability adds up to one, $\langle \mathbb{1} \rangle_\psi = \langle \psi | \psi \rangle = \int |\psi(\mathbf{r})|^2 \mathrm{d}\mathbf{r} = 1$, and we can define the *expectation values* of measurements:

$$\begin{aligned} \langle \hat{\mathbf{x}} \rangle_\psi = \langle \psi | \hat{\mathbf{x}} | \psi \rangle &= \iint \psi(\mathbf{r}_0)^* \psi(\mathbf{r}_1) \langle \mathbf{r}_0 | \, \hat{\mathbf{x}} \, | \mathbf{r}_1 \rangle \, \mathrm{d}^N\mathbf{r}_0 \mathrm{d}^N\mathbf{r}_1 \\ &= \int \mathbf{r}_0 |\psi(\mathbf{r}_0)|^2 \mathrm{d}\mathbf{r}_0. \end{aligned} \tag{2.10}$$

Note how, by using the orthogonality of position eigenstates $\langle \mathbf{r}_0 | \hat{\mathbf{x}} | \mathbf{r}_1 \rangle = \mathbf{r}_1 \langle \mathbf{r}_0 | \mathbf{r}_1 \rangle = \mathbf{r}_1 \delta(\mathbf{r}_0 - \mathbf{r}_1)$, we recovered the formula for the average over the probability distribution $P(\mathbf{r})$.

Canonical quantization includes one final prescription that makes the algebra of operators and states consistent with the classical limit of these equations. We replace everywhere the Poisson brackets for classical variables with the commutator between the respective observables $\{A, B\} \to -i[\hat{A}, \hat{B}]/\hbar$, where $[\hat{A}, \hat{B}] = \hat{A}\hat{B} - \hat{B}\hat{A}$. For the isolated particle, this prescription transforms $\{x_n, p_m\} = \delta_{nm}$ into

$$[\hat{x}_n, \, \hat{p}_m] = i\hbar\delta_{nmj}. \tag{2.11}$$

If our space of positions is continuous and contained in the region $\Omega \subset \mathbb{R}^N$, we can build our Hilbert space using integrable functions $\psi(\mathbf{r}) \in L^2(\Omega)$, associating position and momentum with operators $\hat{x}_n \psi(\mathbf{r}) = r_n \psi(\mathbf{r})$ and $\hat{p}_n \psi(\mathbf{r}) = -i\hbar \partial_{r_n} \psi(\mathbf{r})$ that satisfy the commutation relations (2.11).

Irrespective of the implementation of our Hilbert space, (2.11) implies that canonically conjugate operators are incompatible: They cannot share a common basis of eigenstates. Thus, a state with a well-defined position $|x_0\rangle$ cannot have a well-defined value of the momentum. This leads to the *Heisenberg uncertainty principle*:

$$\Delta x_n \Delta p_m \geq \frac{1}{2} \hbar \delta_{nm}, \tag{2.12}$$

which relates the variances of pairs of observables, $\Delta x_n = \sqrt{\langle \hat{x}_n^2 \rangle - \langle \hat{x}_n \rangle^2}$ and Δp_m. As we will see later, this uncertainty has also a physical manifestation in the world of quantum circuits, where position and momenta are replaced by voltage and intensity, quantities that cannot be measured simultaneously with absolute precision.

2.1.3 Unitary Evolution

We have introduced quantum states and observables as two mathematical objects that together predict the statistics of measurement outcomes. This information is bound to change in time, as observables and states evolve. In canonical quantization, the identification of Poisson brackets with commutators translates the classical equation (2.6) into the *Heisenberg equation*:

$$\frac{\mathrm{d}\hat{O}}{\mathrm{d}t} = -\frac{i}{\hbar}[\hat{O}, \hat{H}] + \frac{\partial \hat{O}}{\partial t}. \tag{2.13}$$

In this model, the dynamics is generated by a Hamiltonian operator that results from replacing the canonical variables with the corresponding observables $\hat{H} = \frac{1}{2m}\hat{\mathbf{p}}^2 + V(\hat{\mathbf{x}})$. In the Heisenberg picture, observables change in time starting from a well-known initial condition $O(t_0) = O_0$. The states $|\psi_0\rangle$ remain stationary, and they are regarded as objects that map the changing observables to their expectation values $\bar{O}(t) = \langle \hat{O}(t) \rangle = \langle \psi_0 | \hat{O}(t) | \psi_0 \rangle$.

The Heisenberg equation is rather inconvenient: We have to work with big and complex operators, and extracting the measurement statistics becomes a very convoluted process. In many situations, we would rather work with an equation that determines how states evolve from, say, an initially localized configuration $\psi_0(x) = \delta(x - x_0)$, spreading to other meaurement outcomes. This information is provided by the *Schrödinger equation* or *Schrödinger picture*, whereby observables have an immutable representation, but wavefunctions change in time $\bar{O}(t) = \langle \hat{O} \rangle_{\psi(t)} = \langle \psi(t) | \hat{O}_0 | \psi(t) \rangle$, with

$$i\hbar\partial_t |\psi(t)\rangle = \hat{H} |\psi(t)\rangle, \quad \text{with } |\psi(t_0)\rangle = |\psi_0\rangle. \tag{2.14}$$

For our isolated particle in an external potential $V(\mathbf{x})$, using the position representation, where $\hat{\mathbf{p}} = -i\hbar\nabla$, this results into a simple wave equation for the complex amplitude of probability $\psi(\mathbf{x})$:

$$i\hbar\partial_t \psi(\mathbf{x}) = \left[\frac{1}{2m}(-i\hbar\nabla)^2 + V(\mathbf{x})\right]\psi(\mathbf{x}). \tag{2.15}$$

Even if they look very different, the Schrödinger and Heisenberg representations give the same predictions because they are both solved by a common unitary transformation called the *evolution operator*. This operator is the solution of an enlarged Schrödinger equation:

$$i\hbar\frac{\mathrm{d}}{\mathrm{d}t}\hat{U}(t,t_0) = \hat{H}\hat{U}(t,t_0), \quad \text{with } \hat{U}(t_0,t_0) = \mathbb{1}. \tag{2.16}$$

In the case of constant Hamiltonians, the unitary operator U is a Lie rotation in the Hilbert space, generated by the Hamiltonian:

$$\hat{U}(t,t_0) = \exp[-i(t-t_0)H/\hbar]. \tag{2.17}$$

This operator is unitary $UU^\dagger = U^\dagger U = \mathbb{1}$. It can be inverted $\hat{U}(t_2,t_1)^{-1} = \hat{U}(t_1,t_2)$ and solves both the Schrödinger $|\psi(t)\rangle = \hat{U}(t,t_0)\,|\psi_0\rangle$ and the Heisenberg equations $\hat{O}(t) = \hat{U}(t_0,t)\hat{O}_0\hat{U}(t,t_0)$, as mentioned before.

2.2 Two-Level Systems

Not all physical systems have continuous degrees of freedom. We are going to work with smaller systems that only have two or three configurations that are active in a given experiment. These discrete systems have smaller Hilbert spaces, with wavefunctions defined in complex vector spaces. For a quantum system with two possible states $|0\rangle$ and $|1\rangle$, the wavefunctions in the two-dimensional Hilbert space are described by two complex amplitudes:

$$|\psi\rangle = \psi_0\,|0\rangle + \psi_1\,|1\rangle \;\leftrightarrow\; \boldsymbol{\Psi} = \begin{pmatrix}\psi_1\\ \psi_0\end{pmatrix} \in \mathcal{H} = \mathbb{C}^2, \tag{2.18}$$

with the usual normalization $|\psi_0|^2 + |\psi_1|^2 = 1$.

Two-dimensional Hilbert spaces are very common. They are a natural representation of the spin $s = 1/2$ states of an electron, a proton, or a neutron: $|0\rangle$ and $|1\rangle$ correspond to spin down and up along a given direction; they are sometimes used for describing the polarization states of a photon, horizontal versus vertical; and they appear most frequently in quantum optics when modeling atomic transitions – i.e., ground state $|0\rangle$ versus excited state $|1\rangle$ – and light–matter interaction. Nowadays, two-dimensional quantum systems are also called *qubits*, because, in analogy to the classical bit, they represent the minimal quantum object where information can be stored and processed. We show in Chapter 6 that it is possible to build superconducting circuits that are accurately described as qubits, and discuss in later chapters how these circuits are applied to quantum computing and simulation.

The algebra of two-level systems is analyzed using a complete set of observables, called the Pauli matrices:

$$\sigma^x = \begin{pmatrix}0 & 1\\ 1 & 0\end{pmatrix}, \quad \sigma^y = \begin{pmatrix}0 & -i\\ i & 0\end{pmatrix}, \quad \sigma^z = \begin{pmatrix}1 & 0\\ 0 & -1\end{pmatrix}. \tag{2.19}$$

Note that in this representation, $|0\rangle$ and $|1\rangle$ are the two eigenstates of σ_z, and we write $\sigma_z = |1\rangle\langle 1| - |0\rangle\langle 0|$. If we enlarge the set of Pauli matrices to include the identity, $\sigma^\alpha|_{\alpha=0}^3 = \{\mathbb{1}, \sigma^x, \sigma^y, \sigma^z\}$, we have a basis where we can expand any observable in this Hilbert space:

$$\hat{O} = \frac{1}{4}\sum_{\alpha=0}^{3} \mathrm{tr}(\hat{O}\hat{\sigma}^\alpha)\hat{\sigma}^\alpha. \tag{2.20}$$

In particular, a general qubit Hamiltonian has the form

$$\hat{H} = E + B\,\mathbf{n}\cdot\hat{\boldsymbol{\sigma}}, \tag{2.21}$$

where $\mathbf{n}$ is a direction in the three-dimensional space, $\hat{\boldsymbol{\sigma}} = (\hat{\sigma}^x, \hat{\sigma}^y, \hat{\sigma}^z)$ and E and B are constants. For spins, the Hamiltonian is interpreted as the coupling between the qubit's dipole moment $\propto \hat{\boldsymbol{\sigma}}$ and the magnetic field $B\mathbf{n}$ along the given direction.

Interestingly, this form has two important properties. First, because $\mathbf{n}\cdot\hat{\boldsymbol{\sigma}}$ has the properties of a Pauli matrix, the eigenvalues of this Hamiltonian are simply $\lambda_\pm = E \pm B$. Second, and for the same reasons, we can use the Pauli expansion to compute the evolution operator of the qubit:

$$\hat{U}(t,0) = \exp\left(-\frac{it\hat{H}}{\hbar}\right) = e^{-iEt/\hbar}\left[\cos(Bt)\,\mathbb{1} - i\sin(Bt)\,\mathbf{n}\cdot\hat{\boldsymbol{\sigma}}\right]. \tag{2.22}$$

We will use this formula when studying superconducting qubits and the implementation of single-qubit gates in Section 6.1.4.

2.3 Density Matrices

The formalism of unitary evolution for a quantum system assumes that the system under study is perfectly isolated from other quantum or classical objects and subject to error-free control. Even if systematic errors are greatly reduced, no physical system can be perfectly isolated: At the very least, there will always be the omnipresent electromagnetic field, carrying the cosmic background radiation and putting our system in contact with the noisy classical world.

It is therefore safe to say that experiments never prepare pure states $|\psi\rangle$. Instead, real quantum systems must be described using an ensemble operator, also known as *density matrix* $\hat{\rho}$: a nonnegative Hermitian operator $\hat{\rho} = \hat{\rho}^\dagger \geq 0$, with proper normalization $\mathrm{tr}(\hat{\rho}) = 1 (\sim \sum_n \langle n|\hat{\rho}|n\rangle)$, which gives expectation values of operators as $\langle \hat{O}\rangle = \mathrm{tr}(\hat{O}\hat{\rho}) \sim \sum_n \rho_{nn}\,\langle n|O|n\rangle$.

Pure states $|\psi_\chi\rangle$ written as density matrices become projectors $\hat{\rho}_\chi = |\psi_\chi\rangle\langle\psi_\chi|$. General states, however, are *mixed states,* because they can be reconstructed as classical ensembles of pure states created with different classical probabilities $p(\chi)$ (Ballentine, 1970):

$$\hat{\rho} = \int |\psi_\chi\rangle\langle\psi_\chi| \, p(\chi)\mathrm{d}\chi. \tag{2.23}$$

This equation (2.23) describes the output of an experiment where parameters have some uncertainty. Mixed states can also arise when a system enters in contact with another system, called the environment. In principle, we should consider the global wavefunction of the system plus its environment, allowing both to be correlated $|\Psi_{\text{global}}\rangle = \sum_{s,E} \Psi_{s,E} |s\rangle \otimes |E\rangle \in \mathcal{H}_{\text{system}} \otimes \mathcal{H}_{\text{environment}}$. However, since we will not have access to all degrees of freedom of the environment, we must trace out all the information that we ignore, obtaining a much smaller ensemble that only describes our system:

$$\hat{\rho}_{\text{sys}} = \text{tr}_{\text{environment}} |\Psi_{\text{global}}\rangle \langle\Psi_{\text{global}}| = \sum_E \sum_{s,s'} \Psi_{s,E}\Psi^*_{s',E} |s\rangle\langle s'|. \tag{2.24}$$

Density matrices are particularly simple in the case of two-level systems, where they can be expanded in the qubit basis $\rho_{ij} := \langle i|\hat{\rho}|j\rangle$, or as a combination of Pauli operators (2.20):

$$\hat{\rho} = \begin{pmatrix} \rho_{00} & \rho_{01} \\ \rho_{10} & \rho_{11} \end{pmatrix} = \sum_{i,j=0}^{1} \rho_{ij} |i\rangle\langle j| = \frac{1}{2}\mathbb{1} + \frac{1}{2} \sum_{\alpha=x,y,z} S_\alpha \hat{\sigma}^\alpha. \tag{2.25}$$

The set of all physical states with $\mathbf{S} = (S_x, S_y, S_z)$ falls inside the *Bloch sphere* $|\mathbf{S}| \leq 1$. The surface of this sphere is formed by pure state ($|\mathbf{S}| = 1$), and its center is the completely depolarized state $\mathbf{S} = 0$ or $\hat{\rho} = \frac{1}{2}\mathbb{1}$.

The completely depolarized state is an example of *classical state*, diagonal density matrices – $\rho_{10} = \rho^*_{01} = 0$ – which may be constructed as a convex combination $\hat{\rho} = \rho_{00} |0\rangle\langle 0| + \rho_{11} |1\rangle\langle 1|$ of preparing state $|0\rangle$ with probability $P_0 = \rho_{00}$ and preparing state $|1\rangle$ with probability $P_1 = 1 - P_0$. Compare this now with the *superposition state*, $|+\rangle = \frac{1}{\sqrt{2}}(|0\rangle + |1\rangle)$. This state exists only in the quantum model for a two-level system. This particular superposition maximizes the off-diagonal elements ρ_{01} and ρ_{10}, also known as *coherences*, recognized as a true signature of quantumness in states.

Sometimes a quantum system is sufficiently isolated and the timescales of study are short enough that we can approximate its evolution with a Hamiltonian that involves just that system. In that case, the unitary evolution of vectors in the Hilbert space dictates a recipe for updating the density matrix $\hat{\rho}(t) = \hat{U}(t, t_0)\hat{\rho}_0\hat{U}(t, t_0)^\dagger$. More generally, a quantum system will get entangled with its environment, suffering an *incoherent evolution*. Under certain physically reasonable assumptions (cf. Appendix B), the equation that describes this dynamics is the *Lindblad master equation*:

$$\partial_t \hat{\rho} = -\frac{i}{\hbar}[\hat{H}, \hat{\rho}] + \mathcal{L}_t(\hat{\rho}). \tag{2.26}$$

The linear *superoperator* $\mathcal{L}_t(\alpha\hat{\rho}_1 + \beta\hat{\rho}_2) = \alpha\mathcal{L}_t(\hat{\rho}_1) + \beta\mathcal{L}_t(\hat{\rho}_2)$, called the *Lindblad superoperator*, contains the information about the noise or the environment-induced decoherence. There are no general prescriptions to write down or even solve this kind of equation, but in a few cases the coupling is so weak and the environment so big that it instantaneously loses all memory about the system's dynamics. This is the so-called *Markovian limit*, in which $\mathcal{L}$ is independent of time and of the system's initial conditions. This limit provides a very simple and very accurate description of the loss of energy and of quantum coherence for many of the superconducting circuits that we will study – see, for instance, Section 5.5.2 or 6.1.5.

2.4 Measurements

The axioms of quantum mechanics prescribe the behavior of a quantum system under a complete measurement of any observable O, also known as *Von Neumann* or *projective* measurements. Each observable is associated to a different Hermitian operator $\hat{O}$. The eigenvalues that result from diagonalizing this operator o_n correspond to the possible measurement outcomes of the measurement. Let $\hat{P}_n = \sum_m |o_n, m\rangle\langle o_n, m|$ be the projector onto all quantum states for which the observable $\hat{O}$ has the value o_n. This projector is built using eigenstates of the observable $\hat{O}\,|o_n, m\rangle = o_n\,|o_n, m\rangle$, understanding that one measurement outcome may be given by many different quantum states, which differ in other generic quantum properties, here denoted as m. According to quantum mechanics, an *ideal projective measurement* of the observable $\hat{O}$ onto a state ρ will produce the outcome o_n with probability $p(o_n) = \mathrm{tr}(\hat{P}_n\hat{\rho})$. If the measurement is also *nondestructive*, the quantum state after the measurement will be projected onto a new density matrix:

$$\hat{\rho} \to \frac{1}{p(o_n)}\hat{P}_n\hat{\rho}\hat{P}_n. \tag{2.27}$$

We can highlight other properties and types of measurements. First, we may realize that the ideal measurement from (2.27) is an instance of a *quantum non-demolition* (QND) measurement, one which, if repeated twice on the same system, always produces the same expected value $\langle\hat{O}\rangle$. Ideal QND measurements are one of the targets for quantum computing setups. In those experiments, we need to determine the state of the superconducting qubit with certainty, through a projective measurement of the qubit's polarization σ^z, that leaves the qubit in a well-defined state $|0\rangle$ or $|1\rangle$. This way, we can use the outcome of the measurement as input to the following steps of quantum algorithms or to error-correction protocols.

Qubit measurements, such as used in those quantum computers, should ideally be *single shot*. This means that every time we run the experiment, we obtain a real value o_n associated to one measurement outcome, without errors. Note, however, that even if we obtain a meaningful value every time we measure, the estimation of $\langle \hat{O} \rangle$ or the probabilities $p(o_n)$ can still be *quantum limited*, and we may need to repeat the experiments many times to obtain such estimates with high accuracy. For instance, the unbiased estimator of the average using M experimental measurements $\bar{O}_{\text{est}} = \frac{1}{M}\sum_{i=1}^{M} o_{n_i}$ is itself a random variable with a standard deviation that approaches the quantum uncertainty:

$$\Delta \bar{O}_{\text{est}} = \frac{\Delta \hat{O}}{\sqrt{M}}. \tag{2.28}$$

Computing a good estimate means bringing this deviation down to zero, which we do by repeating the experiment again and again, until $\Delta \bar{O}_{\text{est}}$ lays below our desired tolerance.

In experiments, we rarely find direct projective measurements. More generally, experiments are designed so that we measure an auxiliary quantum object that has interacted with and extracted the information from the system we want to measure. The reason to operate this way is to reduce decoherence. If we connect an oscilloscope directly to a microwave resonator, the big classical object will quickly deteriorate the quantum state of the photons that are inside the cavity. It is therefore more convenient to create a setup such as the one in Figure 2.1, in which we perform a weak connection between the resonator and a superconducting waveguide that extracts only a tiny fraction of the photons, which are amplified and fed into a detector.

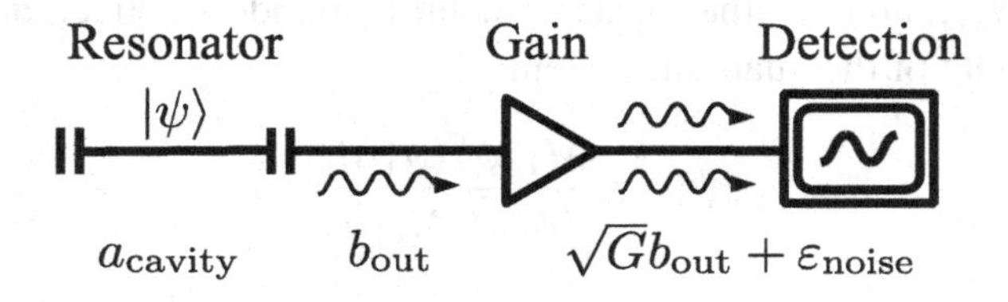

Figure 2.1 A resonator or cavity stores microwave photons whose state we wish to measure. Instead of connecting the measurement apparatus to the cavity, we make a weak connection between the cavity and an outgoing superconducting cable that, after passing through an amplifier, brings the signal to the measurement apparatus. The signal b_{out} is proportional to the cavity signal a_{cavity}, but contains vacuum noise. The measured signal, $Gb_{\text{out}}+\varepsilon$, is amplified with a gain $G > 1$, but contains additional noise ε from the amplifier.

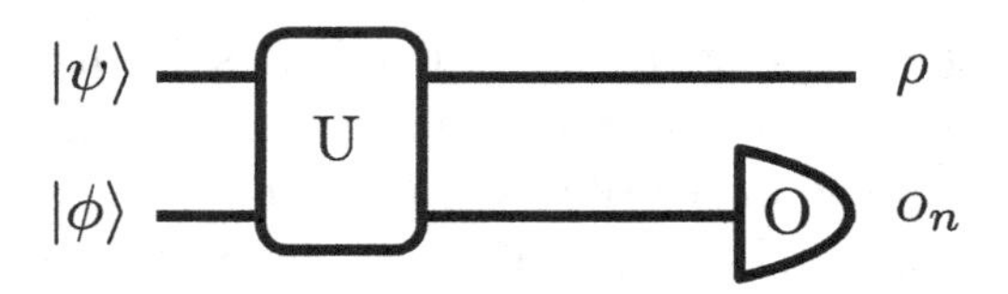

Figure 2.2 Quantum circuit for a generalized quantum measurement.

There are two consequences to using such a setup. First of all, we no longer have a *single-shot* measurement. Amplification introduces noise in the signal, meaning that our oscillator will retrieve a measurement $o_n + \varepsilon$ that is affected by random fluctuating photon noise ε. In absence of systematic errors, those errors average out $\langle\varepsilon\rangle = 0$, and we can still produce meaningful estimates $\bar{O} \simeq \frac{1}{M}\sum_m o_m$, but the presence of large noise prevents us from determining the quantum state of our system after the measurement – i.e., the measurement is no longer quantum limited.

The second consequence of indirect measurements is that we need a broader framework to understand both the measurement statistics and the state of the quantum system after a given measurement outcome. This framework is provided by *generalized quantum measurements*, depicted in Figure 2.2.

(1) The system ψ is put in contact with the auxiliary quantum object ϕ.
(2) Both systems interact through some unitary evolution, $\hat{U}$.
(3) We measure the auxiliary object using a projective measurement $\hat{O}$.

The generalized measurement or positive operator valued measurement (POVM) associates measurement outcomes o_m to operators $\hat{M}_n$ that are no longer projectors, but still satisfy some completeness relation:

$$\sum_n \hat{M}_n^\dagger \hat{M}_n = \mathbb{1}. \tag{2.29}$$

The POVM operators determine the statistics of the measurement outcomes $p(o_m) = \langle \hat{M}_n^\dagger \hat{M}_n \rangle$, and, if the measurement is nondestructive, also the post-measurement state of the quantum system:

$$|\psi\rangle \to \frac{\hat{M}_n |\psi\rangle\langle\psi| \hat{M}_n^\dagger}{\langle\psi|\hat{M}_n^\dagger \hat{M}_n|\psi\rangle}. \tag{2.30}$$

3

Superconductivity

A conductor is a material that can transport electric charge. A typical example of a conductor is a metal, such as copper, iron, or gold. These materials are formed by a somewhat regular lattice of positively charged ions and a sea of mobile electrons. Subject to an external force, the valence electrons will accelerate toward the regions of lowest potential energy, creating an electrical current. As opposing agent to this acceleration, the electrons experience a friction due to the collisions with ions in the lattice. A typical electron will move an average distance, called the *mean free electron path*, before it hits an ion, exchanges kinetic energy with it, and randomly changes both the velocity and direction of motion.

What I described in the last paragraph is the usual microscopic model for Ohmic resistance. At the macroscopic level, it implies that, in order to sustain a current I flowing through a conductor, we have to continously exert a force on the charge carriers, establishing a potential energy difference or *voltage* V between two points of the conductor. The larger the current we wish to establish, the faster the material dissipates energy and the larger the voltage V required. The proportionality constant between current I and voltage V is the device's resistance R:

$$V = I \times R. \tag{3.1}$$

The resistance R is an extensive property that depends on the size of the material through which charge carriers propagate, as well as the rate of collisions that we mentioned before, the so-called *resistivity* ρ.

From our microscopic interpretation of the resitivity ρ, it is obvious that there will be many contributions to this value. Part of the resistivity will be due to *intrinsic* sources. These include lattice defects – dislocations, boundaries between grains in the material, different orientations in the crystal – and impurities or atoms from different materials that penetrate the conductor. In addition to these, we also have to consider the influence of temperature. When the metal heats up, the ions move faster and farther away from their equilibrium positions in the lattice. This increases both

Table 3.1. *Critical temperatures of various superconducting materials.*

Element or compound	Symbol	SC type	T_c	Gap 2Δ
Aluminum	Al	Type I	1.175 K	82.2 GHz
Lead	Pb	Type I	7.20 K	660 GHz
Mercury	Hg	Type I	4.15 K	399 GHz
Niobium	Nb	Type II	9.2 K	738.5 GHz
Yttrium barium copper oxide	$YBa_2Cu_xO_y$	High-T_c	70–90 K	–

the probability of collisions with electrons as well as the amount of energy that they can exchange. When we combine both effects, we typically find an experimental fit of the form

$$\rho \sim \rho_0[1 + \alpha \max\{T - T_0, 0\}], \tag{3.2}$$

with a temperature-dependent contribution and an intrinsic plateau ρ_0 that depends on the material and even on the specific sample.

The study of the temperature-dependent resistance in metals experienced a breakthrough with the discovery of liquid helium by Kamerlingh Onnes in 1908. One of the first applications of liquified helium was to study the conductivity of metals (mercury, tin, and lead) under temperatures $\sim$4.2 K where the thermal contribution to the resistivity should be negligible. In 1911, Kamerlingh Onnes submerged a wire of mercury in helium and observed that the resistivity of the metal suddenly dropped to zero. In other words, the previous plateau disappeared, $\rho_0 \simeq 0$, even though nothing special had been done to improve the material's purity and lattice perfection. After these initial discoveries, further materials were shown to enter this new *superconducting* phase at sufficiently low temperatures, summarized in Table 3.1.

Later studies have shown that a superconducting material cannot be just characterized as a conductor without resistance. As we will soon see, superconductivity is an intrinsically quantum effect that has other unexpected consequences:

- The first one is the possibility of establishing *persistent currents*. If we build a superconducting ring and induce a current, for instance by passing a magnet through the loop or using an external inductor, this current can persist forever without any energy penalty on the material.
- The current that flows around a hole or in a loop can only take certain discrete values that are compatible with the quantization of the magnetic flux that traverses the loop. This *fluxoid quantization* happens in units of the *flux quantum* $\Phi_0 = h/2e$, a universal magnitude governing the operation of SQUIDs (Section 4.7) and superconducting qubits (Chapter 6).

- Superconductors exhibit another surprising phenomenon known as the *Meissner effect*, which is the ability of the superconductor to "expel" magnetic fields. If we take a superconducting object and switch on a magnetic field, the material develops superconducting currents on the surface such that they cancel the magnetic fields inside the object. This phenomenon is similar to how charges on the surface of a conductor arrange to cancel all *electric* fields inside the bulk and lays at the heart of different levitation experiments with magnets and superconductors.
- Finally, superconductivity also leads to counterintuitive behavior in superconductor–insulator–superconductor interfaces. The quantum nature of the superconducting charge carriers allows them to tunnel through thin insulating barriers, establishing currents that behave in unexpected ways in presence of dc and ac potentials. This *Josephson effect* makes it possible to develop *Josephson junctions*, a nonlinear inductor that makes superconducting circuits useful for quantum technology applications.

This phenomenology, which is universal across the board of all superconducting materials, can be explained with a mesoscopic theory that combines quantum mechanics and electromagnetism. However, not all superconducting materials are identical: They can be distinguished by other properties, such as the conditions under which superconductivity appears or is destroyed.

- *Type I superconductors* is the denomination for the first family of superconducting materials that were discovered. This includes most elementary metals, such as mercury, tin, or aluminum. Superconductivity manifests at low *critical temperatures,* between 1–4 K, and is destroyed at relatively low critical magnetic fields. Above such fields, the Meissner effect abruptly disappears and the material experiences a first-order phase transition into an ordinary metal.
- *Type II superconductors* includes niobium and a plethora of alloys with larger critical temperatures and superconducting gaps. These materials survive stronger magnetic fields through the creation of magnetic vortices: thin "tubes" of ordinary metal that allow the field to cross the material, while the rest of the electrons remain in a superconducting state.
- *High-Tc superconductors* is yet another family of superconducting materials, discovered late in the twentieth century. These superconductors are rare-earth ceramic alloys with a complex, quasi-two-dimensional structure that allows superconductivity at temperatures above nitrogen's boiling point (70 K). Since we must work at lower temperatures to engineer microwave quantum circuits (see Section 4.1.1), and since these ceramic materials are difficult to fabricate, they are not very interesting for superconducting quantum information technology.

After this brief overview of superconducting materials, we will introduce both a microscopic interpretation of superconductivity, as well as a macroscopic theory – the London theory or macroscopic wavefunction model – that explains much of the superconducting phenomenology. A distilled version of this model will be the basis in Chapter 4 to develop an effective theory of superconducting circuits.

3.1 Microscopic Model

Which mechanism allows some metals to become superconductors? This is a Nobel Prize–winning question that was answered by physicists John Bardeen, Leon Cooper, and John R. Schrieffer in 1957 (Bardeen et al., 1957a,b). Together, they explained many of the superconducting phenomena mentioned previously, as well as other many-body properties of the superconducting materials, which include the following:

- The evidence of a phase transition and of some type of energy gap to break superconductivity, as suggested by the existence of a critical temperature T_c and a critical magnetic field above which superconductivity disappears.
- The exponential decrease of the superconductor's heat capacity with temperature $C \propto \exp(-1.5T_c/T)$. This property was consistent with a many-body theory in which there is an energy gap, a minimum excitation energy per particle of order $\simeq 1.5k_BT_c$.
- Further evidence of some excitation gap, as provided by the electromagnetic absorption spectrum: the minimum photon energy required to locally excite or break the superconducting state lays somewhere around $2 \times 1.5 \times k_BT_c$.
- Finally, the magnetostatic properties of the superconductor, including superconducting currents and the Meissner effect could be explained by introducing a *coherence length* – i.e., quantum correlations at short distances. As it is now well known in condensed matter physics, the existence of finite coherence lengths is usually an indicator of a gapped model.

The answer by Bardeen and collaborators is known as the *BCS theory*. This theory proposes that the superconductor is actually a Bose–Einstein condensate of charge carriers. As it was already known from studies of Bose–Einstein condensation and early models of ^{4}He superfluidity, a weakly interacting Bose–Einstein condensate can support superfluid currents that never stop and that are immune to small imperfections, impurities, and collisions that do not carry too much energy. In order to justify the existence of Bose–Einstein condensation, the BCS theory introduces an effective attraction between the metal's valence electrons with opposite spin. This attraction is mediated by the phonons of the crystalline structure that forms the metal. At low temperatures, it gives rise to the BCS instability, in which the

Fermi theory breaks down, and electrons join into stable bound particles called *Cooper pairs* (Cooper, 1956). Because of the spin-statistics connection, the pair of two bound electrons is a particle with a bosonic statistics, which condense into a superfluid state.

The BCS theory provides a very elegant and also very straightforward framework for studying the superconductor, with only a few parameters that account for all the physics. The main parameter is the *BCS or superconducting gap*, usually denoted by $\Delta(\mathbf{k})$. The gap is the binding energy of every two electrons that form a pair. It depends mildly on the electron's momenta, and the smallest value $\Delta(0)$ explains many quantiative properties of the superconducting phase and the phase transtion. For instance, as we increase the temperature, we can expect that processes in which pairs are broken become relevant, destroying superconductivity. The quantitative answer is a bit more complicated, as the gap itself depends on temperature $\Delta(T) \simeq 1.74\Delta(0)\sqrt{1 - T/T_c}$, and the critical temperature $T_c \simeq \Delta(0)/1.76k_B$ is the point at which pairing becomes energetically trivial.

The superconducting gap also explains some features in the interaction of the superconductor with electromagnetic fields. A superconductor can absorb photons through two different mechanisms. Low-energy microwaves in the range of 1–20 GHz can excite plasmons of the charged superfluid. These processes create quantum excitations that behave very much like photons (see Chapter 5) or like artificial atoms (see Chapter 6), and which we can route, confine, and operate using superconducting circuits. The second mechanism involves stealing a Cooper pair from the condensate and breaking it into two separate electrons. The energy required for this is $\hbar\omega \simeq 2\Delta(0) \simeq 3.52k_BT$. For the case of the widely used material in superconducting circuits, aluminum, this energy lays in the range of 100 GHz. Therefore, we can suppress this type of event by sufficiently cooling our circuits and isolating them from the environment, with filters that prevent the injection of highly energetic photons.

The BCS theory has other important consequences, including studies of heat capacity, impurities, quasiparticle excitations, Andreev states and normal superconductor interfaces. Overall, this theory applies very well to type I superconductors, and to some extent to type II, but it does not explain high-temperature superconductivity.

Fortunately, we are not so much interested in complex superconducting materials or sophisticated excitations. Rather, we would like an effective model of the superfluid condensate in the simple materials, Al or Nb, which are used in the quantum circuit experiments. As explained by Gor'kov (1959), the BCS model of superconductivity predicts an effective nonlinear theory for the condensate order parameter. This is the Ginzburg–Landau model or, in the simplified linear version that we introduce here, the macroscopic wavefunction model

(Orlando, 1991). Without many complications, this intuitive model will provide a solid and approachable foundation to the engineering of quantum circuit in later chapters.

3.2 Macroscopic Quantum Model

The BCS model for superconductivity proposes that electrons group into a larger unit, the Cooper pair, that has the properties of both being a charged particle, with charge $q = -2e$, as well as being a boson. The bosonic nature of the particle is what makes it possible for all Cooper pairs to condense, sharing a common superfluid state that is insensitive to defects in the material and any other drag force. The theory that we are about to explain is very similar to other models that have been put forward and successfully used, for instance, in the study of weakly interacting Bose–Einstein condensates of alkali atoms (Pitaevskii and Stringari, 2016), and even BCS superfluids built from fermionic atoms. Our formulation of the theory follows closely Orlando (1991), a book we encourage you to read for a better understanding of superconducting properties, magnetostatics, and other interesting phenomenology.

The macroscopic wavefunction theory is based on the assumption that the many-body state is described by a collective wavefunction that is a product state of the same wavefunction for each of the N Cooper pairs:

$$\Xi(\mathbf{x}_1, \mathbf{x}_2 \ldots \mathbf{x}_N; t) = \xi(\mathbf{x}_1, t)\xi(\mathbf{x}_2, t) \cdots \xi(\mathbf{x}_N, t). \tag{3.3}$$

This type of macroscopic accumulation of particles into the same quantum state is what we expect from a Bose–Einstein condensate well below its critical temperature. However, we are also allowing for this accumulation, which is typically a property of a ground state, to also describe the dynamics of the collective system in time, as it reacts to external perturbations from electromagnetic fields, currents, etc. This is a conceptual extension that is only justified by the agreement with experiments and the exact simulations of small systems.

The macroscopic wavefunction theory leads us to introduce new fields n_s and θ, which respectively describe the charge density and, as we will soon see, the flow of particles:

$$\psi(\mathbf{x}, t) = \sqrt{N}\xi(\mathbf{x}, t) \simeq \sqrt{n_s(\mathbf{x}, t)}e^{i\theta(\mathbf{x}, t)}. \tag{3.4}$$

We typically assume a constant and approximately uniform density of carriers throughout most of the material, $n_s(x, t) \simeq \bar{n}_s$. This assumption is approximately valid in many situations because matter tends toward charge neutrality. It will not apply when we consider the charge trapped on a capacitor or in a superconducting island. Those deviations will be studied as perturbative corrections to the background of superconducting particles in the macroscopic theory.

Assuming that the macroscopic wavefunction is a viable model, we now postulate a very general model for its dynamics:

$$i\hbar\partial_t\psi = \left[\frac{1}{2m_s}(-i\hbar\nabla - q_s\mathbf{A})^2 + q_s v(\mathbf{x},t)\right]\psi. \tag{3.5}$$

This model is inspired by the Schrödinger equation for a charged particle moving in an electromagnetic field with scalar and vector potentials $v(\mathbf{x},t)$ and $\mathbf{A}(\mathbf{x},t)$, respectively. The model introduces two effective parameters m_s and q_s to describe the mass and charge of the Cooper pair. As we have seen, the charge is precisely known:

$$q_s = -2e = -2 \times 1.60217662 \times 10^{-19}\ \text{C}. \tag{3.6}$$

However, $m_s = 2m_e^*$ contains the effective mass of the electrons moving through the solid lattice m_e^*, which depends on the band structure and has to be determined experimentally for each material.

3.3 Superfluid Current

From the previous equation, we can already obtain two important properties that we need for studying real superconducting circuits. The first property is the charge distribution, which is given by

$$\rho(\mathbf{x},t) = q_s|\psi(\mathbf{x},t)|^2. \tag{3.7}$$

This superfluid charge includes a very large background that compensates the charge of the ions structuring the lattice of the metal or alloy. From the point of view of circuit theory, it is more interesting to work with the *superfluid current*, a vector field $\mathbf{J}(\mathbf{x},t)$ describing the flow of charges. The evolution of the electric charge Q confined in a volume Ω, is related to the supercurrent flowing across the boundary of that volume $\partial\Omega$:

$$\frac{\mathrm{d}}{\mathrm{d}t}Q = \int_\Omega \partial_t\rho \mathrm{d}^3x = -\int_{\partial\Omega} \mathbf{J}\cdot \mathrm{d}\mathbf{n}. \tag{3.8}$$

Here $\mathbf{n}$ is the unit vector normal to the surface $\partial\Omega$ at each point of the boundary. The same physics is described by the continuity equation:

$$\partial_t\rho = -\nabla\cdot\mathbf{J}. \tag{3.9}$$

As Fritz London conjectured, the superfluid current may be derived from the Schrödinger equation (3.5) as a combination of the macroscopic wavefunction current and the electromagnetic field:

$$\mathbf{J} = q_s \times \mathrm{Re}\left\{\psi^* \left[\left(-i\frac{\hbar}{m_s}\nabla - \frac{q_s}{m_s}\mathbf{A}\right)\psi\right]\right\}. \tag{3.10}$$

This combination explain the fluxoid quantization and the Meissner effect, as described by London et al. (1935).

Note that we can obtain the *electric current intensity* of a circuit I by integrating the charge current $\mathbf{J}(\mathbf{x}, t)$ across any section S of any cable in the circuit:

$$I(t) = \iint_S \mathbf{J}(\mathbf{x}, t) \cdot \mathrm{d}\vec{\mathbf{S}}. \tag{3.11}$$

The current intensity can be different at different points of a large superconducting circuit. However, the conservation of charge – Cooper pairs are not destroyed in our simple, conservative model – implies that the current coming into a superconducting element must balance with the current going out into other circuit elements. This will be key in our analysis of circuits and derivation of quantitative models in Chapter 4.

3.4 Superconducting Phase

The superconducting wavefunction contains information about the charge distribution and the electrical current. We will now argue that most of the information is actually hidden in the phase of the wavefunction. We will also relate this phase to a macroscopically observable quantity, the flux.

The first statement is rather obvious. We have already discussed that the density of charged particles must be a rather uniform property, dependent only on the properties of the material – i.e., how many electrons the atoms donate to the conduction band where Cooper pairs are formed. If we assume that $n_s(\mathbf{x}, t) = |\psi|^2$ is constant and uniform, currents are divergence-free:

$$\nabla \cdot \mathbf{J} = 0, \tag{3.12}$$

and all information about the superconductor must actually reside in the phase of the wavefunction. Indeed, working with (3.10), we obtain

$$\mathbf{J} = q_s n_s \left[\frac{\hbar}{m_s}\nabla\theta - \frac{q_s}{m_s}\mathbf{A}\right]. \tag{3.13}$$

The second statement is more subtle. Let us assume the *Coulomb gauge* $\nabla \cdot \mathbf{A} = 0$, and specialize the macroscopic quantum model (3.5) for a wavefunction with uniform density. The Coulomb gauge implies $\Delta\theta = 0$, and

$$-\hbar\partial_t\theta \simeq \frac{1}{2n_s}\Lambda\mathbf{J}^2 + q_s v. \tag{3.14}$$

Notice the new constant, the isotropic London coefficient[1] $\Lambda = m_s/q_s^2 n_s$.

In the absence of currents, $\mathbf{J} = 0$, (3.14) becomes the so-called *phase-voltage* relation

$$\partial_t\theta \simeq -\frac{q_s}{\hbar}v. \tag{3.15}$$

This relation is an obvious consequence of unitary evolution. For quasi-stationary states, the wavefunction remains constant up to a global phase, determined by the energy of the system $\psi(\mathbf{x},t) = \exp(-iEt/\hbar)\psi(\mathbf{x},0)$. Since the energy of the charged particle in a potential is $E = q_s v$, we obtain $\partial_t\theta = -E/\hbar = -q_s v/\hbar$. The problem with relation (3.15) is that it is not gauge invariant – it is only valid in the Coulomb gauge – and it has been derived under the condition of no persistent currents, $\mathbf{J} = 0$. We have to complete our derivation to regard more general conditions!

3.5 Gauge-Invariant Phase

In order to correct (3.15), we will separate the superconducting phase into a term that is always the same, and a contribution that depends on our choice of electromagnetic gauge. The *gauge-invariant phase* $\varphi(\mathbf{x},t)$ is defined by removing the contribution of the vector potential, taking as reference one (arbitrary) location of the superconductor $\mathbf{x}_0$:

$$\theta(\mathbf{x},t) - \theta(\mathbf{x}_0,t) = \varphi(\mathbf{x},t) - \varphi(\mathbf{x}_0,t) + \frac{q_s}{\hbar}\int_{\mathbf{x}_0}^{\mathbf{x}} \mathbf{A}(\mathbf{r},t)\cdot d\mathbf{l}. \tag{3.16}$$

In the definition of $\varphi(\mathbf{x},t)$, the choice of path from $\mathbf{x}_0$ to $\mathbf{x}$ is arbitrary, but (i) it must be unique for each point $\mathbf{x}$, (ii) it must be continuous, (iii) all paths must remain in the superconductor, and (iv) they must not cross each other.[2] Except for a set of points of zero measure – the discontinuities of φ – we can define a gauge-invariant wavefunction:

$$\psi_{\mathrm{GI}}(\mathbf{x},t) = e^{-i\frac{q_s}{\hbar}\int_{\mathbf{x}_0}^{\mathbf{x}}\mathbf{A}(\mathbf{r},t)\cdot d\mathbf{l}}\psi(\mathbf{x},t) = e^{i\varphi(\mathbf{x},t)}\sqrt{n_s(x,t)}, \tag{3.17}$$

[1] As mentioned before, superconducting currents shield magnetic fields out of the material. The London coefficient is related to the penetration depth of magnetic fields into the superconductor, a fact that can also be derived from this theory (Orlando, 1991).

[2] This choice of path is one of the essential steps in working with any superconducting circuit, as we will see in Chapter 4. However, in that chapter we will assume quasi-one-dimensional structures, where it is easy to believe that such paths do exist, at least in the form $\mathbf{x}_l = \mathbf{x}l + \mathbf{x}_0(1-l)$ for $l \in [0,1]$.

which satisfies a Schrödinger equation without vector potential:

$$i\hbar\partial_t\psi_{\text{GI}} = \left[\frac{1}{2m_s}(-i\hbar\nabla)^2 + q_s v(\mathbf{x},t)\right]\psi_{\text{GI}}. \tag{3.18}$$

We now reach a gauge-independent relation between the phase and the electric field:

$$\begin{aligned}\frac{\partial}{\partial t}\varphi(\mathbf{x},t) &= \frac{q_s}{\hbar}\int_{\mathbf{x}_0}^{\mathbf{x}}\left(-\nabla v - \frac{\partial}{\partial t}\mathbf{A}\right)\cdot \mathrm{d}\mathbf{l} \\ &= \frac{q_s}{\hbar}\int_{\mathbf{x}_0}^{\mathbf{x}}\mathbf{E}(\mathbf{r},t)\cdot\mathrm{d}\mathbf{l} = \frac{q_s}{\hbar}\left[V(\mathbf{x}_0) - V(\mathbf{x})\right].\end{aligned} \tag{3.19}$$

This gauge-invariant phase is related to the voltage difference V across the superconducting circuit,[3] defined as the energy required to transport a unit of charge from $\mathbf{x}_0$ to $\mathbf{x}$. It is convenient to refer the voltage to $V(\mathbf{x}_0) := 0$ and work instead with the electric *flux*, defined up to an irrelevant offset as

$$\phi(\mathbf{x},t) = \int_0^t V(\mathbf{x},\tau)\mathrm{d}\tau. \tag{3.20}$$

Introducing also the *magnetic flux quantum*,

$$\Phi_0 = \frac{h}{|q_s|} = \frac{h}{2e} \simeq 2.067833758(46)\times 10^{-15}\,\text{Wb}, \tag{3.21}$$

we identify the gauge-invariant phase with the electric flux

$$\partial_t\varphi(\mathbf{x},t) = \frac{2\pi}{\Phi_0}\partial_t\phi(\mathbf{x},t), \tag{3.22}$$

As we will see in Chapter 4, this identity makes the electric flux one of the two preferred variables when working with superconducting circuits, the other one being the charge q that accumulates on a superconducting element. Another important property of the gauge-invariant phase is that it determines the superconducting current:

$$\mathbf{J} = q_s n\left[\frac{\hbar}{m_s}\left(\nabla\varphi + \frac{q_s}{\hbar}\mathbf{A}\right) - \frac{q_s}{m_s}\mathbf{A}\right] = \frac{\hbar q_s n_s}{m_s}\nabla\varphi. \tag{3.23}$$

This makes sense, because the superconducting current, just like the circuit voltage, cannot depend on the choice of gauge.

[3] V and v are not the same observable. Only the former is gauge independent.

3.6 Fluxoid Quantization

There is a third consequence of the phase-current relation. This one relates to the allowed values of the current and of the magnetic flux trapped inside superconductors. Let us multiply (3.13) by $\Lambda = m_s/n_s q_s^2$ and integrate it around a closed-loop C around a simply connected region, $C = \partial S$:

$$\oint_C (\Lambda \mathbf{J}) \cdot \mathbf{dl} + \oint_C \mathbf{A} \cdot \mathbf{dl} - \oint_C \frac{\hbar}{q_s} \nabla\theta \cdot \mathbf{dl} = 0. \tag{3.24}$$

We use Stokes's theorem to transform the second line integral into a surface integral over the region S. The contour integral of the phase must produce a multiple of $2\pi \times \hbar/q_s = -\Phi_0$, as otherwise the wavefunction ψ would be discontinuous. This results in the quantization equation

$$\oint_C (\Lambda \mathbf{J}) \cdot \mathbf{dl} + \int_S \mathbf{B} \cdot \mathbf{dS} + \Phi_0 \times m = 0, \quad \text{with } m \in \mathbb{Z}. \tag{3.25}$$

This equation states that the magnetic flux trapped in a loop

$$\Phi_{\text{loop}} = \int_S \mathbf{B} \cdot \mathbf{dS}, \tag{3.26}$$

plus the flux due to the induced supercurrents must be an integer multiple of the magnetic flux quantum (3.21). This is a very powerful result, used by Deaver and Fairbank (1961) to estimate Φ_0 and demonstrate that the superconducting particles $q_s = -2e$ are formed by two electrons.

There is a simpler derivation of fluxoid quantization that is of more interest to us, and which is based on the gauge-invariant phase (3.16). This definition can be reformulated as

$$\nabla\theta = \nabla\varphi + \frac{q_s}{\hbar}\mathbf{A}. \tag{3.27}$$

Integrating around a closed loop, we find

$$\oint_C \nabla\varphi \cdot \mathbf{dl} = 2\pi \times m + \frac{2\pi}{\Phi_0}\Phi_{\text{loop}}. \tag{3.28}$$

Using the phase-voltage relation, this can be rewritten as a condition for the flux differences along different segments of a superconducting circuit, or *fluxoid quantization*:

$$\oint_C \nabla\phi \cdot \mathbf{dl} = \Phi_0 \times m + \Phi_{\text{loop}}. \tag{3.29}$$

This condition is one of the constituent relations in the theory of superconducting circuits from Chapter 4. Note that unlike the wavefunction, the flux and the gauge-invariant phase need not be continuous, as their definition depends on a choice of paths that we make before studying the circuit. However, our derivation reveals that all choices are consistent, and that the existence of loops in a circuit reduces the number of independent variables, because fluxes are not completely independent from each other.

3.7 Josephson Junctions

The last topic that we cover in this chapter is a simple, yet very powerful device that was postulated by Josephson (1962) and verified experimentally shortly thereafter by Anderson and Rowell (1963). The device in question is called a *tunnel* Josephson junction.[4] It is a superconductor–insulator–superconductor "sandwich," where the insulating area is so thin that it allows quantum tunneling of Cooper pairs. As Josephson predicted, the tunneling of pairs creates unexpected relations between the applied voltage (dc or ac) on the superconducting leads and the intensity that circulates through the junction. We can use the macroscopic quantum model to derive those relations.

Figure 3.1b shows the schematics of a Josephson junction, with three separate regions: two superconducting leads of arbitrary size and an insulating barrier. Our toy model for the junction is a one-dimensional[5] potential barrier of height U_0 and width d that hinders the propagation of the macroscopic wavefunction $\psi(x,t)$.

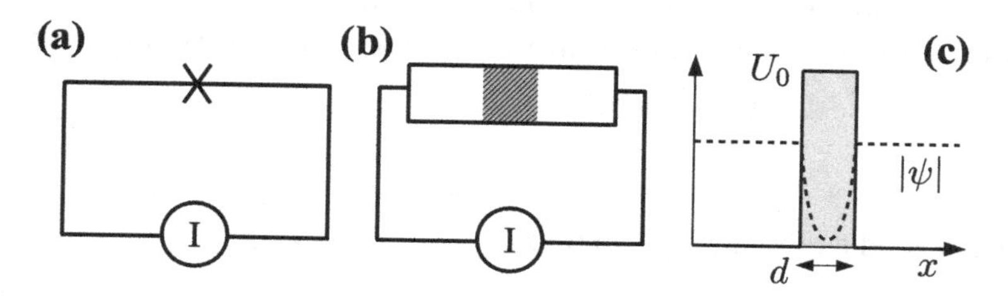

Figure 3.1 (a) Circuit schematics for a Josephson junction connected to an intensity source. (b) The Josephson junction is made of two superconducting leads (white) separated by a thin insulating barrier (gray). (c) We model the junction as a barrier energy U_0, which is thin enough, d, that it allows some quantum tunneling of Cooper pairs.

[4] We can obtain a similar physics through other physical devices, such as constrictions, point contacts, and normal interfaces. However, the theoretical picture is simpler in this case.

[5] The one-dimensional model is sufficient for describing the type of small junctions that appears in typical superconducting circuits. A more detailed model that takes into account the transverse dimensions is found in Orlando (1991).

We seek stationary solutions of the Schrödinger equation for the gauge-invariant wavefunction

$$E\psi_{\mathrm{GI}}(x) = \left[\frac{1}{2m_s}(-i\hbar\partial_x)^2 + v_0(x)\right]\psi_{\mathrm{GI}}(x), \tag{3.30}$$

with the potential barrier

$$v_0(x) = \begin{cases} 0, & |x| > d/2 \\ U_0, & \text{otherwise.} \end{cases} \tag{3.31}$$

As a boundary condition, we impose a current intensity I flowing through the junction. Thanks to the relation between current intensity and superconducting current, our boundary condition can be written as

$$I = \int_S \mathbf{J} \cdot d\mathbf{S} = J \times A, \tag{3.32}$$

where the proportionality constant A is the area perpendicular to the junction – i.e., the cross-section size.

What do the solutions of the Schrödinger equation look like with these conditions? Since far away from the insulator the current is fixed, we expect plane-wave solutions such that $\partial_x\varphi \propto J$. More precisely, we write

$$\psi_{\mathrm{GI}}(x) \propto \sqrt{n}e^{ikx}, \quad |x| > d/2. \tag{3.33}$$

The sign and magnitude of the superconduction current may be derived from the momentum k of the wavefunction as $J = q_s n_s \hbar k/m_s$. The momentum also k determines the energy of the solution

$$E = \frac{\hbar^2 k^2}{2m_s} = \frac{1}{2n}\Lambda^2 J^2. \tag{3.34}$$

When the current is very small $J \simeq 0$, this energy will lay well below the barrier U_0. Inside the insulator, we will have an equation of the form

$$\partial_x^2\psi_{\mathrm{GI}} = \frac{m_s}{\hbar^2}(U_0 - E)\psi_{\mathrm{GI}} = \frac{1}{\xi^2}\psi_{\mathrm{GI}}, \quad |x| \leq d/2, \tag{3.35}$$

which can only be satisfied with exponentially decreasing or increasing solutions, $\psi_{\mathrm{GI}} \propto \exp(\pm x/\xi)$. The final solution then reads

$$\psi_{\mathrm{GI}}(x) = \alpha_+ \cosh(x/\xi) + \alpha_- \sinh(x/\xi), \quad |x| \leq d/2, \tag{3.36}$$

with the parameters

$$\alpha_\pm = \sqrt{n_s}\,\frac{e^{i\varphi_L(-d/2)} \pm e^{i\varphi_R(d/2)}}{e^{d/2\xi} \pm e^{-d/2\xi}}. \tag{3.37}$$

The superfluid current is uniform across the circuit. Along the insulator

$$J = \frac{q_s}{m_s}\mathrm{Re}(-\psi_{\mathrm{GI}}^* i\hbar\partial_x\psi_{\mathrm{GI}}) = J_c \sin[\varphi_R(d/2) - \varphi_L(-d/2)], \tag{3.38}$$

it is proportional to the Josephson junction's critical current

$$J_c = -\frac{q_s\hbar}{m_s\xi}\frac{n_s}{\sinh(d/\xi)} \geq 0. \tag{3.39}$$

Currents above this value cannot be captured by the exponential solution, but instead correspond to plane waves with $E > U_0$.

The previous derivation has shown us that small values of the current have a nonlinear relation with the gauge-invariant phase jump $\delta\varphi = \varphi_{\mathrm{R}} - \varphi_{\mathrm{L}}$ across the junction:

$$I = I_c \sin(\delta\varphi). \tag{3.40}$$

As clearly explained by Orlando (1991), the first Josephson relation still holds in presence of magnetic and electric fields, provided we still work with the gauge-invariant phase.

Assume now that we establish a voltage difference V among the junction's leads. Since we work with metals, the voltage will be approximately uniform along each lead. We then expect a flux difference on both sides of the insulator $\delta\phi(t) = \phi(d/2) - \phi(-d/2) \simeq \int_0^t V(\tau)\mathrm{d}\tau$. Using the connection between flux and phase (3.22), we obtain the second Josephson relation:

$$\delta\varphi = \frac{2\pi}{\Phi_0}\frac{dV}{dt}. \tag{3.41}$$

The Josephson relations combine into an equation connecting flux and current:

$$I = I_c \sin\left(\frac{2\pi}{\Phi_0}\delta\phi\right). \tag{3.42}$$

The dc Josephson effect is a consequence of this equation: a constant voltage bias V induces an oscillating current due to the linearly growing flux:

$$I(t) = I_c \sin\left(\frac{2\pi}{\Phi_0}Vt + \delta\varphi(0)\right). \tag{3.43}$$

However, (3.42) has more general implications, as a constituent equation, allowing us to include Josephson junctions in the general theory of superconducting circuits from Chapter 4. For instance, we can derive the inductive energy stored in the junction as we switch on the voltage – and the flux – to a finite value:

$$E = \int_0^t I(t)V(t)\mathrm{d}t. \tag{3.44}$$

With the change of variables $V(t)\mathrm{dt} = \mathrm{d}\phi$, we can substitute the expression (3.42) and integrate between the initial and final value of the flux

$$E = \int_0^{\delta\phi} I_c \sin\left(\frac{2\pi}{\Phi_0}\phi\right) \mathrm{d}\phi = -\frac{I_c \Phi_0}{2\pi} \cos\left(\frac{2\pi}{\Phi_0}\delta\phi\right). \tag{3.45}$$

From the point of view of superconducting circuit theory, the junction behaves as a nonlinear inductive element. The inductance may be derived from the current–voltage relation or from an expansion of the inductive energy just derived. In both cases, we obtain a similar expression:

$$L_J = \frac{V}{\frac{\mathrm{d}I}{\mathrm{d}t}} = \frac{\Phi_0}{2\pi} \frac{1}{I_c \cos(2\pi\,\delta\phi/\Phi_0)}. \tag{3.46}$$

This nonlinear inductance and the constituent equation (3.42) will be repeatedly used in Chapter 4 when studying superconducting qubits, dc-SQUIDs, and rf-SQUIDs.

4
Quantum Circuit Theory

4.1 Introduction

This chapter introduces a quantum mechanical theory of electrical circuits, developed in the spirit of the original work by Yurke and Denker (1984), and building on the more pedagogical explanations by Devoret (1995). The chapter begins discussing the temperature and power requirements needed to observe quantum phenomena in electrical circuits. Section 4.2 recalls the three lumped circuit elements – the capacitor, the inductor, and the Josephson junction that we use to build most superconducting circuits, describing the classical equations that govern them. Section 4.3 introduces an effective quantum theory of circuits, formulated as a set of rules to transform a circuit network into the quantum Hamiltonian that describes its dynamics. Sections 4.4–4.8 apply this theory to the most important circuits in this book – a simple LC resonator, a transmission waveguide, a nonlinear resonator, and different types of Josephson junction loops. You are strongly advised to work out these examples and exercises on your own, to build a solid foundation for later chapters where we use these circuits to engineer photons (Chapter 5), superconducting qubits (Chapter 6), and light–matter interfaces (Chapter 7).

4.1.1 What Makes a Circuit Quantum

A circuit is "quantum" when we need a quantum mechanical theory to describe it. Quantum circuits exhibit phenomena that are exclusive of quantum physical systems, such as energy quantization, reversible unitary evolution, quantum superpositions, and entanglement. In a quantum circuit, electric current, voltage, charge, and flux are quantum observables. Measurements of these quantities result in probabilistic outcomes, governed by uncertainty relations that limit how much information we can obtain in a single experiment or multiple experiments. Lastly, quantum circuits are modeled with quantum mechanical equations, such as the Schrödinger equations for pure states, or the master equations for mixed states. All this affects the way we design, control, and operate superconducting quantum circuits.

Like all solid-state devices, electrical circuits are built from myriads of quantum systems – a sea of electrons floating through a lattice of ions, covalently bound molecules, oxidized metal, etc. A circuit will exhibit quantum phenomena depending both on how it is built and how it is operated. Common wisdom states that there are three causes for a system to act in a purely classical way, hiding its quantum nature:

(1) When we combine very large number of quantum objects into a many-body system, the properties of the macroscopic object tend to be well defined, with negligible quantum uncertainties. These objects can be safely described with classical equations.
(2) Phenomena such as superpositions and entanglement are destroyed by interaction with the environment – e.g., by surrounding electromagnetic fields and particles. The typical *decoherence time* in which this happens tends to be shorter than the duration of common experiments.
(3) Preparing good quantum states requires cooling the system to a point in which thermal fluctuations do not obscure quantum fluctuations. The temperature at which this happens varies with the physical model.

Let us discuss these challenges in greater detail:

Macroscopicity. To understand the first obstacle, consider the Heisenberg uncertainty principle for two complementary observables such as position and momentum $\Delta x \Delta p \geq \hbar/2$. Learning the position of a particle with a very low uncertainty Δx leads to a large uncertainty in the momentum Δp. However, since the Planck constant is very small, $\hbar \sim 1.055 \times 10^{-34}$ J · s, the actual uncertainty in the *velocity* of a macroscopic object can be very small. Realistically, $\Delta x \sim 800$ nm is limited by the wavelength of the light we use to observe objects; assuming a small particle of 1 gram, then $\Delta v = \Delta p/m \sim \hbar/m\Delta x \sim 10^{-25}$ m/s, which is unobservable for us humans.

A solution to this problem is to work with selected degrees of freedom that can be brought down to a "microscopic" limit. In this book, we study electrical circuits; these are macroscopic objects – with sizes that range between millimeters and meters – but within these objects we manipulate the charge or current degrees of freedom, manifesting their quantum properties. We may create quantum states such as the superposition $\frac{1}{\sqrt{2}}(|N\rangle + |N+1\rangle)$, in which a small superconducting island has either N or $N + 1$ Cooper pairs – a microscopic difference on top of a macroscopic, classical background. Similarly, we can prepare superpositions of left- and right-moving currents, involving a small number of Cooper pairs that rotate along a superconducting loop. Probing these degrees of freedom and small differences is difficult, but we have the technology to do it.

Thermal fluctuations. Once we have selected a degree of freedom out of a macroscopic object, the next obstacle is to prepare it in a state that reveals quantum phenomena. If we do nothing special to a quantum object, its quantum superpositions or entangled states will progressively deteriorate, until the object achieves a thermal state in equilibrium with its surrounding environment.[1] Thermalization brings our quantum system progressively closer to a diagonal mixed state $\rho = \sum_n p_n |n\rangle\langle n|$ with the classical probability distribution $\{p_n\}$ of configurations (see Exercise 4.1). As the quantum system approaches a termal state and *thermal fluctuations* begin to dominate the experiment, they completely mask the quantum uncertainties and correlations that reveal the quantum nature of our system.

We quantify the effect of thermal fluctuations using statistical physics. Let us consider a quantum system with eigenstates $|n\rangle$ that have regularly spaced eigenenergies $E_n \simeq \hbar\omega n$. This is the case of superconducting qubits – two-level systems where n can be 0 or 1; see Chapter 6 – and of microwave photons – where n can be any nonnegative integer $n \in \{0, 1, 2, 3, \ldots\}$; see Chapter 5. In a typical experiment we want to first cool the system as close to the ground state $|0\rangle$ as possible. This is a reference state from which other states and superpositions can be created in a predictable way. If we let our system thermalize, it will have some overlap with the ground state, but it will also spread over other slightly more energetic configurations $|1\rangle, |2\rangle \ldots$. A perfectly thermalized state will follow the Boltzmann distribution, $p_n \propto \exp(-\hbar\omega n/k_b T)$. As shown in Figure 4.1, the probability of *being in any excited state* $p_{\text{excit}} = 1 - p_0 = \exp(-\bar{n})$ decreases exponentially with the ratio $\bar{n} = \hbar\omega/k_B T$ both for photons and two-level systems.

Thermal fluctuations are strongly suppressed once the temperature drops well below a value $\hbar\omega/k_B$ dictated by the characteristic energy scale of excitations. Table 4.1 shows relevant energy scales and temperatures for various experimental platforms. For a typical circuit that works at microwave frequencies $\omega = 2\pi \times 8$ GHz, the temperature reference[2] is $T = \hbar\omega/k_B \simeq 0.38K$. For temperatures well below this limit, of 400, 100, 50, and 10 mK, we predict an excitation probability of 38%, 2%, 0.04%, and $2 10^{-15}$%. Thus, placing the electrical circuit in a good dilution refrigerator that operates in the 20–50 mK guarantees that thermal fluctuations will be negligible.

[1] There is always a reservoir in contact with our system, even if that reservoir is just the electromagnetic field at the temperature of the cosmic background radiation!

[2] A good rule of thumb to remember this calculation is the equivalence between 20 GHz and 1 K, which makes the excitation probability $e^{-1} \sim 36\%$. Every time we halve the temperature, we divide this probability by ~ 3.

Table 4.1. *Equivalence between the frequencies needed to excite a quantum system and the temperature at which thermal excitations become relevant.*

10^{15} Hz $\sim$ 50 000 K	Ultraviolet
300 THz $\sim$ 15 000 K	Near infrarred
1 THz $\sim$ 50 K	High-energy microwaves
1–20 GHz $\sim$ 50 mK–1 K	Microwaves

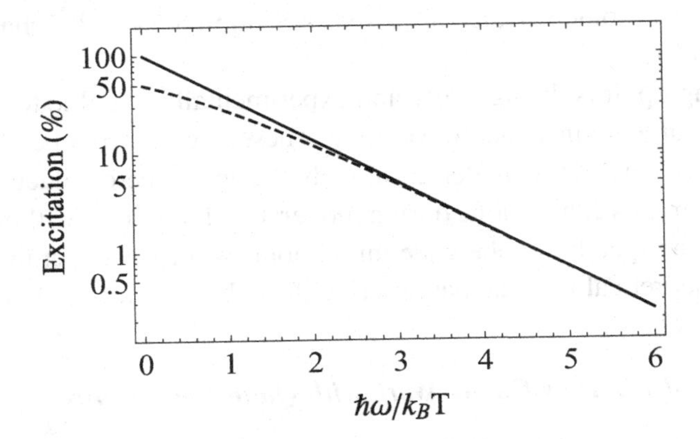

Figure 4.1 Thermal fluctuations in a quantum device. We study a device with either two possible configurations separated by an energy $\hbar\omega$ (dashed), or a ladder of states with energies $\{0, \hbar\omega, 2\hbar\omega, \ldots\}$ (solid). Introducing the ratio $\bar{n} = \hbar\omega/k_B T$, we plot the probability that the system is found in an excited state. The solid line follows the law $\exp(-\bar{n})$.

Environmental influence. Even if we ensure a low-temperature surrounding, our circuit will still couple to the environment: a mixture of surrounding electromagnetic fields, control cables, antennaes, materials and impurities of the substrate, etc. The sample will interact with this environment, radiate energy, absorbe fluctuations, and experience jitter in its eigenenergies. We identify these three generic processes with the terms *dissipation*, *heating*, and *dephasing*. They all lead to a generic phenomenon known as *decoherence*, which is the destruction of quantum coherence or quantum fluctuations in our system, steering it toward a classical state.

There are trade-offs to be made when dealing with decoherence. The materials forming our circuits carry uncontrollable sources of errors that limit our total coherence time. For instance, in ordinary metals, impurities, defects, phonons, etc., cause some electric resistance that cannot be neglected, even at low temperatures. This is one intrinsic source of error that can be eliminated by working with superconductors, but many others persist: trapped charges in defects of a material, two-level fluctuators affecting the inductance of Josephson junctions, radiation to free space, dirt and fabrication defects, etc. On top of this, we must consider the coupling of our quantum systems to measurement devices and controls: this coupling is also a source of decoherence, which we can diminish but never completely suppress.

Altogether, environment and imperfections both limit the time during which we can operate a circuit coherently. In times shorter than this value, we may prepare a quantum state, perform quantum gates, create entanglement, and measure quantum properties with enough accuracy. Past this time, evolution will deviate from a pure dynamics, quantum superpositions will deteriorate, and the whole experiment – unless corrected in some way, as discussed in Section 8.5 – will behave as a noisy classical system.

So, summing up, it is theoretically and experimentally possible to develop electrical circuits that work in a quantum regime. These circuits will operate at very low temperatures, at which some materials already become superconductors. The operation of those circuits can be done during moderately long times without significant errors, preserving quantum coherence throughout the process. Let us now discuss what are the theoretical models that describe the coherent part of evolution.

4.1.2 How Do We Work with Quantum Circuits?

An electrical circuit is a network of metallic, insulating, and semiconducting elements, such as capacitors, inductors, resistors, and transitors, which, combined with appropriate power soures, control the behavior of current I and voltage V. Classical network theory is a systematic framework to analyze complex electrical circuits, finding out the independent variables that describe the circuit – e.g., current or voltage on select nodes of the network – and producing a *set of ordinary differential equations* that describes the dynamics of those variables. If the circuit is conservative – it has no resistance in any element – the theory can be brought to a *Hamiltonian* formulation, quantifying the *total energy of the system* as a function of those variables. As explained in Section 2.1.1, both descriptions contain the same information: We obtain the Hamiltonian from the differential equations, and we can recover the dynamical equations from the Hamiltonian through differentiation.

Let us illustrate this with the help of an LC resonator, a model circuit that will be used frequently throughout this book. A minimal LC resonator, sketched later in Section 4.3, consists of a capacitor C connected in parallel to an inductor L. The pair of variables (I, V) define a phase space similar to that of position and momentum (x, p). They satisfy the ordinary differential equations:

$$\frac{\mathrm{d}V}{\mathrm{d}t} = \frac{I}{C}, \quad \text{and} \quad \frac{\mathrm{d}I}{\mathrm{d}t} = -\frac{V}{L}. \tag{4.1}$$

Alternatively, we can write a second-order differential equation:

$$\frac{\mathrm{d}^2 V}{\mathrm{d}t^2} = -\omega^2 V. \tag{4.2}$$

From your classical mechanics course, you may remember this as the equation for a harmonic oscillator. It describes the periodic oscillations of the voltage (and of the current), with a frequency $\omega = 1/\sqrt{LC}$ that is defined by the capacitance C and the inductance L of the circuit. The total energy of the circuit is conserved (see Exercise 4.4):

$$H = \frac{1}{2}CV^2 + \frac{1}{2}LI^2. \tag{4.3}$$

It is the sum of the capactive and inductive energy stored in the circuit, and also as the Hamiltonian associated to (4.1), where V and I play the role of position and momentum x and p.

Quantum circuits are really not that different. Because we are interested in suppressing thermal fluctuations and decoherence, these circuits are built from superconductors – alluminum and niobium, typically – and operated at temperatures of tens of milli-Kelvin. We design these circuits to only include capacitors, inductors, and Josephson junctions. There are no transistors, and the resistance of the junctions is usually neglected.

As before, the circuit elements induce relations between observables, which result in two possible formulations of a *quantum network theory* for a circuit, both of which rely on the notions from Section 2.1.3. We can construct a *set of ordinary differential equations*. These are Heisenberg equations for the *quantum observables* of current and voltage, which become operators with associated expectation values, uncertainties, and correlations that can be probed in experiments. Alternatively, we can build a *Hamiltonian*, a quantum operator that is a function of those observables. This new operator can be used to recover the same Heisenberg equations, or to define a *Schrödinger equation* for the wavefunction of the circuit.

In the Heisenberg formulation, we convert (4.1) or (4.2) to equations for the infinite-dimensional operators $\hat{V}$ and $\hat{I}$:

$$\frac{\mathrm{d}^2}{\mathrm{d}t^2}\hat{V}(t) = -\omega^2 \hat{V}(t). \tag{4.4}$$

Interestingly, we can compute the expectation value $\langle\psi_0|\hat{V}(t)|\psi_0\rangle$ of the voltage $\hat{V}(t)$ at time t, for a circuit that is initially in quantum state $|\psi_0\rangle$. We find that it reproduces the dynamics of a classical circuit:

$$\frac{\mathrm{d}^2}{\mathrm{d}t^2}\left\langle \hat{V}(t)\right\rangle = -\omega^2 \left\langle \hat{V}(t)\right\rangle. \tag{4.5}$$

In the Schrödinger formulation, we build a quantum Hamiltonian $\hat{H} = \frac{1}{2}C\hat{V}^2 + \frac{1}{2}L\hat{I}^2$ using (4.3) and replacing variables with now static operators. The Hamiltonian defines a Schrödinger equation for the quantum state of the resonator circuit $|\psi(t)\rangle$, as

$$i\hbar\frac{\mathrm{d}}{\mathrm{d}t}|\psi\rangle = \hat{H}|\psi\rangle, \tag{4.6}$$

with initial condition $|\psi(0)\rangle = |\psi_0\rangle$. As explained in Section 2.1.3, when we work in the Schrödinger picture, observables do not change in time, but states do. Now the expectation value of the voltage across the capacitor will be $\langle\psi(t)|\hat{V}|\psi(t)\rangle$, which consistently follows the same equation (4.5).

We leave the study of Schrödinger or Heisenberg equations for later chapters. In this one, the goal is to take any quantum circuit and derive equations such as (4.2) or (4.4), together with their Hamiltonians. This requires first some basic notions of electrical network theory. In Section 4.2, we provide definitions and conventions for voltage and current, as well as the relations between them imposed by the circuit elements. Section 4.3 adds Kirchoff's laws to the mix, formulating the complete quantum network theory.

4.2 Circuit Elements

Our study of quantum circuits is a generalization of the lumped-element circuit theory. This is a very basic model for electrical circuits where we disregard the internal structure of capacitors, inductors, or nonlinear devices, treating them as point-like elements. Each element is described by two *branch variables*: the *electric current intensity* I flowing through the element, and the *voltage drop* V experienced by the charged particles that cross it. The lumped-element approximation is justified when voltage and current oscillate at frequencies whose wavelength is much longer than the size of the circuit elements. Under such conditions, V and I have no spatial dependence and are related by the constitutive equations of the element under study.

This point of view is illustrated in Figure 4.2. A generic circuit element has two connection points, A and B. When placed in a circuit network, a certain amount of current I passes through the element. We assign a direction to the element and associate the positive sign of the current $I > 0$ to currents that flow along that direction. In this particular drawing, a positive current implies a flow of *positively charged particles* from B to A. In ordinary circuits, the current originates from a potential difference between the elements' leads. Positive charges flow from a region of high potential energy to a region of lower potential energy. This sets the convention that when there is current going from B to A, there will be an associated voltage difference $V = V_B - V_A > 0$ growing in the opposite direction.

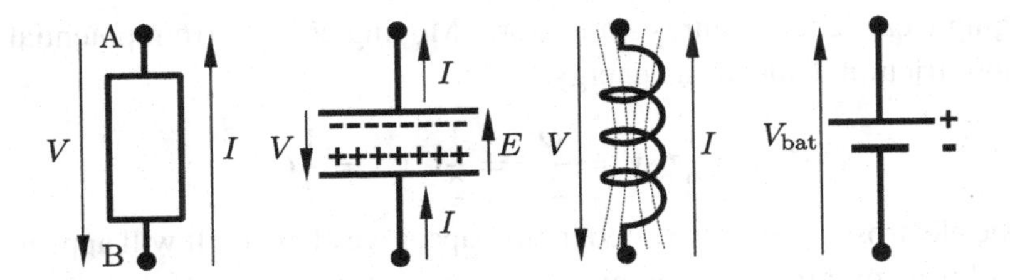

Figure 4.2 Circuit elements, together with the usual convention for current and voltage. From left to right: generic element, capacitor, inductor, and battery. Note that particles flow from the point of the circuit with the largest potential energy to the area with the lowest potential energy. Hence, V and I have opposite directions.

Note that these conventions still work when the current is formed by negatively charged particles, such as our Cooper pairs. Negatively charged particles move along the direction of growing electric potential, from A to B, because their total energy $E = -2e \times V$ is therefore decreased. However, if we understand the current intensity as the time derivative of the charge $I = \mathrm{d}Q/\mathrm{d}t$, a flow of negative particles from A to B is equivalent to a flow of positively charged particles from B to A.

Using these conventions, we will now analyze two linear elements, the capacitor and the inductor, and then move on to the nonlinear Josephson junction. For all three circuits, we will establish the voltage-current relations that will be used in Section 4.3 to build the circuit's dynamical equations.

4.2.1 Capacitor

We model a generic capacitor using two parallel plates, as shown in Figure 4.2. We charge the capacitor by establishing a (positive) current that drains charges from plate A and induces a positive charge in the opposite plate. As we know from our physics textbooks, the charge imbalance creates a uniform electric field E directed from the positive to the negative plate. This electric field is the gradient of a potential $E = -\partial V/\partial x$ that decreases from the positively charged plate B down to the plate A. Since E is proportional to the charge stored in the plates, we have a proportionality relation:

$$V = V_B - V_A = \frac{Q}{C} = \frac{1}{C}\int_0^t I(\tau)\mathrm{d}\tau, \tag{4.7}$$

where the constant $C \propto$ (area$\times$plate distance) is the total *capacitance* of the circuit. We frequently use this relation in its differential form:

$$I = C\frac{\mathrm{d}V}{\mathrm{d}t} = C\ddot{\phi}. \tag{4.8}$$

Charging a capacitor requires some work. Moving charges from potential V_A to V_B without friction demands an energy

$$E_{\text{cap}} = \frac{1}{2}QV = \frac{Q^2}{2C} = \frac{1}{2}CV^2 = \frac{1}{2}C\dot{\phi}^2. \tag{4.9}$$

This is the electrostatic energy stored in the capacitive element. It will appear in this form in all sorts of circuits, from microwave resonators to Josephson junctions and superconducting qubits.

If the capacitor is made of a superconducting material, the transfer of n extra Cooper pairs between the plates will change the energy by an amount

$$E_{\text{cap}} = 4E_C n^2, \quad \text{with } E_C = \frac{e^2}{2C}. \tag{4.10}$$

Typical superconducting circuit elements will have small capacitances, ranging from the picofarad to the femtofarad – 1 pF = 10^{-12} F and 1 fF = 10^{-15} F in SI units – with a *charging energy*[3] E_C taking values between 10 μeV and 10 meV – or frequencies from 40 MHz to 40 GHz. As with many other properties of a circuit, the values the capacitances in a circuit will be typically fixed, and very accurately predicted by the geometry of the metallic elements and dielectrics in the circuit.

4.2.2 Inductor

The inductance is the inertia that a conductor exhibits when we try to change the current that traverses it. As sketched in Figure 4.2, a change in the current I passing through a cable induces a change in the magnetic field $\mathbf{B}$ around it. By Maxwell's equations, an electric field will appear that opposes the change in magnetic field $\nabla \times \mathbf{E} = -\partial \mathbf{B}/\partial t$. This electric field is equivalent to a potential drop parallel and proportional to the time variation of the electric current:

$$V = L\frac{\mathrm{d}I}{\mathrm{d}t}, \quad \text{or} \quad \phi = LI. \tag{4.11}$$

The proportionality constant L is called the *inductance*, and (4.11) is the constitutive relation for this lumped-element circuit.

Similar to the capacitance, the inductance relates the magnetic energy stored in a circuit to the current intensity that passes through it. In an linear inductor L, we need to spend an energy

$$E_{\text{ind}} = \frac{1}{2}LI^2 = \frac{1}{2L}\phi^2 \tag{4.12}$$

to raise the current intensity – or the electric flux – from zero to I.

[3] Remember the factor 4 in the definition of E_C, which originates from the fact that the charging energy is a concept that is also found in quantum dots and other electrical objects where the charge unit is an electron, not a Cooper pair.

The inductance of a circuit element depends on its geometry and the material with which it is constructed. In the case of a superconducting material, we may also need to consider the *kinetic inductance*. This accounts for how the current flow is affected by microscopic details, such as the penetration depth of the material and the actual thickness of the superconductor film (Orlando, 1991). This source of inductance is particularly interesting when we wish to enhance the magnetic interaction between circuit elements (see Section 7.3.4).

4.2.3 Josephson Junctions or "Nonlinear Inductors"

So far we have discussed circuit elements that appear in classical circuits. We now introduce the Josephson junction, a circuit that is unique to the superconducting world. If you haven't done so yet, I recommend that you read Chapter 3 and in particular Section 3.7. There we explain that a junction consists of a thin insulator sandwiched in between two superconducting leads. The quantum tunneling of Cooper pairs along the junction establishes a nonlinear relation between the current I flowing through the junction and the flux difference ϕ between the leads:

$$I = I_c \sin\left(\frac{2\pi}{\Phi_0}\phi\right) = I_c \sin\left(\frac{\phi}{\varphi_0}\right). \tag{4.13}$$

The constant I_c is the critical current of the junction, and we have introduced the flux-phase relation $\varphi_0 = \Phi_0/2\pi$.

For small fluxes, a Josephson junction behaves as a linear inductor. Expanding (4.13) to first order in the flux difference $\phi \ll \Phi_0$, we obtain $\phi \simeq L_J I$ with the Josephson inductance

$$\frac{1}{L_J} = \frac{I_c}{\varphi_0}. \tag{4.14}$$

In general, however, the Josephson junction is a nonlinear device that makes quantum circuits more interesting, departing from traditional circuit theory, and enabling new devices such as superconducting qubits and SQUIDs.

The nonlinear inductance adds new contributions to both the inductive and the capacitive energies:

$$E_{\mathrm{JJ}} = \frac{q^2}{2C_J} - E_J \cos\left(\frac{2\pi}{\Phi_0}\phi\right). \tag{4.15}$$

The Josephson junction energy $E_J = I_c/\varphi_0 \simeq L_J^{-1}$ will be the dominant term, but we will often see also small capacitive contribution, mediated by charge differences between the superconducting leads and their interaction through the insulating dielectric. To account for this, junctions are usually drawn in parallel with a fake capacitor C_J that captures this physics (see Section 4.6 and the lumped-element circuits therein). Keep in mind, however, that this is just a notational convenience: The capacitor and the nonlinear inductor are just one entity, and the loop formed by these two lumped elements does not exist.

The values of the Josephson critical current and of the Josephson energy depend very much on the actual application. For instance, the work by van der Wal et al. (2000) uses three junctions with critical currents in the range $I_c \sim 400-600$ nA. This implies Josephson energies $E_J \sim 13.2-19.8$ nJ or $E_J/\hbar \sim 1\,250-1\,870 \times 2\pi$ GHz. However, you will also find many qubits with Josephson energies that have lower values, in the 100s of GHz.

4.2.4 Other Elements

General circuits also include voltage and current sources. We can treat these elements as boundary conditions that fix either the voltage $V(t)$ or the current $I(t)$, leaving the other variable free to satisfy any other dynamical constraints in the circuit. Figure 4.2 sketches a battery, a constant voltage source. The potential energy is larger in the positive electrode A, and lower in the negative one, and the difference between them is approximately constant, so that $V_A - V_B = V_{\text{bat}}$ is independent of time. An ideal battery sets no limit how much current it can supply to the circuit: The electric current is determined by the dynamical equations of the whole circuit. We will also find dc- and ac-current sources that provide either a constant $I = I_0$ or oscillating current $I = I_0 \cos(\omega t + \phi)$, leaving the potential V free to accommodate all other circuit restrictions.

General circuits also contain resistive elements. Establishing a current I through such element demands a potential difference $V = RI$ that compensates for the loss of energy across the resistor R. There may appear some resistance when we combine superconducting circuits with metallic parts, or as part of the Josephson junction equations. In general, we try to minimize these resistances until they become negligible, because of two problems. First, they remove energy in an incoherent way. This prevents them from fitting into the unitary evolution formalism from Section 4.3. And second, such resistive elements may have other negative effects, such as transforming the absorbed energy into heat that affects the whole sample.

4.3 Quantization Procedure

Circuit quantization is a mathematical process aimed at building a quantum mechanical model for a superconducting quantum circuit formed by linear elements – capacitors, inductors, Josephson junctions, and voltage and current sources. As discussed before, the theory applies at low temperatures, and also for excitation energies well below the superconducting gap. The quantization is a multistep process that constructs classical equations for the electrical circuit, then derives a

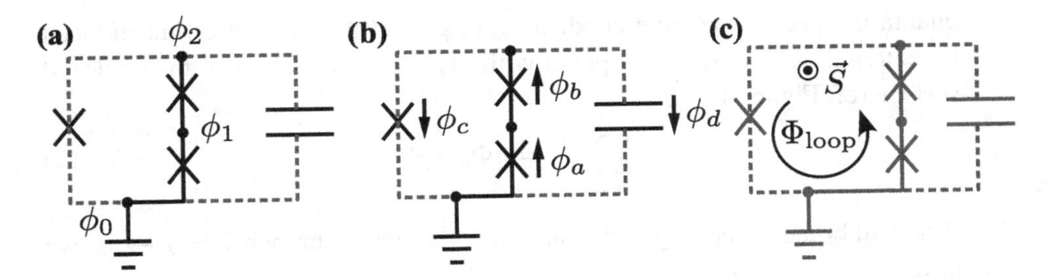

Figure 4.3 Quantization steps. (a) Open tree and node fluxes; (b) identify all oriented branch fluxes; (c) define loop fluxes and use flux quantization to find independent branches. The direction of the loop defines the orientation of the enclosed surface $\vec{S}$ with which we compute the flux (3.26).

classical Hamiltonian associated to those equations, and concludes with a quantum version of the model. The process is similar to the LC resonator example in Section 4.1.2 and to the canonical quantization of moving particles from Section 2.1. I will present you with a version of the theory that is very similar to the quantization rules described by Devoret (1995), discussing a few subtleties that appear in the application and interpretation of the rules.

Assuming that we start from a lumped-element circuit scheme, such as the one in Figure 4.3, the quantization steps are as follows:

(1) **Labels:** Identify all distinct nodes in the circuit and associate a flux variable to each of them ϕ_j (cf. Figure 4.3a). A node is an intersection where two or more circuit elements connect. All points in the same line are equivalent, and a line can only have one distinct node.

(2) **Circuit tree:** Select a node – e.g., a ground plane – and build an open and directed tree of branches T that runs through all nodes, without any loops. Create a different set of closing branches that complete the loops and the circuit C. In Figure 4.3b, $T = \{a, b\}$, and $C = \{c, d\}$.

(3) **Branch fluxes:** Associate a flux difference

$$\phi_x = \phi_j - \phi_i, \quad x = i \to j \in T, \tag{4.16}$$

to each branch x that includes a circuit element (cf. Figure 4.3b). This variable is the difference between the fluxes at the end (j) and beginning (i) of the branch. For convenience, we can consider redundant orientations, using the convention $\phi_{-x} = \phi_{j\to i} = -\phi_{i\to j} = -\phi_x$.

(4) **Flux quantization:** For all branch elements that close loops $x' \in C$, determine the flux differences $\phi_{x'}$ using fluxoid quantization to set their value (see Section 3.6). Given a series of directed branches $\{x\}$ that form a properly oriented loop, the sum of flux differences is equal to the number n of flux

quanta trapped in the superconducting ring[4] and the magentic flux induced by external fields Φ_{loop}, computed with the right orientation of the enclosed surface (cf. Figure 4.3c):

$$\sum_{x\in\text{loop}} \phi_x = n\Phi_0 + \Phi_{\text{loop}}. \tag{4.17}$$

When all branches belong to the main tree, except for branch $x' = i \to j$, we have

$$\phi_{x'} = n\Phi_0 + \Phi_{\text{loop}} + (\phi_j - \phi_i), \quad x' = i \to j \in C, \tag{4.18}$$

to be compared with (4.16). Note that capacitors actually implement a physical discontinuity in the conductor. When capacitive elements close a loop, $n = 0$, and there will be no trapped fluxes.

(5) **Branch currents:** Compute the currents traversing each circuit element. For a circuit on a branch $b = i \to j$, the current will flow from j to i with a value I:

Capacitor	$I_{j\to i} = C_b \dot{V}_b = C_b \ddot{\phi}_b$	See Section 4.2.1
Inductor	$\dot{I}_{j\to i} = V_b / L_b \Rightarrow I = \phi_b / L_b$	See Section 4.2.2
Junction	$I_{j\to i} = I_{J,b} \sin(\phi_b / \varphi_0)$	See Section 4.2.3

The branch flux ϕ_b is either (4.16) or (4.17), depending on whether $b \in T$ or $b \in C$. The capacitance C_b, inductance L_b, or critical current $I_{J,b}$ are determined by the circuit element and do not depend on the branch orientation.

(6) **Current conservation:** Write down one equation per node, establishing that the sum of currents coming in and out of the node add up to zero:

$$\sum_j I_{j\to i} = 0. \tag{4.19}$$

This will produce one ordinary differential equation per node i in the tree, with a general form

$$\sum_{b\in\text{capacitor}} s_i^b C_b \ddot{\phi}_b + \sum_{b\in\text{inductor}} s_i^b \frac{1}{L_b} \phi_b + \sum_{b\in\text{junction}} s_i^b I_b \sin\left(\frac{\phi_b}{\varphi_0}\right) = 0, \tag{4.20}$$

where $s_b = \partial\phi_b / \partial\phi_i = \pm 1$ is a sign determining whether the branch b leaves or enters the ith node.

(7) **Holonomic constraints:** Circuits are typically connected to ground planes, batteries, or microwave sources. These fix either the voltage of a node $\dot{\phi}_k = V(t)$ or the current along a branch. Use those constraints to eliminate variables and equations that are no longer needed.

[4] This value should be ideally zero, if the system is properly annealed. Note that discrete changes in the trapped fluxes along an experiment may cause decoherence in the quantum state of the circuit.

(8) **Lagrangian:** Rewrite the differential equations (4.20) in the form

$$\frac{\mathrm{d}}{\mathrm{d}t}\frac{\partial \mathcal{L}}{\partial \dot{\phi}_i} = \frac{\partial \mathcal{L}}{\partial \phi_i}, \tag{4.21}$$

and identify the Lagrangian $\mathcal{L}$ from which they are derived. If we follow the preceding conventions, the Lagrangian will be a function of the node fluxes $\boldsymbol{\phi} = (\phi_0, \phi_1, \ldots)^T$, containing the capacitive and inductive energies

$$\mathcal{L}(\boldsymbol{\phi}, \dot{\boldsymbol{\phi}}_i) = E_{\text{cap}}(\boldsymbol{\phi}, \dot{\boldsymbol{\phi}}_i) - E_{\text{ind}}(\boldsymbol{\phi}, \dot{\boldsymbol{\phi}}), \text{ with} \tag{4.22}$$

$$E_{\text{cap}}(\boldsymbol{\phi}, \dot{\boldsymbol{\phi}}_i) = \sum_{b \in \text{capacitor}} \frac{1}{2} C_b \dot{\phi}_b^2, \text{ and}$$

$$E_{\text{ind}}(\boldsymbol{\phi}, \dot{\boldsymbol{\phi}}_i) = \sum_{b \in \text{inductor}} \frac{-\phi_b^2}{2L_b} + \sum_{b \in \text{junction}} E_{J,b} \cos(\phi_b/\varphi_0). \tag{4.23}$$

We keep the branch fluxes ϕ_b because they make the expression more compact, but these values should ultimately be replaced with the independent node variables ϕ_i and $\dot{\phi}_i$. Note also that, in order to have a proper identification of energies, we cannot rescale (4.20) or introduce new variables before deriving the Lagrangian.

(9) **Canonical variables:** Define the node charges as the canonically conjugate momenta of the node fluxes:

$$q_i = \frac{\partial \mathcal{L}}{\partial \dot{\phi}_i}. \tag{4.24}$$

Use these variables and a Legendre transform to construct a Hamiltonian

$$H(\mathbf{q}, \boldsymbol{\phi}) = \sum_i q_i \dot{\phi}_i - \mathcal{L}(\boldsymbol{\phi}, \dot{\boldsymbol{\phi}}), \tag{4.25}$$

where we use (4.24) to write flux derivatives as functions of charges and fluxes, $\dot{\boldsymbol{\phi}}(\mathbf{q}, \boldsymbol{\phi})$. With this Hamiltonian, the evolution equations read

$$\frac{\mathrm{d}\boldsymbol{\phi}}{\mathrm{d}t} = \frac{\partial H}{\partial \mathbf{q}}, \qquad \frac{\mathrm{d}\mathbf{q}}{\mathrm{d}t} = -\frac{\partial H}{\partial \boldsymbol{\phi}}. \tag{4.26}$$

(10) **Quantization:** As in Section 2.1, replace the canonical variables (ϕ_i, q_i) with operators $(\hat{\phi}_i, \hat{q}_i)$ that satisfy commutation relations

$$[\hat{\phi}_i, \hat{q}_j] = i\hbar\delta_{ij}. \tag{4.27}$$

Define a Hamiltonian operator $\hat{H}$ using (4.25) with the flux and charge operators as variables. Due to the separation between capacitive and inductive energies, this operator will be Weyl-ordered and Hermitian. As explained in Section 2.1.3, this operator can be used to derive the *Heisenberg equations* for charge and flux, or to study the *Schrödinger equation* for the circuit wavefunction (4.6).

In this whole treatment, we have worked with fluxes ϕ_i on the nodes of the tree because they are the canonical variables and because one can derive the general form of the Lagrangian (4.22) from the current equations on these nodes (4.20). However, this is not a unique way to work with the circuit: Any change of coordinates that satisfies the constraints of our system will lead to a valid quantization procedure. We can work with node charges, branch charges,[5] or branch fluxes as variables instead. We will discuss this last case because it illustrates the structure of the resulting Hamiltonians and because it is convenient for some Josephson junction–based circuits.

Nodes and tree branches are largely interchangeable. On every connected tree there will be $N+1$ nodes and N branches. When the tree includes voltage sources or ground planes, the number of independent variables is identical in both representations. Each node flux can then be written as a reference potential plus various branch fluxes from on the tree. The change of coordinates from branch variables $\boldsymbol{\phi}_b$ to node fluxes $\boldsymbol{\phi} = A\boldsymbol{\phi}_b$ is performed by a matrix of ± 1's and 0's. We use this matrix to express the Lagrangian (4.22) as a function of the branch fluxes $\mathcal{L}(\boldsymbol{\phi}_b, \dot{\boldsymbol{\phi}}_b)$. In doing so, we must remember that closing branches are sums of branch variables on the tree, due to flux quantization $\phi_c = n\Phi_0 + \Phi - \sum_b \phi_b$. We may also need to reduce the number of variables in $\boldsymbol{\phi}_b$ to account for voltage or current constraints. The inductive terms remain similar in structure and do not need to be discussed. The capacitive terms adopt a quadratic form:

$$E_{\text{capacitive}}(\dot{\boldsymbol{\phi}}_b) = \frac{1}{2}\dot{\boldsymbol{\phi}}_b^T C \dot{\boldsymbol{\phi}}_b + C_g \mathbf{V}^T \dot{\boldsymbol{\phi}}. \tag{4.28}$$

This includes a capacitance matrix $C_{bb'}$ and a voltage source term $\mathbf{V}$ connected to the circuit through the coupling capacitances C_g. The capacitive term fixes the definition of branch charges $\mathbf{Q_b} = \partial\mathcal{L}/\partial\dot{\boldsymbol{\phi}}_b = C\dot{\boldsymbol{\phi}} - C_g\mathbf{V}$, and gives the Hamiltonian

$$H(\mathbf{Q_b}, \boldsymbol{\phi}_b) = \frac{1}{2}(\mathbf{Q_b} - C_g\mathbf{V})^T C^{-1}(\mathbf{Q_b} - C_g\mathbf{V}) + \sum_{b\in\text{inductor}} \frac{\phi_b^2}{2L_b} - \sum_{b\in\text{junction}} E_{J,b}\cos(\phi_b/\varphi_0). \tag{4.29}$$

This Hamiltonian is mathematically identical to the one derived for $\boldsymbol{\phi}$. We can now adopt canonical commutation relations for the new flux and charge operators $[\hat{\phi}_b, \hat{Q}_{b'}] = i\hbar\delta_{bb'}$, and work with them in the same way as with the node operators.

[5] These are, for instance, the variables used in the works by Yurke and Denker (1984) or Blais et al. (2004) to derive equations for transmission lines and resonators.

It is important to remark how the capacitance matrix C appears in the final Hamiltonian. The original equivalent circuit was based on a series of nearest-neighbor interactions, given by the capacitive energy across different circuit elements, $\frac{1}{2}C(\dot{\phi}_i - \dot{\phi}_j)^2$. This typically means that C and C_g are both sparse matrices and vectors reproducing a simple, typically planar topology. However, once we derive the canonical variables, we find that the inversion of the matrix C^{-1} establishes long-range, dipolar-type interactions between arbitrary pairs of charges. On the one hand, this is one of the sources of *cross-talk* that has to be taken care of when building complex quantum circuits – e.g., in large-scale quantum computers with hundreds of qubits and resonators, we would like to have no residual interaction between faraway elements. On the other hand, such long-range interactions are very similar to what we find in many-body systems, and they can be harnessed for the purpose of simulating complex condensed matter or theoretical physics models in the lab!

4.4 LC Resonator

Let us apply the quantization formalism to the LC circuit shown in Figure 4.4a, which consists of a superconducting resonator coupled to a constant voltage source. The four lumped elements in this circuit are the resonator's (i) capacitor C and (ii) inductor L, (iii) the voltage source, and (iv) the *gate* capacitor that connects it to the resonator, C_g. Using this information, please try to develop the quantum Hamiltonian yourself, checking the result with the following derivation – one of many equivalent ones.

In Figure 4.4b, we have constructed one possible circuit tree. The illustration marks the three different nodes $\{\phi_0, \phi_1, \phi_2\}$, three tree branches $\{\phi_a, \phi_b, \phi_c\}$, and a

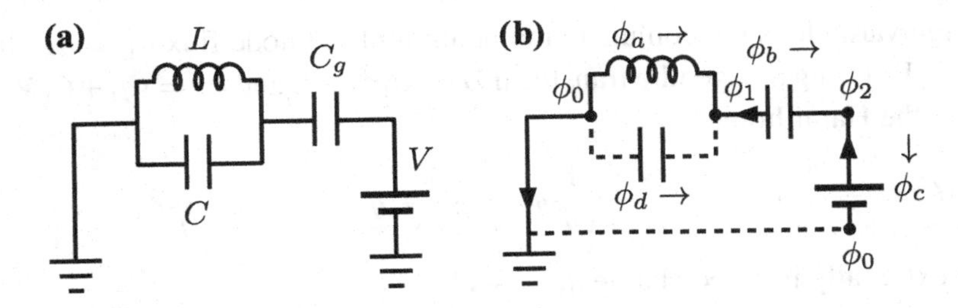

Figure 4.4 (a) Equivalent circuit for an LC resonator consisting on an inductor in parallel with a capacitor, subject to an external potential $V(t)$. (b) Nodes, node fluxes, and branch fluxes for the circuit description and quantization.

closing branch ϕ_d. From left to right, the nodes include a ground plane – common to the inductor, the capacitor, and the battery – the intersection between the inductor and the two capacitors ϕ_1, and the cathode ϕ_2. We have also chosen directions for all branches, defining the sign of potential differences:

$$\phi_a = \phi_1 - \phi_0, \tag{4.30}$$

$$\phi_b = \phi_2 - \phi_1, \tag{4.31}$$

$$\phi_c = \phi_2 - \phi_0. \tag{4.32}$$

The ground plane is a constant potential reference, which allows us to eliminate one variable $\phi_0 = 0$. The battery fixes the branch fluxes $\dot{\phi}_c = \dot{\phi}_2 - \dot{\phi}_1 = V$, which implies $\dot{\phi}_2 = V$ and $\dot{\phi}_b = V - \dot{\phi}_1$. Since the LC cannot support a supercurrent, the voltage difference along the inductor L (solid line) and along the capacitor C (dashed line) are identical, $\phi_a = \phi_d = \phi_1$. This leaves only one independent variable, ϕ_1.

With the conventions of voltage and current from Section 4.2, the flux difference ϕ_a is associated to the left-moving currents $\dot{\phi}_a/L$ and $C\ddot{\phi}_a$ along the capacitor and inductor. These currents equal the current coming from the battery through C_g,

$$C\ddot{\phi}_1 + \frac{1}{L}\phi_1 = C_g(\dot{V} - \ddot{\phi}_1). \tag{4.33}$$

Introducing $C_\Sigma = C + C_g$,

$$\frac{\mathrm{d}}{\mathrm{d}t}\left(C_\Sigma\dot{\phi}_1 - C_g V\right) = -\frac{1}{L}\phi_1. \tag{4.34}$$

This equation is conservative and derived via (4.21) from the Lagrangian

$$\mathcal{L} = \frac{1}{2}C_\Sigma\dot{\phi}_1^2 - C_g\dot{\phi}_1 V - \frac{1}{2L}\phi_1^2. \tag{4.35}$$

The charge variable is the conjugate momentum of the node flux $q_1 = \partial\mathcal{L}/\partial\dot{\phi}_1 = C_\Sigma\dot{\phi}_1 - C_g V$. Using a Legendre transform $H = q_1\dot{\phi}_1 - \mathcal{L}$ and $\dot{\phi}_1 = (q_1 + C_g V)/C_\Sigma$, we obtain the Hamilltonian

$$H = \frac{1}{2C_\Sigma}(q_1 + C_g V)^2 + \frac{1}{2L}\phi_1^2 = \frac{1}{2C_\Sigma}(q_1 - q_g)^2 + \frac{1}{2L}\phi_1^2, \tag{4.36}$$

with the externally induced charge $q_g = -C_g V$.

4.5 Transmission Line

We will now study a model for circuits that transport microwave energy without significant losses. Many physical objects can do this: coaxial cables that bring the

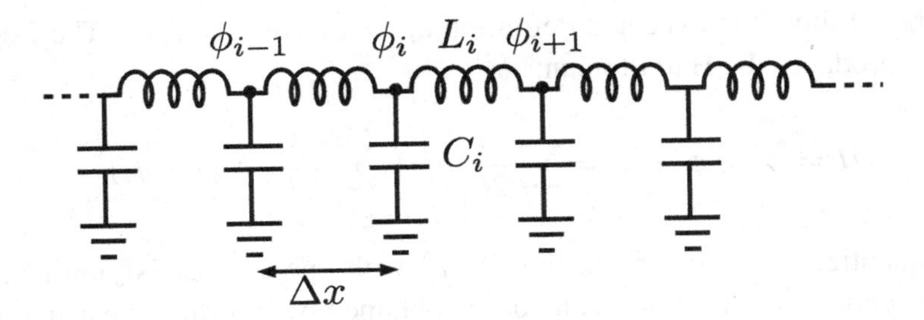

Figure 4.5 Equivalent circuit for a transmission line or a coplanar waveguide. The line is discretized as a set of coupled resonators, each of size Δx and with capacitances and inductances given by (4.37). The tree starts in the ground plane and spans as a star to all nodes through the vertical capacitors.

TV signal to your home, superconducting lines printed on a two-dimensional chip, or copper and aluminum tubes. Most of these objects are described with the equivalent circuit from Figure 4.5. This lumped-element model is an array of coupled LC resonators. The capacitors account for the electrostatic energy stored in the waveguide, while the inductors model the inertia against changes in the electric current. The circuit is a discretized version of the guide where each capacitor and inductor accounts for a small segment Δx that is much smaller than the guided wavelengths. The properties of these elements depend on the capacitance and inductance per unit length:

$$C_i = c_i \Delta x, \qquad L_i = l_i \Delta x. \tag{4.37}$$

We will assume that the quantities $c_i = c(x_i)$ and $l_i = l(x_i)$ are smooth functions of the position along the waveguide x_i.

The circuit tree is an open star with the ground plane at the center ϕ_0, and branches along the capacitors that connect the ground to the individual node fluxes ϕ_i. As in the LC, we can take $\phi_0 = 0$ as reference. The branch fluxes along the inductors are leftward oriented and $\phi_{i+1\to i} = \phi_{i+1} - \phi_i$. Current conservation on the nodes balances the flow of charges between inductors with the current flowing in and out of the ground plane:

$$C_i \ddot{\phi}_i = \frac{1}{L_i}(\phi_{i+1} - \phi_i) - \frac{1}{L_{i-1}}(\phi_i - \phi_{i-1}). \tag{4.38}$$

This equation can be derived from the Lagrangian:

$$\mathcal{L} = \sum_{i=1}^{N} \frac{1}{2} C_i \dot{\phi}_i^2 - \sum_{i=1}^{N-1} \frac{1}{2L_i}(\phi_{i+1} - \phi_i)^2. \tag{4.39}$$

The charges follow the expected definition $q_i = \partial\mathcal{L}/\partial\dot{\phi}_i = C_i\dot{\phi}_i$. The Legendre transform produces the Hamiltonian:

$$H = \sum_{i=1}^{N} q_i\dot{\phi}_i - \mathcal{L} = \sum_{i=1}^{N} \frac{1}{2C_i}q_i^2 + \sum_{i=1}^{N-1} \frac{1}{2L_i}(\phi_{i+1} - \phi_i)^2. \tag{4.40}$$

We can quantize the theory replacing (ϕ_i, q_i) with operators satisfying $[\hat{\phi}_i, \hat{q}_j] = i\hbar\delta_{ij}$. The continuum limit of this theory is obtained by inserting the actual size of the discretization:

$$H = \sum_{i=1}^{N} \frac{\Delta x}{2c(x_i)} \left(\frac{\hat{q}_i}{\Delta x}\right)^2 + \sum_{i=1}^{N-1} \frac{\Delta x}{2l(x_i)} \left(\frac{\hat{\phi}_{i+1} - \hat{\phi}_i}{\Delta x}\right)^2, \tag{4.41}$$

defining the charge density $\hat{\rho}(x_i) = \hat{q}_i/\Delta x$, and replacing sums with integrals to obtain a quadratic field theory:

$$H = \int \mathrm{d}x \left\{ \frac{1}{2c(x)}\hat{\rho}(x)^2 + \frac{1}{2l(x)}[\partial_x\hat{\phi}(x)]^2 \right\}. \tag{4.42}$$

These fields satisfy new commutation relations $[\hat{\phi}(x), \hat{\rho}(y)] = i\hbar\delta(x - y)$.

4.6 Charge and Transmon Qubits

We will now study a circuit that, depending on the parameter regime, describes either a charge qubit and a transmon qubit. The circuit, shown in Figure 4.6, is a simple nonlinear version of the LC circuit: It consists of a superconducting island coupled to a ground plane via a Josephson junction and externally controlled by a voltage source. We model this qubit using four lumped elements: (i) a junction with critical current I_c; (ii) the total capacitance C between the island and the ground

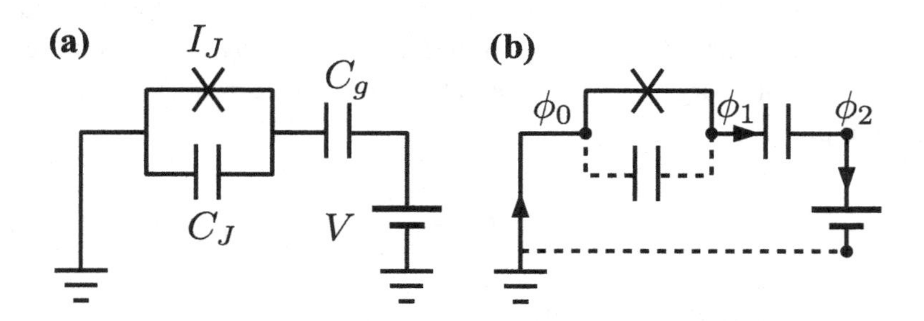

Figure 4.6 (a) Equivalent circuit for the charge qubit, with (b) the actual network on the right. The superconducting island is the region contained between the two capacitors, C_J and C_g. The nonlinear element E_J represents the channel by which pairs can tunnel into the ground plane.

plane – both direct and mediated by the junction; (iii) a gate capacitor C_g; and (iv) voltage source V, which accounts for any external control on the island.

The quantization procedure copies the one in Section 4.4, using the same independent variables and constraints. We simply replace the linear inductor with the junction's nonlinear current-voltage relation

$$\frac{\mathrm{d}}{\mathrm{d}t}\left[C_\Sigma\left(\dot{\phi}_1 - \frac{C_g}{C_\Sigma}\dot{V}\right)\right] = -I_c \sin(\phi_1/\varphi_0). \tag{4.43}$$

The Lagrangian's potential energy is now the nonlinear inductive energy of the junction

$$L = \frac{1}{2}C_\Sigma\dot{\phi}_1^2 - C_g\dot{\phi}_1 V + E_J\cos(\phi_1/\varphi_0), \tag{4.44}$$

with $E_J = I_c\varphi_0$. The charge operator has the same expression and the Hamiltonian becomes a nonlinear function

$$H = \frac{1}{2C_\Sigma}(q - q_g)^2 - E_J\cos(\phi_1/\varphi_0), \tag{4.45}$$

where the external potential sets the equilibrium charge $q_g = -C_g V$ in the ground state of the qubit.

4.7 SQUIDs

We will now discuss the circuits for two different superconducting quantum interference devices (SQUIDs). These are superconducting circuits that make use of flux quantization and Josephson tunneling. This feature makes the SQUID so sensitive to magnetic fields that they can be used as high-precision magnetometers in biology, medicine, or geology, among other fields. In a way, SQUIDs are the peak of superconducting circuit technology, a vast field of research[6] that predates and enables modern-day quantum circuits. In this book, we do not have space to go deep into this field, but will frequently rely on the SQUIDs to implement superconducting qubits, add tuneability to circuits, and build quantum amplifiers.

4.7.1 rf-SQUID

The rf-SQUID consists of a Josephson junction that is shunted by a linear inductor, as shown in Figure 4.7. The device is called a radio-frequency SQUID because it relies on the ac Josephson junction effect. It is a cheap magnetometer and also

[6] See, for instance, *The SQUID Handbook* by Clarke and Braginski (2004).

one of the earliest superconducting qubits, which still finds applications in modern quantum annealers (see Section 9.4.1).

Compared to previous devices, the rf-SQUID introduces new formal challenges that were not discussed in circuit quantization: (i) We have a loop without reference potential; (ii) the SQUID couples to external current sources via a mutual inductance M; and (iii) the external currents induce a net magnetic flux in the loop $\Phi \neq 0$, which affects the dynamical equations.

The equation for the node variable ϕ_0 balances the currents on the junction and the linear inductor. The current in the inductor is partially caused by the self-inductance of the branch ϕ_L/L, and partially by the mutual inductance. According to our sign conventions, the downward external current I induces an upward moving current on the SQUID $-MI$, so as to reduce the total magnetic interaction between both circuits. Hence,

$$C_J\ddot{\phi}_J + I_c \sin(\phi_J/\varphi_0) = \frac{1}{L}\phi_L - MI. \tag{4.46}$$

We replace the branch fluxes ϕ_J and ϕ_L with their definitions in terms of flux nodes, using the fluxoid quantization. Assuming no trapped vortices and with our choice of tree, we have $\phi_L = \phi_1 - \phi_0$ and $\phi_J = -\Phi + (\phi_0 - \phi_1) = -(\Phi + \phi_L)$. Note that the total magnetic flux Φ is created both by the external currents I and by the current supported by the loop itself ϕ_L/L. Far away from the regime of bistability, we can neglect this last contribution and approximate $\dot{\Phi} = 0$. The equation on the lower node

$$\frac{\mathrm{d}}{\mathrm{d}t}\left[C_J(\dot{\phi}_0 - \dot{\phi}_1)\right] = -I_c \sin\left[\frac{\phi_0 - \phi_1 - \Phi}{\varphi_0}\right] + \frac{\phi_1 - \phi_0}{L} - MI \tag{4.47}$$

is identical to that of the one on the upper node, up to a global sign. Both equations have the form 4.21 with the Lagrangian

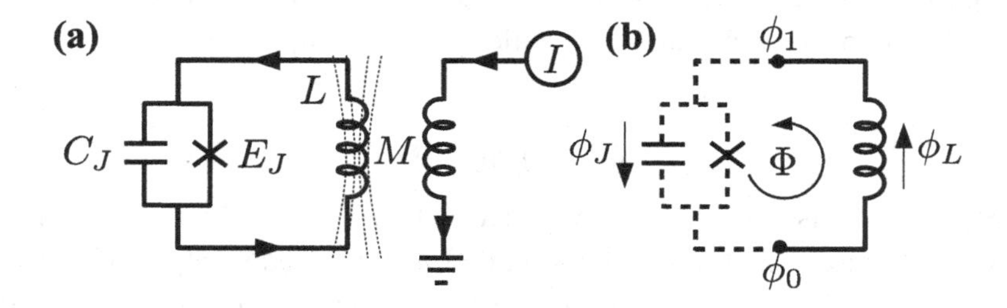

Figure 4.7 Equivalent circuit for the rf-SQUID qubit in full version (a) or restricted to the SQUID (b). A Josephson junction (C_J, E_J) is shorted by a large inductor L that may couple to an external magnetic field generated by a current source. The mutual inductance M is introduced in the text.

$$\mathcal{L} = \frac{1}{2}C_J(\dot{\phi}_1 - \dot{\phi}_0)^2 - \frac{1}{2L}(\phi_1 - \phi_0)^2 - M(\phi_1 - \phi_0)I - E_J \cos\left[\frac{\phi_0 - \phi_1 - \Phi}{\varphi_0}\right], \tag{4.48}$$

and the Josephson energy $E_J = I_c\varphi_0$.

These equations do not set any value for $\phi_0 + \phi_1$, which is a constant of motion $\frac{d}{dt}(\phi_0 + \phi_1) = 0$ decoupled from the flux difference ϕ_L. Following Section 4.3, we use ϕ_L as independent variable. Expressing the Lagrangian in terms of this variable, we obtain the conjugate momentum $Q_L = C_J\dot{\phi}_L$ and the Hamiltonian

$$H \simeq \frac{1}{2C_J}Q_L^2 + \frac{1}{2L}\phi_L^2 - MI\phi_L + E_J \cos\left[\frac{1}{\varphi_0}(\phi_L + \Phi)\right]. \tag{4.49}$$

If we look at the inductive terms, the role of MI is to shift the flux ϕ_L at which the minimum of energy is reached. However, this shift MIL will have to compete with the tendency of the nonlinear potential to trap the flux around $\phi_L \propto 2\pi\varphi_0 + \Phi$. In a regime of maximum frustration, where $\Phi/\varphi_0 \sim \pi$, we can obtain an energy landscape with two local minima that can be applied to create a qubit, as discussed in Section 6.4.

4.7.2 dc-SQUID

The "interference" in the word SQUID is revealed by circuits with multiple junctions, such as the dc-SQUID from Figure 4.8a. In this lumped-element model, there are two junctions connected in parallel, forming a loop. The incoming supercurrents split into two distinct paths, interfering at the exit port. The outcome of this interference and the total dc current that passes through depend on the magnetic flux Φ injected through the SQUID.

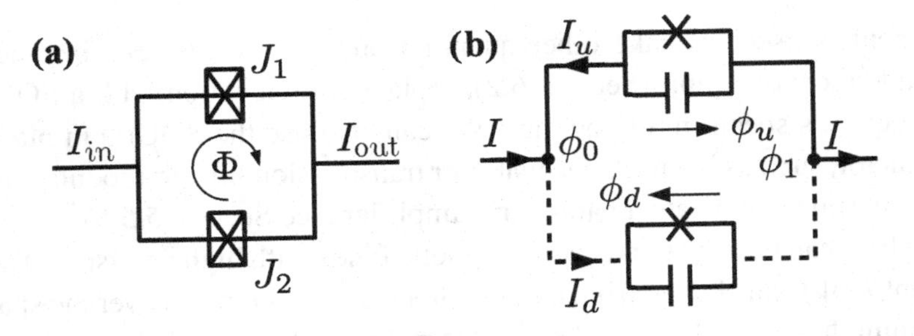

Figure 4.8 (a) A dc-SQUID is a device consisting on two junctions in parallel, threaded by some magnetic flux. (b) A more detailed version of the circuit must take into account the capacitive and inductive energy of the junctions if the pair is unbalanced.

In a conservative scenario, where the resistance of the junctions is negligible, the total incoming and outgoing currents must be the same and must split equally on both nodes: $I = I_d - I_u = I_{\text{in}} = I_{\text{out}}$. As in the rf-SQUID, there is only one independent equation that summarizes the conservation of current:

$$C_{J2}\ddot{\phi}_d + I_{J2}\sin(\phi_d/\varphi_0) = C_{J1}\ddot{\phi}_u + I_{J1}\sin(\phi_u/\varphi_0) + I.$$

Using flux quantization and our choice of main tree, the upper and lower branch fluxes are $\phi_u = \phi_1 - \phi_0$ and $\phi_d = \Phi + \phi_0 - \phi_1$, with the total external and trapped fluxes Φ. We assume identical junctions and use the identity $\sin(a) \pm \sin(b) = 2\sin\left(\frac{a\pm b}{2}\right)\cos\left(\frac{a\mp b}{2}\right)$ to write

$$C_J(\ddot{\phi}_d - \ddot{\phi}_u) = I - 2I_c \sin\left(\frac{\phi_d - \phi_u}{2\varphi_0}\right)\cos\left(\frac{\phi_d + \phi_u}{2\varphi_0}\right)x. \tag{4.50}$$

Let us make a second change of variables $\phi_\pm = \frac{1}{2}(\phi_d \pm \phi_u)$. The sum $\phi_+ = -\Phi/2$ is the circulating flux in the SQUID, while the difference ϕ_- defines the total current passing through the device. The effective Lagrangian and corresponding Hamiltonian are

$$\mathcal{L} = \frac{1}{2}(2C_J)\dot{\phi}_-^2 + E_J(\Phi)\cos\left(\frac{\phi_-}{\varphi_0}\right) - I\phi_-, \tag{4.51}$$

$$H = \frac{Q_-^2}{4C_J} + I\phi_- - E_J(\Phi)\cos\left(\frac{\phi_-}{\varphi_0}\right). \tag{4.52}$$

This Hamiltonian describes a new type of circuit element: the *tunable Josephson junction*. This is an effective junction with twice the original capacitance $2C_J$, where the Josephson energy can be adjusted between $-2I_c\varphi_0$ and $+2I_c\varphi_0$ using an external magnetic field Φ:

$$E_J(\Phi) = 2I_c\varphi_0\cos(\Phi/2\varphi_0) = E_J(0)\cos(\Phi/2\varphi_0). \tag{4.53}$$

This element is used to make other quantum circuits tunable. For instance, we can upgrade a charge qubit (Section 6.2), replacing its junction with a SQUID to adjust the qubit's spectrum of energies. We can also use the SQUID in place of a linear inductor, such as in an LC resonator or transmission line, to modulate the LC resonance in time and create a parametric amplifier (see Section 5.5.6).

dc-SQUIDs can be used to measure magnetic fields with high precision. The critical current $I_c(\Phi)$ can change from 0 to hundreds of nanoamperes over the span of a flux quantum $\Phi_0 \sim 2\times 10^{-15}\ \mathrm{T}\cdot\mathrm{m}^{-2}$. This means that if we can detect changes in $I_c(\Phi)$, we can use a SQUID to measure a magnetic field (or flux) with sensitivities close to attoteslas. In order to measure I_c, we can rely on the dynamics of a current-biased SQUID:

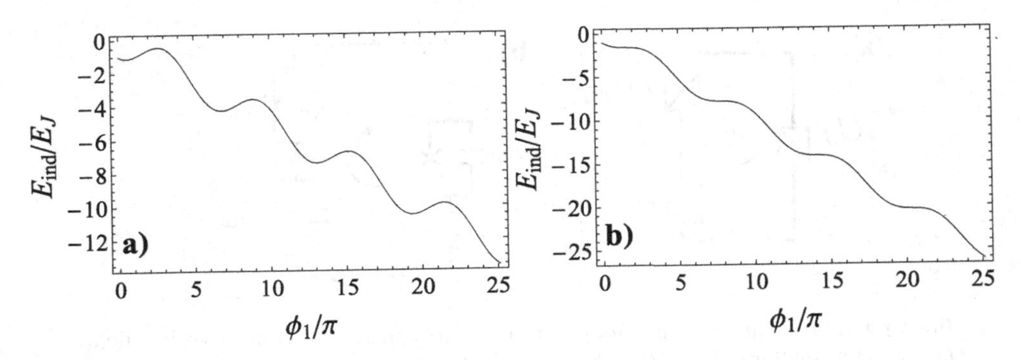

Figure 4.9 Washboard potential of the dc-SQUID (4.54) (a) below and (b) on the verge of switching to a voltage state.

$$E_{\text{ind}} = -E_J(\Phi)\cos\left(\frac{\phi_-}{\varphi_0}\right) - I\phi_- \tag{4.54}$$

For small biases, this function has the shape of a washboard potential (cf. Figure 4.9a). The infinitely many local minima are metastable configurations of the flux ϕ_- with $V \propto \dot{\phi}_- = 0$. When the bias current I exceeds the critical value $I_c(\Phi) = 2I_c\cos(\Phi/2\varphi_0)/\varphi_0$, all such minima disappear (cf. Figure 4.9a), and the flux "flows" down the slope, linearly accelerating in time.[7] This translates into a measurable potential difference between the SQUID's ports, which can be detected.

In the world of quantum circuits, SQUIDs are used to measure the current state of other circuits. For instance, a flux qubit has two states $|0\rangle$ and $|1\rangle$ that correspond to different orientations of the supercurrent. These states can inject different values of the flux Φ_0 and Φ_1 into a nearby SQUID. We can bias the SQUID to a point in between the critical currents for those two states, say $I_c(\Phi_0) < I < I_c(\Phi_1)$. When doing so, the SQUID will either remain the same or switch to a voltage state, depending on whether the qubit is in the $|1\rangle$ or $|0\rangle$ states. These are almost projective measurements, limited in precision by noise in the SQUID and the finite duration of the measurement.

4.8 Three-Junction Flux Qubit

The last circuit in this chapter is a superconducting loop with three Josephson junctions, shown in Figure 4.10a. The lumped-element circuit has three independent node variables $\phi_{0,1,2}$, which combine to create two independent branch variables

[7] This is a formally ill-defined problem, which stabilizes to an asymptotically stable solution once we consider the resistance of the junctions, or add a resistor in parallel to the circuit.

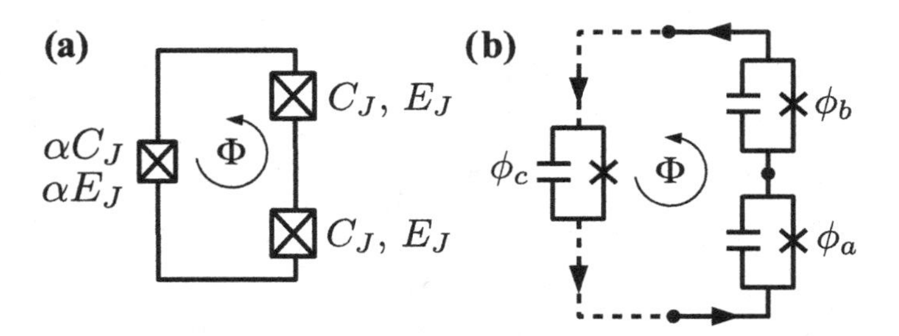

Figure 4.10 (a) The qubit is a loop with three Josephson junctions, two identical (1, 2) and a smaller one (3). (b) The equivalent circuit where we focus only on the branch fluxes.

$\phi_a = \phi_1 - \phi_0$ and $\phi_b = \phi_2 - \phi_1$, plus a closing branch $\phi_c = -\Phi - \phi_a - \phi_b = -\Phi + \phi_0 - \phi_1$. There are two current equations

$$I_{J3}\sin(\phi_c/\varphi_0) + C_{J3}\ddot{\phi}_c = I_{J1}\sin(\phi_a/\varphi_0) + C_{J1}\ddot{\phi}_a \\ = I_{J2}\sin(\phi_b/\varphi_0) + C_{J2}\ddot{\phi}_b, \tag{4.55}$$

which derive from the Lagrangian:

$$\mathcal{L} = \frac{1}{2}(C_{J1} + C_{J3})(\dot{\phi}_a^2 + \dot{\phi}_b^2) + C_{J3}\dot{\phi}_a\dot{\phi}_b \\ + E_{J1}\cos\left(\frac{\phi_a}{\varphi_0}\right) + E_{J2}\cos\left(\frac{\phi_b}{\varphi_0}\right) + E_{J3}\cos\left(\frac{\Phi + \phi_a + \phi_b}{\varphi_0}\right). \tag{4.56}$$

These qubits are designed with two identical junctions $C_{J1} = C_{J2} = C_j$ and $E_{J1} = E_{J2} = E_J$, and a junction with an area α times smaller, so that $E_{J3} = \alpha E_J$, $C_{J3} = \alpha C_J$. As in the rf-SQUID, the sum of node fluxes $\sum_i \phi_i$ is a constant of motion,[8] which leaves only two branch fluxes as independent variables. We now introduce a change of coordinates

$$\phi_\pm = \phi_a \pm \phi_b. \tag{4.57}$$

With the choice of the preceding parameters and the new variables, we find the charges $Q_+ = \frac{1}{2}C_J(1 + 2\alpha)\dot{\phi}_+$, and $Q_- = C_J\dot{\phi}_-$, and the Hamiltonian:

$$H = \frac{1}{2}\frac{Q_+^2}{C_J(\frac{1}{2} + \alpha)} + \frac{1}{2}\frac{Q_-^2}{C_J} \\ - E_J\left[\alpha\cos\left(\frac{\Phi + \phi_+}{\varphi_0}\right) + 2\cos\left(\frac{\phi_-}{2\varphi_0}\right)\cos\left(\frac{\phi_+}{2\varphi_0}\right)\right]. \tag{4.58}$$

This energy functional has a complicated landscape of equilibrium states, depending on the ratio of areas α and the external flux Φ, which we will discuss in Section 6.4.3.

[8] We can set this variable to zero by grounding one of the nodes.

4.9 Number-Phase Representation

In the last sections, we have derived quantum mechanical Hamiltonians for various circuits, specifying no more than a generic commutation relation between flux and charge operators $[\hat{\phi}_b, \hat{q}_{b'}] = i\hbar\delta_{bb'}$. We will now develop two *mathematical representations* of these operators. These are descriptions of the flux $\hat{\phi}$ and charge $\hat{q}$ observables, as operators acting on a Hilbert space of quantum states.

Unfortunately, even though flux and charge satisfy a canonical commutation relation, they are not completely equivalent to position and momentum. First of all, unlike momentum, the charge is a discrete operator that is proportional to the number of Cooper pairs $\hat{q} = -2e\hat{n}$. Second, the flux is related to the superconductor's phase $\hat{\varphi} = 2\pi\hat{\phi}/\Phi_0$. This results in a periodic representation for the phase operator, remarkably different from the usual position operator $\hat{x}$. Both differences imply that we cannot simply use the representation of position and momenta in terms of plane waves, and we must derive a specific *phase-number* representation.

There exist two representations of the phase-number or charge-flux commutation relations[9] that are operationally useful and can be applied to solve most problems:

Phase representation: We construct the Hilbert space using the eigenstates of the flux operator $\hat{\phi}$ or the equivalent phase operator $\hat{\varphi} = 2\pi\hat{\phi}/\Phi_0$. In this representation, the circuit wavefunction

$$|\Psi\rangle = \int_0^{2\pi} \mathrm{d}\varphi_1 \cdots \int_0^{2\pi} \mathrm{d}\varphi_N \Psi(\varphi_1, \ldots, \varphi_N) |\varphi_1, \ldots, \varphi_N\rangle$$

expands over states with a well-defined value of the node phases $|\varphi_i\rangle$. This continuous representation transforms the Schrödinger equation into a partial-differential equation for $\Psi(\varphi_1, \ldots, \varphi_N)$, which can be solved analytically in some important cases, such as the transmon qubit (Section 6.3).

Number representation: We construct the Hilbert space using the eigenstates of the charge operator $\hat{q}$, which are states with a given number of excess Cooper pairs $\hat{q} = -2e\hat{n}$. These states form a discrete basis $\{|n\rangle, n \in \mathbb{Z}\}$. Operators such as $\hat{n}, \hat{q}$, or the Hamiltonian H are represented as infinitely large matrices, which in most cases can be truncated to a reasonable cutoff $|n| \leq n_{\max}$, either for analytical approximation – see the charge qubit in Section 6.2 – or for some numerical diagonalization method – see the three-junction flux qubit in Section 6.4.3.

In each representation, we have one operator that is diagonal and simple, but we ignore the conjugate observable. In the number representation, the charge

[9] Our notation is consistent with Pegg and Barnett (1989) and with the original definition of the phase operator by Susskind and Glogower (1964).

operator is a sum of projectors onto well-defined numbers of Cooper pairs, $\hat{q} = \sum_n (-2en) |n\rangle\langle n|$ for $n \in \mathbb{Z}$. We know that all states in the Hilbert space can be expressed as linear superpositions of number states $|\Psi\rangle = \sum_n \Psi_n |n\rangle$, but we ignore how to write flux $\hat{\phi}$ and phase $\hat{\varphi}$ in this basis.

In order to find a faithful representation of the flux and phase, we must look at the Lie algebra they generate. We use the canonical commutation relations between $\hat{\phi}$ and $\hat{q}$ to find an alternative equation:

$$e^{i\hat{\phi}/\varphi_0}\hat{q} = (\hat{q} - 2e)e^{i\hat{\phi}/\varphi_0}. \tag{4.59}$$

Note how the phase operator $\hat{\varphi} = \hat{\phi}/\varphi_0$ is the generator of displacements in the space of charges. Using the number representation, we recover a dimensionless version of this identity:

$$\hat{n}e^{i\hat{\varphi}} = e^{i\hat{\varphi}}(\hat{n} - 1), \text{ or} \tag{4.60}$$

$$e^{i\hat{\varphi}} |n\rangle = |n - 1\rangle, \tag{4.61}$$

and conclude that the exponential of the phase is a ladder operator:

$$e^{i\hat{\varphi}} = \sum_n |n - 1\rangle\langle n|. \tag{4.62}$$

The Josephson energy combines two exponentials, allowing processes in which a Cooper pair tunnels in or out of the superconducting island:

$$\cos(\hat{\varphi}) = \frac{1}{2}\sum_n |n + 1\rangle\langle n| + \frac{1}{2}\sum_n |n\rangle\langle n + 1|. \tag{4.63}$$

We can perform a similar calculation in the basis of eigenstates of the phase operator. Now we have to study how the phase and number operators act on a generic wavefunction $\Psi(\varphi)$:

$$\hat{n}\left[e^{i\hat{\varphi}}\Psi(\varphi)\right] = e^{i\varphi}\left[(\hat{n} - 1)\Psi(\varphi)\right]. \tag{4.64}$$

Notice that we have safely replaced the $\hat{\varphi}$ operator with its representation, the phase degree of freedom φ. The previous equation is solved by the identification $\hat{n} = i\partial_\varphi$ and we can derive the phase eigenstates from the equation

$$\langle m|e^{i\hat{\varphi}}|\varphi_c\rangle = \langle m + 1|\varphi_c\rangle = e^{i\varphi_c}\langle m|\varphi_c\rangle, \tag{4.65}$$

which implies

$$|\varphi\rangle = \frac{1}{\sqrt{2\pi}}\sum_{m\in\mathbb{Z}} e^{i\varphi_c m} |m\rangle. \tag{4.66}$$

This discussion completes a dictionary for working with flux and charge, summarized in Table 4.2. Using these tools, we can take the Hamiltonian of a Josephson

Table 4.2. *Mathematical representations of the flux and charge operators.*

	Continuous	Discrete
Variable	$\varphi \in \mathbb{R}$	$n \in \mathbb{Z}$
Charge	$q = -i2e\partial_\varphi$	$q = -2e \sum_n n \vert n\rangle\langle n\vert = -2e\,\hat{n}$
Flux	$\hat{\phi} = \frac{\Phi_0}{2\pi} \times \varphi$	$e^{i2\pi\hat{\phi}/\Phi_0} = \sum_n \vert n-1\rangle\langle n\vert.$

junction (4.45) and write it in different representations. In the phase variables, the state is given by a periodic wavefunction $\Psi(\varphi,t)$ that evolves in time with a partial-derivative equation:

$$i\hbar\frac{\partial\Psi}{\partial t} = -4E_C\frac{\partial^2\Psi}{\partial\varphi^2} - E_J\cos(\varphi)\Psi. \tag{4.67}$$

In Chapter 6, we solve analytically a stationary version of this equation, but in practical applications with two or more variables, or with additional terms, it is not possible to do it. We could in principle address those complicated examples using a method such as finite differences. However, a simpler approach is to write the same Hamiltonian in the number representation. For instance, in the Josephson junction case (4.67) we have

$$H = \sum_n \left[4E_C n\,|n\rangle\langle n| - E_J(|n+1\rangle\langle n| + |n\rangle\langle n+1|)\right]. \tag{4.68}$$

In many real-world applications, where $E_J/E_C \leq 100$, the probability that a state with large number of Cooper pairs is occupied is very low. This means we can reduce the complexity of the Hamiltonian, truncating the matrices and vectors to a smaller subspace, of $|n| \leq n_{\max} \sim 10$ or less. In some cases, such as the charge qubit 6.2, this truncation is even more dramatic, and we end up in a two-dimensional subspace, such as $n \in \{0, 1\}$.

Exercises

4.1 Discuss the differences between the thermal state $\rho_\infty = \frac{1}{2}|0\rangle\langle 0| + \frac{1}{2}|1\rangle\langle 1|$, and the quantum superposition $|\psi\rangle = \frac{1}{\sqrt{2}}(|0\rangle + |1\rangle)$. What is the density matrix ρ_q associated to the latter state? How can we distinguish both states experimentally if we have access to measurements of the σ^z and σ^x operators?

4.2 Using the commutation relation $[\phi_n, q_m] = i\hbar\delta_{nm}$, prove (4.59). Hint: Write down the Taylor expansion of the exponential, and use the commutator on each term.

4.3 Reproduce the derivations of all the circuit models and effective Hamiltonians from Sections 4.4 to 4.8.

4.4 Using (4.1), show that the energy of the LC circuit $H = \frac{1}{2}CV^2 + \frac{1}{2}LI^2$ is conserved, that is, $\mathrm{d}H/\mathrm{d}t = 0$. What happens if you add a resistor in parallel to the circuit?

4.5 Draw a circuit of two capacitively coupled LC resonators. The circuit should be the result of replacing the battery in Figure 4.4 with another resonator. Find out the quantum model for this circuit. Find out the canonical variables $\tilde{Q}_n$ and $\tilde{\phi}_n$ that diagonalize the Hamiltonian, and bring it to a form

$$H_{\text{eff}} = \sum_n \frac{\tilde{Q}_n^2}{2C_n} + \frac{1}{2}L_n\tilde{\phi}_n^2. \tag{4.69}$$

4.6 Similar to Problem 4.5, derive the circuit and the Hamiltonian for two identical charge qubits connected via a capacitor. Show that, in the limit of weak coupling, the interaction between both qubits is modeled by the product of charges $\propto q_1q_2$ from both qubits.

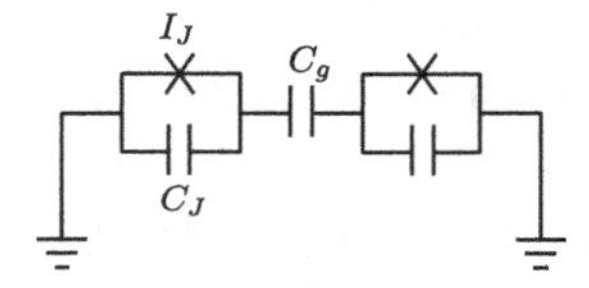

4.7 Compute analytically the value of the bias current at which the washboard potential from Figure 4.9 loses all its local minima.

4.8 Think how you would write the Hamiltonian for the 3-JJ flux qubit (4.58) as a matrix in the number-phase representation. Remember to use the decomposition of trigonometric functions into exponentials that have a simple matrix representation $\cos(\varphi) = \frac{1}{2}(e^{i\varphi} + e^{-i\varphi})$. Also realize that $e^{i\varphi_+}e^{i\varphi_-}$ will translate into a Kronecker or tensor product of two matrices $A \otimes B$, each representing the action of the exponential on a different Hilbert space – such as A for $e^{i\varphi_+}$ and B for $e^{i\varphi_-}$.

5

Microwave Photons

In this chapter, we begin a journey through the practical applications of superconducting quantum circuits. Our first stop is linear circuits, such as the LC resonators and transmission lines introduced in Sections 4.4 and 4.5. We discuss how the energy of these circuits is quantized in units that we call *photons* – more precisely, *microwave photons* – and how this reflects in the Hamiltonian and the representation of observables. Microwave photons can be stored in LC resonators and microwave cavities for a long time, or they can travel through transmission lines to distant circuits and detectors. Microwave photons may be prepared in very different quantum states, from well-defined number states, to entangled states and sophisticated quantum superpositions. We explain how to create and characterize these states using linear or "Gaussian" operations, which include microwave sources, inductive and capacitive interactions, parametric quantum amplifiers, and devices that meassure voltage and power.

5.1 LC Resonator

5.1.1 Energy Quantization and Photons

Figure 5.1a shows a very small LC circuit made of two superconducting elements: a very visible interdigitated capacitor with many fingers, shunted by a linear inductor in the form of a small superconducting loop. The LC circuit is so small that the lumped-element circuit approximation is well justified. To be more precise, we can tune global properties, such as the total charge, flux, voltage, or intensity, but we cannot control these properties' distribution across the circuit, because this "structure" or *mode* is fixed in a state of minimum energy. If we want to alter the mode structure, we need to introduce changes over distances $d \sim 100\ \mu\text{m}$. The energies associated to these tiny details is hundreds of GHz, well above the superconducting gap and far from our experiments' operation regime.

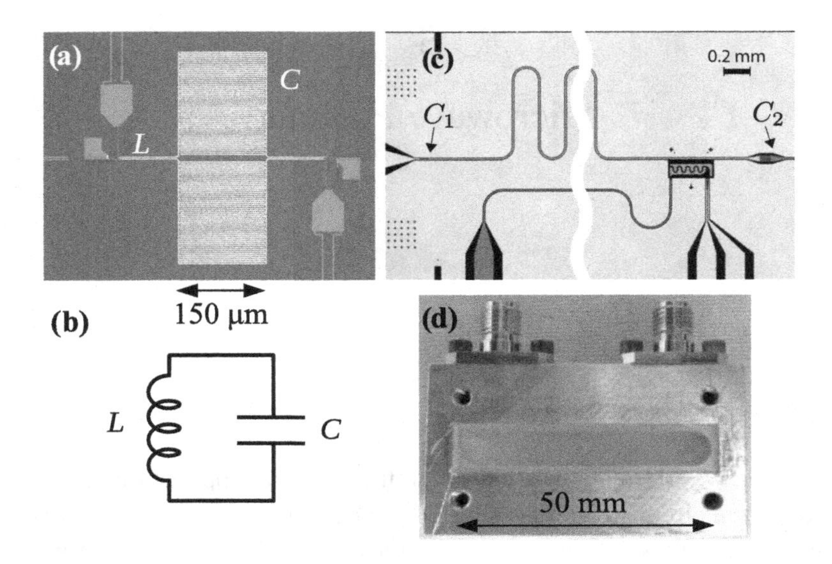

Figure 5.1 (a) Zero-mode LC resonator formed by an interdigitated fingers capacitor and a short line that acts as an inductor. Courtesy of Pol Forn-Díaz. (b) Equivalent circuit. (c) $\lambda/2$ microwave resonator formed by a waveguide intersected by two capacitors, C_1 and C_2. Reprint from Pechal et al. (2014). (d) Three-dimensional microwave cavity formed by carving two aluminum blocks that will be joined together, forming a closed hull. Reprinted figure with permission from Paik et al. (2011). Copyright (2011) by the American Physical Society.

The resonator in Figure 5.1a not only a very good example of the *lumped-element* approximation, it even resembles the graphical representation of the LC circuit, shown in Figure 5.1b, which we introduced in Section 4.4. As discussed then, charge, flux, voltage, and current all follow a similar oscillatory equation, $\ddot{\phi} = -\omega_{LC}^2\phi$, solved as $\phi(t) \sim \cos(\omega t + \varphi)$ with a single frequency that depends on the electrical properties of the circuit:

$$\omega_{LC} = \frac{1}{\sqrt{LC}}. \tag{5.1}$$

Because this circuit only supports a fixed spatial distribution of charge and flux, it only has a single mode or frequency, which is why it is also called a *zero mode resonator*.

The LC resonator is conceptually similar to the ideal oscillators that Planck used in his model of the black-body's electromagnetic radiation (see Section 2.1). Following Planck's and Einstein's ideas, we could propose that our superconducting resonator is a quantum mechanical object that only exchanges discrete units of energy with the environment – our *quanta* or *microwave photons*. The states of the resonator would then be labeled by the number of photons that it stores $|n\rangle$, and the total energy would be a sum of quanta $E_n \sim \hbar\omega \times n$, $n \in \{0, 1, 2, \ldots\}$.

This qualitative reasoning is essentially correct. It can be justified using Dirac's quantization and the quantum circuit models from Section 4.3. The "photons" that we find in the superconductor have a mixed nature: They are excitations of the electromagnetic field that are accompanied by waves of charge and current on the surface of the metal. This hybrid nature would make it more appropriate to call them *microwave plasmons*, but we keep the name "photon" to distinguish these plasmonic excitations from others that have a more anharmonic spectrum, which we use to build qubits.

The microwave photons stored in a resonator do not interact with each other, at least not until we reach high powers that excite nonlinear contributions to the inductance and capacitance. Also, when placed on extended larger systems, such as a superconducting waveguides (cf. Figure 5.1c), the photonic nature of these plasmons evidences in a propagation without dispersion, at a fraction of the speed of light.

However, the microwave photons do have a matter-like component. This component allows us to control the properties of the photon – frequency, speed, wavelength, etc. – by a proper design of the circuit: capacitance, inductance, thickness of the material, etc. Moreover, the matter component in the wave enables the strong interaction between these microwave photons and other circuits, atoms, molecules, NV-centers, etc., opening regimes of extreme light–matter interactions that were previously unreachable in quantum optics.

Unfortunately, working with microwave photons has some drawbacks. In particular, the small energy contained in a microwave photon makes it very difficult to build a microwave photodetector or photon counter. The lack of such devices prevents us from directly verifying the quantization of the cavity's excitations $E_n = \hbar\omega n$, a fact that only becomes evident when we introduce other, more matter-like nonlinear circuit elements, such as qubits.

5.1.2 Hamiltonian Diagonalization

We already have a quantum mechanical model for the LC resonator (4.36):

$$H = \frac{1}{2C_\Sigma}(\hat{q} - q_g)^2 + \frac{1}{2L}\hat{\phi}^2. \tag{5.2}$$

This model is analogous to the problem of a quantum particle in a harmonic potential. In absence of charge bias $q_g = 0$, we identify

$$\begin{aligned} \tfrac{1}{2C}\hat{q}^2 + \tfrac{1}{2}\hat{\phi}^2 \quad &\Leftrightarrow \quad \tfrac{1}{2m}\hat{p}^2 + \tfrac{1}{2}m\omega^2\hat{x}^2, \\ [\hat{\phi}, \hat{q}] = i\hbar \quad &\Leftrightarrow \quad [\hat{x}, \hat{p}] = i\hbar. \end{aligned}$$

The capacitance is the oscillator's mass $C \sim m$, while the inductance acts as a restoring potential, $\frac{1}{2L} \sim \frac{1}{2}m\omega^2$. Together they produce the natural frequency of the resonator $\omega = (LC)^{-1/2}$.

This dictionary also maps the textbook methods[1] for bringing the harmonic oscillator into diagonal form $\hat{H} = \sum_n E_n |n\rangle\langle n|$, with a complete set of eigenstates $|n\rangle$. Specifically, we know that the oscillator eigenstates are Fock states $|n\rangle$ labeled with all nonnegative integers[2] $n \in \{0, 1, 2, \ldots\}$. The excitations in these states are created and annihilated by the bosonic Fock operators $\hat{a}$ and $\hat{a}^\dagger$:

$$\hat{a}\,|n\rangle = \sqrt{n}\,|n-1\rangle\,, \qquad \hat{a}^\dagger\,|n\rangle = \sqrt{n+1}\,|n+1\rangle\,,\ \forall n \in \{0, 1, 2, \ldots\}. \tag{5.3}$$

These operators satisfy bosonic commutation relations $[a, a^\dagger] = 1$ and together define the number operator $\hat{n} = \hat{a}^\dagger a$. Using this representation, we write the Hamiltonian and the flux and the charge operators as

$$\hat{H} = \hbar\omega\left(\hat{a}^\dagger\hat{a} + \tfrac{1}{2}\right) = \hbar\omega\left(\hat{n} + \frac{1}{2}\right), \text{ with } \begin{cases} \hat{\phi} = \sqrt{\frac{\hbar Z}{2}}(\hat{a} + \hat{a}^\dagger), \\ \hat{q} = \sqrt{\frac{\hbar}{2Z}}i(\hat{a}^\dagger - \hat{a}). \end{cases} \tag{5.4}$$

The spectrum of eigenenergies of $\hat{H}$ is discrete and numerable, $E_n = \hbar\omega(n + \frac{1}{2})$, and each energy level corresponds to a different number of photons n, as anticipated. The diagonalization gives expressions for the flux and charge operators in the Fock space, but also for voltage and intensity:[3]

$$\begin{aligned} \hat{V} &= \frac{\mathrm{d}}{\mathrm{d}t}\hat{\phi} = -\frac{i}{\hbar}[\hat{\phi}, \hat{H}] = \omega\sqrt{\frac{\hbar Z}{2}}i(\hat{a}^\dagger - \hat{a}) = \frac{\hat{q}}{C}, \\ \hat{I} &= \frac{\mathrm{d}}{\mathrm{d}t}\hat{q} = -\frac{i}{\hbar}[\hat{q}, \hat{H}] = -\omega\sqrt{\frac{\hbar}{2Z}}(\hat{a} + \hat{a}^\dagger) = -\frac{\hat{\phi}}{L}. \end{aligned} \tag{5.5}$$

In these formulas, the *impedance* $Z = \sqrt{L/C}$ plays the role of an oscillator wavepacket size. This parameter defines a unit of flux $\phi_{\text{vac}} = \sqrt{\hbar Z}$, and the "strength" of quantum fluctuations of observables in the vacuum state $|0\rangle$. Due to the canonical commutation relations $[\hat{\phi}, \hat{q}] = i\hbar$, those fluctuations are also constrained by the Heisenberg uncertainty principle (2.12):

$$\Delta\hat{V}\Delta\hat{I} \propto \Delta\phi\Delta q \geq \frac{1}{2}\hbar. \tag{5.6}$$

In other words, the more precisely we measure $\hat{\phi}$ (or $\hat{q}$), the more uncertain is our estimate of complementary observable $\hat{q}$ ($\hat{\phi}$).

[1] See, e.g., chapter 6 from Ballentine (1998).
[2] Even though we use the same symbol as in Section 4.9, here n represents the quanta of excitation, and is only indirectly related to the number of Cooper pairs.
[3] These relations, derived with the Heisenberg equation, hold also in the Schrödinger picture!

5.1.3 Phase Space Dynamics

We can solve the Schrödinger equation for a time-independent Hamiltonian $\hat{H}$ as $|\psi(t)\rangle = e^{-i\hat{H}t/\hbar}\,|\psi(0)\rangle$. In the case of an LC circuit, we write

$$|\psi(t)\rangle = \sum_n \exp(-iE_n t/\hbar)\,\langle n|\psi(0)\rangle \times |n\rangle\,. \tag{5.7}$$

This method no longer works when the Hamiltonian depends on time, such as when the resonator is connected to an oscillating potential $V(t)$ in (5.2). This problem is best analyzed in the Heisenberg picture (2.13).

The Hamiltonian for a resonator coupled to an oscillating potential $V(t)$ reads

$$\hat{H} = \hbar\omega a^\dagger a + i\hbar\Omega(t)(a^\dagger - a) + E_{\text{global}}(t). \tag{5.8}$$

The linear displacement $\hbar\Omega(t) = \sqrt{\hbar/2Z}C_g V(t)/C_\Sigma$ contains the external potential. The energy shift $E_{\text{global}}(t) \propto V(t)^2$ will be ignored, since it commutes with all operators. The Heisenberg equations for the canonical and Fock operators are linear, first-order differential equations:

$$\left.\begin{aligned} \tfrac{\mathrm{d}}{\mathrm{d}t}\hat{\phi} &= \tfrac{1}{C_\Sigma}[\hat{q} - C_g V(t)], \\ \tfrac{\mathrm{d}}{\mathrm{d}t}\hat{q} &= -\tfrac{1}{L}\hat{\phi}, \end{aligned}\right\} \Leftrightarrow \frac{\mathrm{d}\hat{a}}{\mathrm{d}t} = -i\omega a + \Omega(t). \tag{5.9}$$

The Fock operators experience a phase and a complex displacement $\alpha(t)$:

$$\hat{a}(t) = e^{-i\omega t}\hat{a}(0) + \int_0^t e^{-i\omega(t-\tau)}\Omega(\tau)\mathrm{d}\tau = e^{-i\omega t}\hat{a}(0) + \alpha(t), \tag{5.10}$$

and provide the solution for flux and charge:

$$\hat{\phi}(t) = \sqrt{\frac{\hbar Z}{2}}(e^{+i\omega t}a^\dagger + \text{H.c.}) + \phi_{\text{ext}}(t), \qquad \hat{q}(t) = \sqrt{\frac{\hbar}{2Z}}(ie^{+i\omega t}a^\dagger + \text{H.c}) + q_{\text{ext}}(t), \tag{5.11}$$

with $\alpha = [\phi_{\text{ext}}(t) + iZq_{\text{ext}}(t)]/\sqrt{\hbar Z}$.

In absence of external drive, the solutions in the phase space $(\hat{\phi}, \hat{q})$ are elliptical orbits with angular frequency ω:

$$\hat{\phi}(t) = \cos(\omega t)\hat{\phi}(0) + \sin(\omega t)Z\hat{q}(0), \tag{5.12}$$

$$\hat{q}(t) = \cos(\omega t)\hat{q}(0) + \sin(\omega t)\frac{1}{Z}\hat{\phi}(0). \tag{5.13}$$

As in the classical theory, the impedance is the ratio between the positive (or negative) frequency components of the flux and charge, or voltage and current intensity. This can be seen from a Fourier transform, or directly, taking the positive frequency terms from (5.11):

$$\frac{\langle\tilde{\phi}(+\omega)\rangle}{\langle\tilde{q}(+\omega)\rangle} = \frac{\sqrt{\hbar Z/2}\,\langle a^\dagger(0)\rangle}{\sqrt{\hbar/2Z}\,\langle a^\dagger(0)\rangle} = \frac{\tilde{V}(+\omega)}{\tilde{I}(+\omega)} = -iZ. \tag{5.14}$$

Here and throughout this book, we often write the expected value of a non-Hermitian operator $\langle a^\dagger \rangle$. Even though $a^\dagger$ is not an observable, it can be decomposed as the addition of two measurable quantities:

$$a^\dagger(t) = \sqrt{\frac{2}{\hbar Z}}\hat{\phi} - i\sqrt{\frac{2Z}{\hbar}}\hat{q}. \tag{5.15}$$

Therefore, the expectation value of $a^\dagger$, a, $a^\dagger a$, or even higher moments $a^2 a^\dagger$, etc., can be inferred from the composition of two separate measurements. This will be relevant in later sections, where we study the amplification of light and the simultaneous measurement of quadratures (Section 5.5.7).

5.1.4 Are There Real Photons?

Chapter 2 mentioned direct evidences of light quantization in the photelectric effect. In these experiments, light carries enough energy – 0.1 – 10 eV in the infrarred to UV range – that can excite a short-lived but measurable electric current. The discreteness of the detection events and the distribution of energies in the generated currents are evidences that light is absorbed by the material as individual quanta of energy, the photons.

Unfortunately, we cannot produce similar evidence using microwave photons. In a typical frequency range of 1–10 GHz, light particles carry a very small amount of energy, 4—40 μeV, insufficient to trigger a conventional photodetector.[4] We must therefore seek other ways to explore the quantization of light and of observables in superconducting resonators.

One approach is to stick to linear circuits and devices, extending them with amplifiers and voltage-current measurement devices (cf. Section 5.5.7). Such a setup allows us to measure the quadratures of the field, including the energy operator itself $\hat{a}^\dagger \hat{a}$. Unfortunately, these are noisy processes that do not yet discriminate individual photons and only provide good ensemble estimates of the field operators. This is enough for many tasks, including state tomography, entanglement witnesses, or squeezing, but not for a single-shot readout of photons.

The other approach is to combine resonators with nonlinear devices. As explained in Chapter 6, there are superconducting circuits with discrete, highly anharmonic spectra, called *qubits*. When paired with a resonator, these saturable devices can only emit or absorb a single quantum of energy at a given frequency. Using qubits, we can create photons one by one, engineering complex superpositions of one-, two-, and more-photon states predictably. We can also do accurate tomography

[4] Note the word *conventional*. A microwave photon can still trigger other elements, such as superconducting circuits Romero et al. (2009a,b), as a basis for microwave photodetection.

of the number of photons and reconstruct the wavefunction of the resonator or in any other bosonic mode. This kind of protocol, introduced later in Chapter 7, is also single-shot: Each experiment provides a measurement outcome that does not destroy the photon state and can be use for feedback and control of the photon's state.

5.2 Transmission Lines or Waveguides

The zero-mode resonators are a theorist's dream: They have a simple design, they can be described by a one pair of variables $(\hat{\phi}, \hat{q})$, and the two parameters L and C can be accurately engineered to produce the desired resonance ω_{LC}. However, zero-mode resonators are rarely used in quantum experiments. They are necessarily small, so that it becomes difficult to pack the antennas needed to control them, together with other circuits such as superconducting qubits. The limitation is not so much on the fabrication part, but on the operation part. Increasing the density of components leads to errors in the form of undesired crosstalks between circuits, bad isolation, etc.

Photons can also be injected and stored in other linear circuits that are generically called *waveguides*. Like the strings of a guitar, they accommodate many different *photonic modes*, in the form of stationary waves with distinct wavelengths and frequencies. In an extreme limit, the work by Sundaresan et al. (2015) resolved spectroscopically 75 photonic modes within the same waveguide. However, when so many modes participate in the dynamics, it is preferable to disregard the notion of standing waves and work with propagating photons, as discussed in Chapter 7.

There are various elements that transport photons without distortion or significant losses. The simplest one is the coaxial cable that brings the TV signal to our receivers. These cables have a conducting core surrounded by a grounded metallic shell. The electromagnetic waves move longitudinally in the free space among conductors, dragging with them charge and current excitations on both surfaces.

The equivalent of a coaxial cable in a 2D superconducting microchip would be either a stripline or a coplanar waveguide. The stripline consists of two lines of superconducting material sitting above and below the substrate, and one of them connected to ground, as seen in Figure 5.2b (bottom). The field moves in between both lines, confined in the substrate. The coplanar waveguide is an alternative and more popular design where two ground planes surround a thin superconducting line. Now the electromagnetic field runs also outside the substrate, making it a bit more robust under perturbations of the material.

All 2D superconducting waveguides are typically modeled with the same lumped-element circuit, discussed in Section 4.5. The circuit includes a capacitive term for the interaction between the core conductor and the ground planes, and an

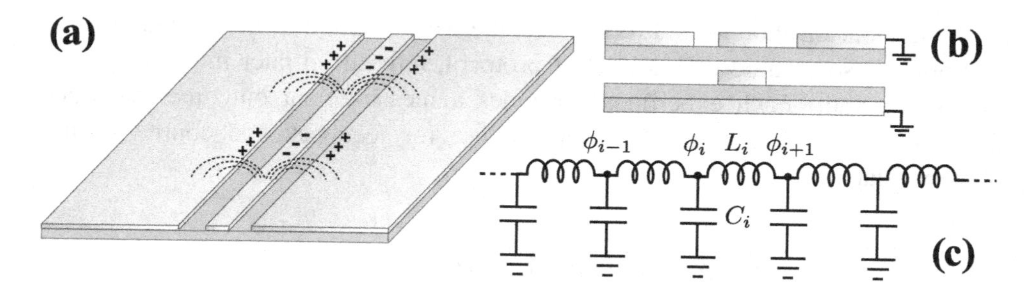

Figure 5.2 (a) A coplanar waveguide. The left and right superconducting planes are grounded to a fixed potential while the charge and electromagnetic waves run through the middle section. (b) Other planar waveguides, such as the stripline (top), differ from the coplanar waveguide (bottom) on the position of ground planes relative to the main conductor. (c) Equivalent circuit for the waveguide (see Section 4.5).

inductive energy describing the "inertia" of the conductors to change the current. Introducing the capacitance and inductance per unit length, $c(x)$ and $l(x)$, we derived two alternative models: a discrete theory (4.40) and a continuous field theory (4.42). We will now diagonalize and study these models in three different scenarios with different boundary conditions and line terminations.

5.2.1 Periodic Boundary Conditions

Let us take the discrete model for a transmission line (4.40):

$$H = \sum_{i=1}^{N} \frac{1}{2C}\hat{q}_i^2 + \sum_{i=1}^{N} \frac{1}{L}\left(\hat{\phi}_{i+1}^2 - \hat{\phi}_{i+1}\hat{\phi}_i\right). \tag{5.16}$$

The line has length d and is divided into N segments with the same capacitance and inductance, $C = c\Delta x$ and $L = l\Delta x$. We assume that the line closes onto itself, so that $\hat{\phi}_{N+1} = \hat{\phi}_1$.

The second term in this model is a quadratic form, $\boldsymbol{\phi}^T B\boldsymbol{\phi}/2L$, with a nonnegative, symmetric matrix B (cf. Section A.1.1). Following the mathematical tools from Appendix A.2, we diagonalize the matrix B and obtain N normal modes to write the Hamiltonian as a sum of N independent oscillators:

$$\hat{H} = \sum_k \hbar\omega_k \left(\hat{b}_k^\dagger \hat{b}_k + \frac{1}{2}\right). \tag{5.17}$$

Each collective degree of freedom is associated to a pair of Fock operators $[\hat{b}_k, \hat{b}_{k'}^\dagger] = \delta_{kk'}$, with a frequency and a momentum:

$$\omega_k = \frac{1}{\sqrt{CL}}\sqrt{2 - 2\cos(k\Delta x)}, \qquad k_n = \frac{2\pi}{d} \times n \in \left(-\frac{\pi}{\Delta x}, \frac{\pi}{\Delta x}\right]. \tag{5.18}$$

There are a total of N modes, which we identify with left- and right-moving waves, $k < 0$ and $k > 0$, labeled by the integer numbers $n \in (-N/2, N/2]$. We can reconstruct the flux and charge fields along the waveguide's positions $x_m = \Delta x \times m$ as

$$\hat{\phi}(x_m) = \sum_k \sqrt{\frac{\hbar}{2c\omega_k}} \left[u_k(x)\hat{b}_k + u_k(x)^* \hat{b}_k^\dagger\right], \tag{5.19}$$

$$\hat{q}(x_m) = \Delta x \sum_k \sqrt{\frac{\hbar c \omega_k}{2}} \left[i u_k(x)^* \hat{b}_k^\dagger - i u_k(x)\hat{b}_k\right],$$

using the normalized eigenmode wavefunctions $u_x(x) = e^{ikx_m}/\sqrt{d}$.

For small momenta, the dispersion relation is approximately linear:

$$\omega_k \simeq v|k|, \quad v = \frac{1}{\sqrt{c_0 l_0}}, \quad |k| \ll \frac{1}{\Delta x}. \tag{5.20}$$

The dispersion relation, the speed of light v, and the operators $\hat{\phi}$ and $\hat{q}/\Delta x$ are all defined in terms of intensive quantities and scale-independent normal modes. As anticipated in Section 4.5, this facilitates taking the limit $\Delta x \to 0$, where flux and charge are replaced with continuous fields $[\hat{\phi}(x), \hat{\rho}(y)] = i\hbar\delta(x - y)$. The new Hamiltonian

$$H = \int_0^d \mathrm{d}x \left\{ \frac{1}{2c}\hat{\rho}(x)^2 + \frac{1}{2l}\left[\partial_x \hat{\phi}(x)\right]^2 \right\} \tag{5.21}$$

corresponds to a massless Klein–Gordon theory. These fields satisfy a wave equation with periodic boundary conditions $\hat{\phi}(x + d, t) = \hat{\phi}(x, t)$:

$$\left(\partial_t^2 - \frac{1}{c(x)l(x)}\partial_x^2\right)\hat{\phi}(x, t) = 0, \tag{5.22}$$

whose stationary solutions $(\omega_k^2 - \partial_x^2)u_k(x) = 0$ are the continuous version of the discretized eigenmodes that we found before.

5.2.2 $\lambda/2$ and $\lambda/2$ Microwave Cavities

Transmission lines are often cut to form a one-dimensional open cable of finite length d, coupled to the environment through two capacitors C_1 and C_2. As shown in Figure 5.1b, photons are confined between those two capacitors, which act as partially reflective mirrors and create the microwave equivalent of a Fabry–Perot cavity.

Instead of working with the discrete model, it is more convenient to start from the continuous theory (5.21) and change the boundary conditions. Initially we neglect the coupling to the outer waveguides $C_{1,2} = 0$ and focus on the eigenmodes of an

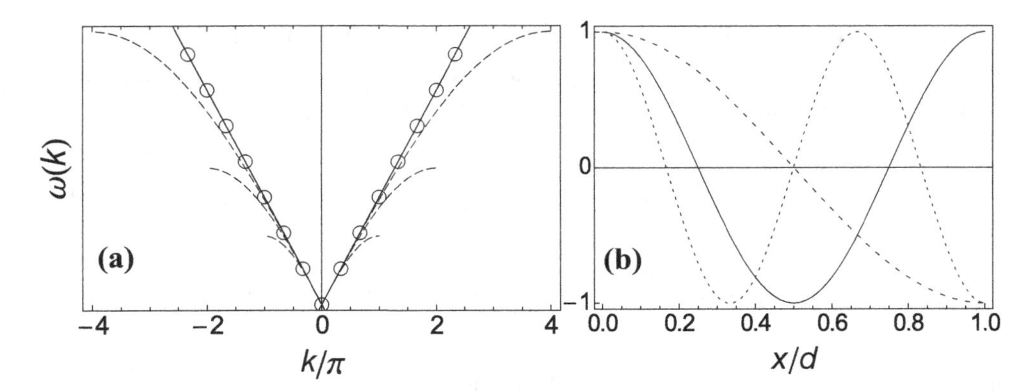

Figure 5.3 (a) Dispersion relations for a transmission line. The dashed line plots the relations for $\Delta x = 1, 1/2, 1/4$, and the solid line plots the limiting relation for $\Delta x = 0$. When the transmission line has a finite length $d = 6$, the momenta and frequencies are discrete (circles). (b) First flux modes for an open trasmission line of finite length.

isolated line. In this limit, no charge can escape the line and the currents must vanish at the boundaries. The zero-current property translates into Neumann-type boundary conditions for the flux $\partial_x \phi|_{x=0,d} = 0$. The Hamiltonian (5.21) with these boundary conditions is diagonalized by an infinite set of standing waves:

$$u^{(n)}(x) = \sqrt{\frac{2}{d}} \cos\left(\frac{\pi n}{d} x\right), \; n \in \mathbb{N} = \{1, 2, 3, \ldots\} \tag{5.23}$$

with quasimomentum $k_n = \pi n/d$ and wavelength $\lambda_n = d/2n$. Figure 5.3a shows the first three eigenmodes, all with zero derivative at the boundaries. Note how the length of the resonator is *half the wavelength* of the fundamental mode $d = \lambda_1/2$.

Due to the boundary conditions, the edges of a $\lambda/2$ resonator have no current, but reach a maximum of the charge on all modes:

$$\rho(x) = \sum_n u^{(n)}(x) \sqrt{\frac{\hbar c_0 \omega}{2}} i \left(b_n^\dagger - b_n\right). \tag{5.24}$$

As show in Figure 5.1c, we can use this fact to couple capacitively the resonator to a semi-infinite transmission line. The coupling, whose strength depends on the capacitors C_1 or C_2 used, is our means to control the resonator, injecting photons or measuring its internal state (see Section 5.5).

If we only need one control access, we may want to connect the other boundary to a ground plane, as shown in Figure 5.4b. This connection replaces one of the Neumann boundary conditions with a fixed value for the voltage at that point $\hat{\phi}(0) = 0$,

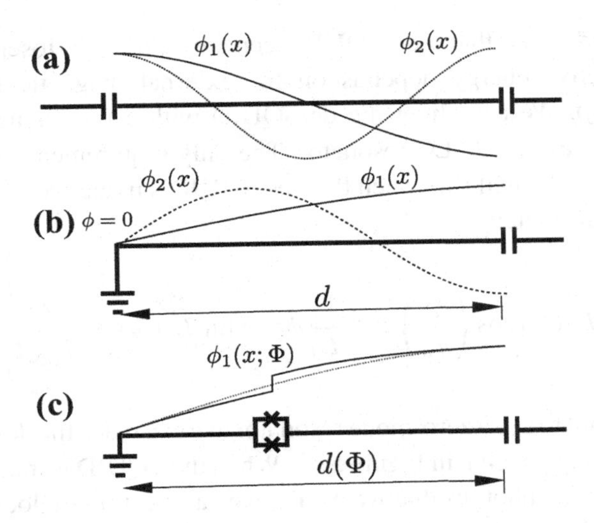

Figure 5.4 First $\phi_0(x)$ and second $\phi_1(x)$ eigenmodes of (a) $\lambda/2$ and (b) $\lambda/4$ transmission line resonators. (c) A transmission line resonator with a SQUID becomes an adjustable resonator. The external flux Φ affects the SQUID's inductance, the lowest energy eigenmodes, and the effective length of the resonator $d(\Phi)$. Note that the drawing is not on scale: A typical SQUID is much smaller than the waveguide.

while the other condition remains the same $\partial_x \hat{\phi}(d) = 0$. The new set of modes that satisfies these boundary conditions are the sinusoidal functions

$$u^{(n)} = \sqrt{\frac{2}{d}} \sin\left(\frac{\pi(2n-1)}{2d} x\right). \tag{5.25}$$

The length of the waveguide is a *quarter of the wavelength* of the fundamental mode, $d = \lambda_1/4$. Such $\lambda/4$ resonators are interesting because close to the edge $x = 0$ the current $\partial_x \phi / l$ is maximal. This is beneficial for coupling inductive elements to the resonator and to place inductive elements that tune the properties of the cavity, as we will now see.

5.2.3 Tunable Cavities

In theory, a microwave resonator can be tuned by changing the capacitance or the inductance of the elements that form it. In practice, adjustable capacitors require mobile parts to change their geometry,[5] which is impractical to operate within a fridge and may introduce noise and heat. Tuning the inductance, on the other hand, is a simpler task.

[5] Does anyone remember old radio receivers with a frequency selection wheel that adjusted a big capacitor? In this device, the tuning wheel changed the opposing surface between two metallic plates, adjusting the capacitance.

We have already seen that a dc-SQUID recreates a tunable Josephson junction, where the inductive energy depends on the external magnetic field $E_J(\Phi) = 2E_J(0)\cos(\Phi/\varphi_0)$. We can shunt the dc-SQUID with a big enough capacitor to create a tunable, zero-mode LC resonator. The only requirement is to operate the circuit in a regime of small flux, such that the SQUID's inductive potential remains approximately harmonic:

$$- E_J(\Phi)\cos\left(\frac{\hat{\phi}}{\varphi_0}\right) \simeq \frac{1}{2L_J}\hat{\phi}^2, \text{ with } L_J(\Phi) = \frac{\varphi_0^2}{E_J(\Phi)}. \tag{5.26}$$

If we do not want to use a zero-mode resonator, we can place the dc-SQUID inside a transmission line, as shown in Figure 5.4c. When the SQUID is much smaller than the wavelength of the photons that we excite, we can treat it as a local perturbation in the Klein–Gordon model (5.21):

$$\Delta\hat{H}_{\text{ind}} \simeq \int_0^d \frac{1}{2L_J(\Phi)}\hat{\phi}(x)^2\delta(x - x_0)\mathrm{d}x + \mathcal{O}(\phi^4). \tag{5.27}$$

This term affects the eigenfrequencies and eigenmodes of the guide. In particular, the normal modes become discontinuous around the SQUID: The flux jumps by an amount $\delta\phi \sim \partial_x\phi \times \delta x$, which would correspond to extending the waveguide by an additional length $\delta x = L_J/l_0$ – a ficticious length that remains "folded" inside the SQUID.

The inductance of a junction can be much larger than that of typical coplanar waveguide. The change in the length $\delta x/\lambda$ can be comparably large, affecting the location of the nodes and antinodes of the resonator. If we connect other circuits – superconducting qubits, resonators, etc. – to those points, we can use a dc-SQUID to bring them in and out of resonance, coupling and decoupling the resonator from them (Forn-Díaz et al., 2017).

The change in resonator length can also be much faster than the time d/v for a photon traverse the waveguide (Sandberg et al., 2008). In this regime of driving, it is as if the "mirrors" of the resonator moved at close to the speed of light. Such conditions simulate a gedanken experiment known as the dynamical Casimir effect, which results in the production of entangled photon pairs out of the vacuum. This experiment would be impractical with real mirrors and cavities, but has been repeatedly demonstrated by Wilson et al. (2011) and later works.

Finally, in Section 5.5.6 we will see that a periodic modulation of the inductance can be used to achieve an effect known as *parametric amplification*. When we send a quantum signal to this resonator, it is reflected back with one or two quadratures that are enlarged or shrunk. Quantum amplification is a very strong and robust mechanism, instrumental for verifying the quantum properties of a photonic field (see Section 5.5.7).

5.3 Three-Dimensional Cavities and Waveguides

Our description of microwave resonators would not be complete without a discussion of three-dimensional resonators. Unlike transmission lines and LC resonators, which are built on a two-dimensional chip, a 3D cavity is an actual cavity inside block of metal. As shown in Figure 5.1c, 3D cavities are fabricated into two or more blocks that join together to fully enclose a three-dimensional vacuum space. The blocks are normally made of aerospace-grade aluminum, but larger cavities and waveguides have been constructed using plain copper. As the transmission line, the 3D cavity is an extended object with a complicated mode structure, where the electromagnetic field propagates in free space, supported by charge and current excitations on the inner surface of the "tube." This produces multiple eigenfrequences and quantized bosonic excitations that we can group together in a single Hamiltonian $H = \sum_n \omega_n \hat{b}_n^\dagger \hat{b}_n$, as we did for the waveguides.

Three-dimensional cavities are no newcomers to the quantum information world. They were used in the pioneering experiments by the Nobel Prize winner Serge Haroche (Raimond et al., 2001) because they can trap photons for very long times. Three-dimensional cavities are usually combined with other 3D circuits. Superconducting qubits, antennas, flux injectors, coaxial waveguides, etc., are all components that can be fabricated independenly and later inserted at different points in the cavity. The result is a kind of superconducting Lego structure that can be adjusted and reused for different purposes over time. This is an interesting paradigm shift from traditional 2D circuits, which can only be fabricated for a single task, with all other circuits permanently attached. Moreover, it must be remarked that 3D waveguides and coaxial cables can transport the photons along relatively long distances. As of this writing, the record was set by the ETHZ group, who extended the setup from Magnard et al. (2020) to connect two separate cryostats by waveguides 30 m long, operating at tens of millikelvins.

Three-dimensional cavities pose a minor theoretical inconvenience. Simulating the behavior of microwaves, reproducing the resonator mode structure and frequencies, or predicting the interaction with other embedded circuits becomes a great numerical challenge. We have to use finite-element methods to solve Maxwell's equations, and apply a sophisticated framework known as *black-box circuit quantization* (Nigg et al., 2012) to reverse-engineer a lumped-element model that describes the cavity with all the embedded components. In this case, the equivalent circuit is really a mathematical convenience: We cannot ascribe the effective capacitances, inductances, and nonlinearities to any particular region of the superconductor – they are just models for the dynamics of the combined system.

5.4 Photon States

Microwave cavities and resonators are used to explore quantum superpositions, macroscopic quantum states, entanglement, or the statistical properties in quantum measurements. The superconducting circuit literature is rich in examples of creation, manipulation, and detection of photonic quantum states. Some great works are the methods by Hofheinz et al. (2009) to engineer arbitrary superpositions states of up to 15 photons, the experiments by Vlastakis et al. (2013) demonstrating long-lived Schrödinger cats with up to 100 photons, or the generation of two-mode squeezing by Mallet et al. (2011). In this section, we discuss families of quantum states that are often used in the literature. We close the section with a mathematical representation of quantum states, the Wigner function, that is both experimentally and theoretically convenient to investigate the photonic field.

5.4.1 Fock States

The Fock states are the eigenstates of the number operator. They are obtained from the *vacuum* $|0\rangle$, climbing upward in the number of photons with the creation operator $|n\rangle = \frac{1}{\sqrt{n!}}a^{\dagger n}|0\rangle$. Fock states can be expanded in the eigenbasis of the dimensionless quadratures:

$$\hat{x} = \frac{1}{\sqrt{2}}(\hat{a} + \hat{a}^\dagger) = \frac{\hat{\phi}}{\sqrt{\hbar Z}}, \qquad \hat{p} = \frac{i}{\sqrt{2}}(\hat{a}^\dagger - \hat{a}) = \frac{\hat{q}}{\sqrt{\hbar/Z}}. \tag{5.28}$$

For instance, in terms of position eigenstates $\hat{x}|x_0\rangle = x_0|x_0\rangle$, the Fock states are Gaussians modulated by Hermite polynomials[6] $H_n(x)$:

$$\langle x_0|n\rangle = \frac{1}{\sqrt{2^n n!\sqrt{\pi}}} H_n(x) e^{-x^2/2}. \tag{5.29}$$

In particular, the vacuum state $|0\rangle$ has a purely Gaussian wavefunction.

The Fock basis is a useful representation for pure $|\psi\rangle = \sum_{n=0}\psi_n|n\rangle$ and for mixed states $\rho = \sum_{nm}\rho_{nm}|n\rangle\langle m|$. In these expansions, $|\psi_n|^2 = |\langle n|\psi\rangle|^2$ and $\rho_{nn} = \langle n|\rho|n\rangle$ are the probabilities of finding n photons in the resonator. These expansions are often truncated at a cutoff $N_{\max}$ consistent with the maximum energy injected in the resonator. This simplifies both the experimental characterization of

[6] We use the definition $H_0(x) = 1$, $H_n(x) = (x - \partial_x)H_{n-1}(x)$.

quantum states – we ignore high-order moments that are hard to measure – as well as the theoretical simulations of practical circuits.

Fock states are simple and useful, but difficult to prepare. If we drive a linear resonator with an external field, the coupling $\Omega(t)(\hat{a}^\dagger + \hat{a})$ spreads the wavefunction over *all* Fock states. To create a state of one photon, we need a *single photon source*: a saturable quantum device that is always excited to the same state, and which always decays emitting just one photon. The word "saturable" means that the device can take a finite amount of energy: typically, it will jump from some ground state $|g\rangle$ to the same excited state $|e\rangle$, and it will stop there. When we stop driving the source, the photon source decays back to $|g\rangle$, emitting a single photon. These nonlinear two-level circuits are the superconducting *qubits* that we introduce in Chapter 6.

Using qubits, we can play some magic tricks. For instance, we can prepare the qubit in a superposition of ground and excited state $\alpha\,|0\rangle + \beta\,|1\rangle$, so that when it exchanges energy with the microwave resonator it implements the operation $(\alpha + \beta\hat{a}^\dagger)$. Multiple iterations of this trick lead to arbitrary superpositions of Fock states, as demonstrated by Hofheinz et al. (2009). Moreover, the *same* qubit can also be used as *photon counter*, to reconstruct the full wavefunction of the microwave resonator – see Section 7.3.2 and works by Schuster et al. (2007) and Hofheinz et al. (2009).

5.4.2 Thermal States

A saying goes that *real experiments do not have pure states:* Since we never have 100% control of all experimental conditions, states in the lab are always *mixed states* – classical superpositions of experiments with slightly varying conditions. This is in particular true when we consider quantum circuits at thermal equilibrium with their environment. These circuits are modeled with a canonical ensemble of quantum states $|n\rangle$, excited with Boltzmann weights $\exp(-E_n/k_BT)$. Applied to the single-mode resonator, this creates a Bose–Einstein distribution of photons:

$$\rho = \frac{e^{-\beta H}}{Z(\beta)} = \sum_{n=0}^{\infty} e^{-\beta\hbar\omega n}(1 - e^{-\beta\hbar\omega})\,|n\rangle\langle n|\,. \tag{5.30}$$

This ensemble gives a nonzero probability of finding photons in the resonator $p_{n>0}$ and a nonzero average occupation $\bar{n}$:

$$p_{n>0} = \sum_{n>0} c\rho_{nn} = 1 - \rho_{00} = e^{-\hbar\omega/k_BT}, \tag{5.31}$$

$$\bar{n} = \langle a^\dagger a\rangle = \frac{1}{e^{\hbar\omega/k_BT} - 1}. \tag{5.32}$$

Recalling that $\hbar \times 2\pi \times 20$ GHz $\simeq k_B \times 1$ K, the average population of photons ranges between 1.5 photons for a 1 GHz resonator at 100 mK, to 10^{-21} photons for a 10 GHz resonator at 10 mK – i.e., from something that may hinder the fidelity of quantum operations to a negligible correction.

For some applications in quantum simulation, it might be desirable to prepare nonequilibrium thermal states at a temperature higher than that of the cryostat. This can be done by pumping the oscillator with narrow-band white noise; this results in a master equation at an artificially high temperature (cf. Section 5.5.4) that equilibrates to the desired state, without "breaking" the sample or the superconducting state.

5.4.3 Coherent States

The coherent states are eigenstates of the annihilation operator $a\,|\alpha\rangle = \alpha\,|\alpha\rangle$. There is one coherent state for every complex eigenvalue $\alpha = \alpha_x + i\alpha_p \in \mathbb{C}$. The vacuum is a coherent state with $\alpha = 0$. The *displacement operator* is a unitary transformation $D(\alpha)^\dagger = D(-\alpha) = D(\alpha)^{-1}$ that shifts the Fock operators:

$$D(\alpha)^\dagger \hat{a} D(\alpha) = \hat{a} + \alpha, \qquad D(\alpha)^\dagger \hat{a}^\dagger D(\alpha) = \hat{a}^\dagger + \alpha^*. \tag{5.33}$$

Hence, all coherent states result from displacements of the vacuum:

$$|\alpha\rangle = D(\alpha)\,|0\rangle = e^{\alpha\hat{a}^\dagger - \alpha^*\hat{a}}\,|0\rangle = \sum_n \frac{\alpha^n}{\sqrt{n!}} e^{-|\alpha|^2/2}\,|n\rangle\,, \tag{5.34}$$

and can be uniquely identified by their displacement $\alpha = \alpha_{\text{re}} + i\alpha_{\text{im}} = \frac{1}{\sqrt{2}}\,\langle\alpha|\hat{x}|\alpha\rangle + \frac{i}{\sqrt{2}}\,\langle\alpha|\hat{p}|\alpha\rangle$, which is their "position" in the phase space defined by $(\langle\hat{x}\rangle, \langle\hat{p}\rangle)$. As they evolve in time $e^{-iHt/\hbar}\,|\alpha\rangle \propto |e^{-i\omega t}\alpha\rangle$, coherent states describe elliptical orbits in phase space, similar to a classical oscillator. Coherent states have Gaussian wavefunctions in position, momentum, or any other direction in phase space. The width of the Gaussian

$$\langle x|\alpha\rangle = \frac{1}{\pi^{1/4}} e^{-(x-\sqrt{2}\alpha_{\text{re}})^2/2 + i\alpha_{\text{im}}x} \tag{5.35}$$

is the same along any direction and saturates the Heisenberg uncertainty relation, that is, $\Delta\hat{x} = \Delta\hat{p} = \frac{1}{\sqrt{2}}$ and $\Delta\hat{\phi}\Delta\hat{q} = \frac{1}{2}\hbar$. For this reason, coherent states are also called *minimum uncertainty states*.

Resonators are prepared in coherent states by feeding them with coherent microwave signals, via the coupling term in (5.8) (see Section 5.5.1). Unlike visible/IR/UV coherent fields, which require lasers, there are many ways to generate coherent microwaves: from high-power masers, magnetrons, and tubes, to low-power solid-state electronic devices. A typical coherent microwave source is formed by a chain of electronic devices. An ultrastable crystal oscillator – referenced to an atomic clock for phase and frequency stability – creates a signal at a fixed frequency around hundreds of MHz. The output signal is frequency-multiplied in a nonlinear circuit, doubling its frequency until it reaches the desired range of GHz. The last fine-tuning is done with additional (digital) electronics that shift the frequency to the desired value, in resonance with the LC circuit.

5.4.4 Schrödinger Cat States

Quantum mechanics predicts the existence of superpositions of "macroscopically" distinguishable states. The textbook example is Schrödinger's cat experiment: a live cat is placed inside a box with a poisoning device that is activated by a decaying radioactive particle. At any given point in time, the state of the cat in the box must be described as a quantum superposition of being alive (with the radioactive atom still active), and being dead (with a decayed atom and free poison). Schrödinger's gedanken experiment challenges the limits of quantum mechanics, setting a boundary between a somewhat loosely defined "macroscopic" world, where we do not experience quantum superpositions, and the realm of coherent quantum mechanics. Scientists have accepted Schrödinger's challenge, exploring how to create large quantum superpositions using microwave resonators.[7]

Photonic Schrödinger cats are non-Gaussian states and require nonlinear operations to be created. As explained in Chapter 7 and in the work by Vlastakis et al. (2013), a qubit with a dispersive coupling can push a microwave resonator into a superposition of two coherent states with opposite displacements:

$$|\Psi_{\text{cat}}(\theta)\rangle = \frac{1}{\sqrt{2}}(|\alpha\rangle + e^{i\theta}\,|-\alpha\rangle). \tag{5.36}$$

In state-of-the-art experiments, the cat has as many as $|\alpha|^2 \sim 100$ photons. At this point, the two states $|\pm\alpha\rangle$ can be macroscopically distinguished through quadrature measurement – i.e., the voltage or current leaking from the resonator. This distinguishability also makes the state all the more fragile, decreasing its lifetime linearly with the number of photons.

Interestingly, it is possible to extend the lifetime of a photonic Schrödinger cat using an error-correcting code that detects if the state has lost any photon. When this happens, the cat state is not immediately destroyed, but simply changes phase, jumping around in the Hilbert space $a\,|\Psi_{\text{cat}}(\theta)\rangle \propto \frac{1}{\sqrt{2}}(|\alpha\rangle - e^{i\theta}\,|-\alpha\rangle) = |\Psi_{\text{cat}}(\theta+\pi)\rangle$. This new state is orthogonal to the original

[7] Also on other circuits, such as superconducting current loops (van der Wal et al., 2000), and on trapped ions, optical lattices, etc.

state: $\langle\Psi_{\rm cat}(\theta)|\Psi_{\rm cat}(\theta+\pi)\rangle \propto \exp(-2|\alpha|^2)$. However, using qubits we can detect those tiny phase changes, either correcting the state or bookkeeping the accumulated phase θ_n and correcting later measurements. This technique can stabilize a cat state to a time that is around the resonator lifetime (Ofek et al., 2016). Further extensions would need to account for dephasing – difference in the relative phases of $|\alpha\rangle$ and $|-\alpha\rangle$ – using a larger space of states and error-detection strategies.

5.4.5 Single-, Two-, and Multimode Squeezed States

Squeezed states are Gaussian states where the uncertainty of one or more quadratures is reduced at the expense of increasing the uncertainty in the conjugate quadratures. There are two important motivations for working with squeezed states: seeking a metrological advantage and creating entanglement. The metrological advantage appears already in the *single-mode squeezed state*. The reduction in the uncertainty of one quadrature, such as $\Delta\hat{x}$, means that an experimental estimate of $\langle\hat{x}\rangle$ will need a lower number of measurements M to achieve the same precision – see (2.28) in our discussion of unbiased estimators.

Single-mode squeezing is a unitary transformation, parameterized by the squeezing strength r and angle θ:

$$U_{\text{sq-1}}(r) = \exp\left(\frac{1}{2}\left(re^{i2\theta}\hat{a}^{\dagger 2} - re^{-i2\theta}\hat{a}^2\right)\right). \tag{5.37}$$

In the Heisenberg picture,[8] this Bogoliubov transformation combines annihilation and creation operators:

$$\hat{a}(r) := U^{\dagger}_{\text{sq-1}}(r)\hat{a}(0)U_{\text{sq-1}}(r) = \cosh(r)\hat{a}(0) - \sinh(r)\hat{a}^{\dagger}(0)e^{i2\theta}. \tag{5.38}$$

Introducing the rotated quadrature $\hat{x}_\phi = \frac{1}{\sqrt{2}}(\hat{a}^{\dagger}e^{i\phi} + \hat{a}e^{-i\phi})$, we find two observables with opposite behavior

$$\hat{x}_\theta(r) = e^{-r}\hat{x}_\theta(0), \qquad \hat{p}_\theta(r) := \hat{x}_{\theta+\pi/2} = e^{+r}\hat{p}_\theta(0). \tag{5.39}$$

The operator $\hat{p}_\theta$ and its uncertainty $\Delta\hat{p}_\theta$ are both amplified with a *gain factor* $\sqrt{G} = e^{+r}$, while the operator $\hat{x}_\theta(r)$ and its variance $\Delta\hat{x}_\theta(r)$ are both contracted by the inverse factor $e^{-r} = 1/\sqrt{G}$.

Experimentally, this means that measuring $\hat{x}_\theta(0)$ on the squeezed vacuum $U_{\text{sq-1}}(r)\,|0\rangle$ produces results with an uncertainty below the standard quantum limit $\Delta\hat{x}_\theta = e^{-r}/\sqrt{2}$. Squeezing is measured in decibels $r_{\rm dB} := 10 \times \log_{10}(\Delta O_{\rm new}/\Delta O_{\rm old})$.

[8] Instead of studying how the states are transformed, we study how the observables that we want to measure are transformed.

A value $r_{\mathrm{dB}} \sim 3$ means a 50% reduction in uncertainty $e^{-r} \sim 0.5$. Not too long ago, such values were considered the state of the art in quantum optics – for instance, the Laser Interferometer Gravitational-Wave Observatory (LIGO) uses between 2.7 and 3.2 dB of squeezing in their interferometer. These days, much larger values of 10–12 dB (0.1–0.06 reductions) are available in the superconducting lab, thanks to the parametric amplifiers described in Section 5.5.6.

The second motivation for engineering squeezed states is to create entanglement and correlations between two or more bosonic modes (Laurat et al., 2005). In the typical *two-mode squeezed state*, we compress a quadrature that is a linear combination of two oscillator modes $b_1 + b_2$, using the general two-mode squeezing Gaussian operation:

$$U_{\text{sq-2}}(r) = \exp\left[re^{i2\theta} b_j^\dagger b_i^\dagger - r^{-i2\theta} b_i b_j\right]. \tag{5.40}$$

This operator mixes the annihilation and creation of different modes, as in

$$\begin{aligned}\hat{b}_1(r) &= \cosh(r)\hat{b}_1(0) - \sinh(r)e^{2i\theta}\hat{b}_2^\dagger(0), \\ \hat{b}_2^\dagger(r) &= \cosh(r)\hat{b}_2^\dagger(0) - \sinh(r)e^{-2i\theta}\hat{b}_1(0).\end{aligned} \tag{5.41}$$

Applying this transformation onto a thermal state, we achieve squeezing and amplification of the joint quadratures $\hat{x}_\theta^\pm = \frac{1}{2}(\hat{b}_1^\dagger \pm \hat{b}_2^\dagger)e^{-i\theta} + \text{H.c.}$ and $\hat{p}_\theta^\pm = \frac{i}{2}(\hat{b}_1^\dagger \pm \hat{b}_2^\dagger) + \text{H.c.}$ similar to before:

$$\hat{x}_\theta^\pm(r) = e^{\mp r}\hat{x}_\theta^\pm(0), \qquad \hat{p}_\theta^\pm(r) = e^{\pm r}\hat{x}_\theta^\pm(0). \tag{5.42}$$

We will discuss the creation of two-mode squeezing with linear amplifiers and these gain/contraction relations in Section 5.5.6.

The two-mode squeezed state is an entangled state because it cannot be written as product of two independent states for each of the modes $\rho_{\text{sq-2}} \neq \rho_1 \otimes \rho_2$. This is qualitatively appreciated when we write the Schmidt decomposition of the state in the Fock basis $|0_r\rangle := U_{\text{sq-2}}(r)|0\rangle = \sqrt{1-\lambda_r}\sum_n \lambda_r^{n/2}|n,n\rangle$, with $\lambda_r = \tanh(r)^2$. We can also compute the reduced density matrix that results from tracing out or ignoring one of the modes $\rho_1 = \mathrm{tr}_2(|0_r\rangle\langle 0_r|) = (1-\lambda_r)\sum_n \lambda_r^n |n\rangle\langle n|$. The von Neumann entropy of this matrix is a measure of the entanglement between the modes $S = -\mathrm{tr}(S\log(S)) \sim 2r$ and diverges with the amount of squeezing r.

It is possible to create entangled states of multiple degrees of freedom. In particular, the transmission line is an example of *multimode squeezed state*. From (5.19), we see that the fields $\{\hat{b}_k, \hat{b}_k^\dagger\}$ are macroscopic superpositions of quadratures at different positions. Thus, while the vacuum state may seem a rather trivial state with no photons $\hat{b}_k|\Omega\rangle = 0$, it is in fact a highly correlated state from the point of view of the field operators along the line, $\{\hat{\phi}(x), \hat{q}(x)\}$. If we try to estimate the uncertainty of the local observables, we will find strong divergences:

$$\langle \hat{\phi}(x)\hat{\phi}(y)\rangle - \langle \hat{\phi}(x)\rangle \langle \hat{\phi}(y)\rangle = \sum_k \frac{\hbar e^{ik(x-y)}}{2c_0 \Delta x \omega_k} \langle b_k b_k^\dagger \rangle, \tag{5.43}$$

both in the infrared and ultraviolet limit.[9] This manifestation of entanglement in the 1D field theory is but one example of other quantum field theory concepts that can be explored using superconducting circuits – particle localization, propagation of entanglement and causality, vacuum energy and Casimir effect, etc.

5.4.6 Wigner Functions and Gaussian States

Coherent, thermal, and squeezed states are representatives of the Gaussian state family. Defining this family requires a tiny bit of quantum information theory.[10] Let us assume a system of N bosonic modes, described with dimensionless position and momenta (5.28). We group all operators into a vector $\hat{\mathbf{R}} = (\hat{x}_1, \ldots, \hat{x}_N, \hat{p}_1, \ldots, \hat{p}_N)^T$, introducing the commutator matrix $i\Omega_{ij} = [\hat{R}_i, \hat{R}_j]$ and the displacement operator:

$$D(\mathbf{R}_0) = \exp\left(i\hat{\mathbf{R}}^T \Omega \mathbf{R}_0\right) = \exp\left(i\left(\hat{\mathbf{x}}^T, \hat{\mathbf{p}}^T\right)\begin{pmatrix} 0 & 1 \\ -1 & 0 \end{pmatrix}\begin{pmatrix} \mathbf{x}_0 \\ \mathbf{p}_0 \end{pmatrix}\right). \tag{5.44}$$

Note that for a single mode

$$D(\mathbf{R}_0) = e^{i(p_0\hat{x} - x_0\hat{p})} = e^{\alpha\hat{a}^\dagger - \alpha^*\hat{a}}, \quad \text{with } \alpha = \frac{1}{\sqrt{2}}(x_0 + ip_0). \tag{5.45}$$

The *Wigner function* is the Fourier transform of the characteristic function $\chi(\mathbf{R}) = \mathrm{tr}\,[\rho D(\mathbf{R})]$:

$$W_\rho(\mathbf{R}) = \frac{1}{(2\pi)^{2N}} \int_{\mathbb{R}^{2N}} e^{i\mathbf{Y}^T\Omega\mathbf{R}} \mathrm{tr}\,[\rho D(\mathbf{R})]\, \mathrm{d}^{2N}\mathbf{y}. \tag{5.46}$$

The marginals of the Wigner function give the probability distributions of the position and momentum operators, $\langle \mathbf{x}|\rho|\mathbf{x}\rangle = \int W_\rho(\mathbf{x}, \mathbf{p})\mathrm{d}^N\mathbf{p}$, and $\langle \mathbf{p}|\rho|\mathbf{p}\rangle = \int W_\rho(\mathbf{x}, \mathbf{p})\mathrm{d}^N\mathbf{x}$. We can also use W_ρ to compute any expected values involving the quadratures:

$$\langle F(\hat{\mathbf{x}}, \hat{\mathbf{p}})\rangle_\rho = \mathrm{tr}[F(\hat{\mathbf{x}}, \hat{\mathbf{p}})\rho] = \int_{\mathbb{R}^{2N}} W_\rho(\mathbf{x}, \mathbf{p}) F(\mathbf{x}, \mathbf{p}) \mathrm{d}^N\mathbf{x}\mathrm{d}^N\mathbf{p}. \tag{5.47}$$

9 Infrared because the denominator approaches zero for $k = 0$, and ultraviolet because the fraction is proportional to $1/n$ for $|n| = 1, \ldots, N/2$, and the sum diverges logarithmically with the discretization size.

10 Notation follows Adesso et al. (2014) and Olivares (2012).

If $W_\rho \geq 0$, the Wigner function represents a true probability distribution. In this case, the experimental measurement outcomes can be classically simulated by sampling this distribution. However, W_ρ in general is just a *quasiprobability distribution*, because it can reach negative values. The states for which this happens are usually referred to as *nonclassical states.*

A *Gaussian state* is one whose Wigner function is a Gaussian:

$$W(\mathbf{R}) = \frac{1}{\pi^N \sqrt{\det(\Gamma)}} \exp\left[-(\mathbf{R}-\mathbf{r})^T \Gamma^{-1} (\mathbf{R}-\mathbf{r})\right]. \tag{5.48}$$

The center of the Gaussian $\mathbf{r} := \langle \hat{R}_i \rangle_\rho$ and its **covariance matrix** Γ

$$\Gamma_{ij} := \langle \{\hat{R}_i - r_i, \hat{R}_j - r_j\} \rangle_\rho = \langle \hat{R}_i \hat{R}_j + \hat{R}_j \hat{R}_i \rangle_\rho - 2 \langle \hat{R}_u \rangle \langle \hat{R}_j \rangle_\rho, \tag{5.49}$$

are uniquely determined by the first and second moments of the quadratures.

As we anticipated, all single- and multimode coherent states $|\alpha\rangle$ are Gaussian states. Their Wigner functions look like displaced vacua: They are centered on $d = \sqrt{2}(\alpha_{\text{re}}, \alpha_{\text{im}})$ and – because they are minimal uncertainty states – they have the same width on all directions. Figure 5.5a and b illustrate the Wigner function for a vacuum state and for a coherent state at $\alpha = 5/2$.

A squeezed state is a Gaussian state with unequal widths along two or more mutually orthogonal directions. Figure 5.5e shows the Wigner function of a single-mode squeezed vacuum state – state $U(r)|0\rangle$ with $\theta = 0$ in (5.37). As expected, the quadrature $\hat{x}$ is reduced by a factor $e^{-r} = 1/2$, while the canonically conjugate momentum is enlarged by the inverse factor. The result is a Gaussian with elliptical equiprobability contours, and a minimum width $\Delta\hat{x} = 1/2\sqrt{2}$ below the Heisenberg limit.

The family of Gaussian states includes also thermal states. Moreover, all Gaussian states can be written as a finite temperature state of one quadratic model, $\rho = e^{-\beta H_{\text{eff}}}/\text{tr}\, e^{-\beta \hat{H}_{\text{eff}}}$ with $\hat{H}_{\text{eff}} = \frac{1}{2} H_{ij}(\hat{R}_i - r_i)(\hat{R}_j - r_j)$. The process of determining the model $\hat{H}_{\text{eff}}$ and the temperature $\beta = 1/k_B T$ is equivalent to finding the simplectic transformation that diagonalizes Γ.

Gaussian operations are those that preserve the Gaussian nature of a state. This includes all linear transformations between modes $\mathbf{R}' = U^\dagger \mathbf{R} U = O\mathbf{R}$ generated by quadratic Hamiltonians $U = \exp(-i H_{\text{eff}} t)$. Examples of those coherent transformations are the single-mode and two-mode squeezing operations from Section 5.4.5. However, Gaussian states are also preserved by incoherent operations that change the area covered by the Wigner function, enlarging the uncertainty of one or more quadrature. We will see examples of those operations when we study the coupling of a Gaussian model to a linear environment in Section 5.5.4.

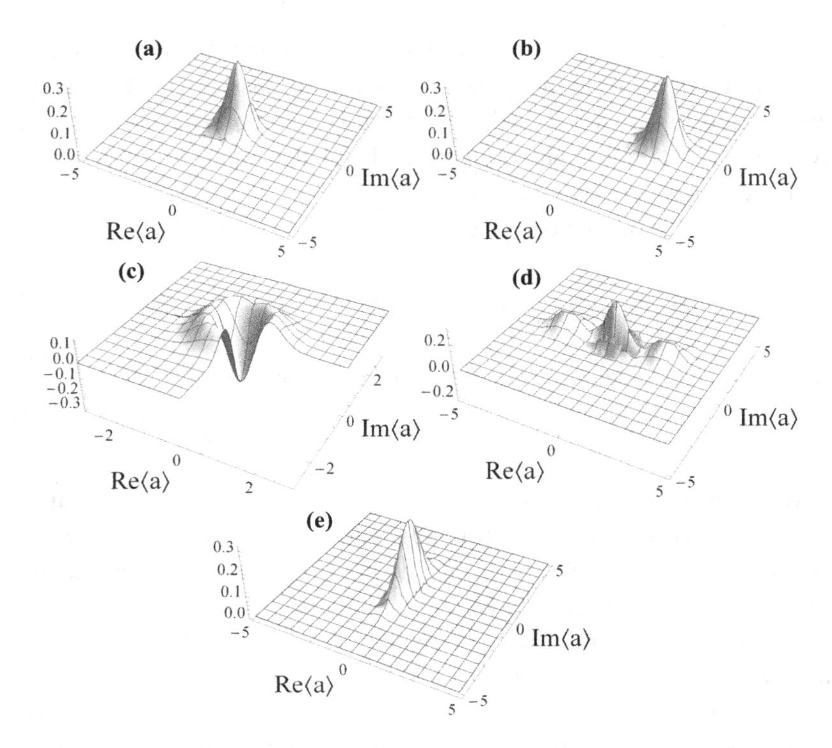

Figure 5.5 Wigner functions of (a) the vacuum state $|0\rangle$, (b) a coherent state $|\alpha = 5/2\rangle$, (c) a Fock state $|1\rangle$, (d) a Schrödinger cat $\propto |5/2\rangle + |-5/2\rangle$, and (e) a squeezed state with $e^{-r} = 1/2$. Note how the Wigner function for the $|0\rangle$ and $|\alpha\rangle$ state look identical, up to a displacement. The Wigner function becomes negative for the "nonclassical" states (c) and (d).

The Wigner function is a useful tool to reconstruct and identify bosonic quantum states, even if they are not Gaussian. Figure 5.5c and d show the Wigner function of a Fock state and of a Schrödinger cat (5.36). The Fock state spreads over a ring of radius $|\alpha|^2 \sim n$, while the Schrödinger cat shows the interference between to coherent states $|\pm\alpha\rangle$.

The Wigner function is an infinite-dimensional object that can only be approximately reconstructed under realistic assumptions. The following are main reconstruction methods:

Gaussian state verification. This method only works for Gaussian states. One measures moments of the quadratures $\langle (a^\dagger)^m a^n \rangle$ up to a certain order $n + m = N_{\text{max}} > 2$. These moments are used to verify that the state is Gaussian, showing that higher-order moments $n + m > 2$ are related to the first and second moments by Wick's theorem (Menzel et al., 2012). Once the Gaussian nature is verified, the covariance matrix Γ and the mean values $\mathbf{r}$ can be derived from the moments with $n + m \leq 2$.

Parity measurements. There is a formula that relates the Wigner function to the parity operator $\Pi = (-1)^{a^\dagger a}$:

$$W_\rho(\alpha) = \frac{2}{\pi}\mathrm{tr}(\Pi\rho_\alpha) = \frac{2}{\pi}\left\langle D(-\alpha)^\dagger \Pi D(-\alpha)\right\rangle_\rho . \tag{5.50}$$

This formula enables the computation of the Wigner function at every point of phase space, through a three step process: (i) prepare the state ρ, (ii) displace the bosonic mode using some external microwave drive to construct $\rho_\alpha := D(\alpha)^\dagger \rho D(\alpha)$, and (iii) measure the parity over this new state $\mathrm{tr}(\Pi\rho_\alpha)$, for instance with the help of a qubit (Hofheinz et al., 2009).

Truncated moments. For states where there is a maximum number of photons, higher-order moments above N_{max} can be vanishingly small. This allows us to reconstruct the Wigner function from the smaller set of nonzero moments (Eichler et al., 2011):

$$W_\rho(\alpha = x + ip) = \sum_{n,m}\int_{\mathbb{C}} \mathrm{d}^2\lambda \frac{(-\lambda^*)^m\lambda^n}{\pi^2 n!\, m!}\left\langle(\hat{a}^\dagger)^m\hat{a}^n\right\rangle_\rho e^{-\frac{1}{2}|\lambda|^2+\alpha\lambda^*-\alpha^*\lambda}. \tag{5.51}$$

Radon transform. This is a method for unbiased tomography with first-order quadrature measurements. It reconstructs $W(x,p)$ from the marginals $P(x_\theta) = \int \mathrm{d}p_\theta W(\cos(\theta)x_\theta + \sin(\theta)p_\theta, \cos(\theta)p_\theta - \sin(\theta)x_\theta)$ along various directions θ in phase space. The statistics of $P(x_\theta)$ is recovered from binning measurements of a rotated quadrature $\hat{Q}_{\omega,\theta} \propto x_\theta$, such as voltage or intensity, with an angle determined by the phase of the reference oscillator (see Section 5.5.7). The inverse Radon transform approximates $W_\rho(x,p)$ from a set of $p(x_\theta)$ over the different angles (Mallet et al., 2011).

5.5 Gaussian Control of Microwave Photons

We have discussed some linear superconducting circuits – microwave resonators, transmission line resonators, and three-dimensional cavities – and the quantum states that they support. We have suggested linear operations – displacements, heating, amplification and squeezing, measurement – that can be formally used to create and reconstruct those states. In this section, we explain how all those *Gaussian operations* are implemented in actual experiments, together with the Hamiltonian models that describe them.

5.5.1 Coherent Drivings and Displacement Operations

The simplest way to control an LC circuit is to connect it to a voltage source, providing a time-dependent potential bias $V(t)$, as in Figure 4.4. The canonical quantization produces the linear, driven harmonic oscillator from (5.8), with the external field $\Omega(t) = \sqrt{\hbar/Z} C_g V(t)/C_\Sigma$. This external drive induces a displacement transformation onto the field $\hat{a} \to D(\alpha)^\dagger \hat{a} D(\alpha) = \hat{a} + \alpha$. If we switch from the Heisenberg to the Schrödinger picture, the same operation interpreted as preparing the resonator, from a vacuum state, into a coherent state with adjustable displacement and phase $|\alpha\rangle = D(\alpha)\,|0\rangle$ (see Problems 5.4 and 5.5).

Let us study an oscillating microwave field $\Omega(t) = \Omega_0 \cos(\omega_d t + \varphi)$ that resonates with the circuit for a brief period of time. The solution of (5.10) predicts a displacement of the resonator with two parts:

$$\hat{a}(t) = e^{-i\omega t}\left[\hat{a}(0) + \alpha_{\text{RWA}}(t) + \alpha_{\text{non-RWA}}(t)\right]. \tag{5.52}$$

The *rotating term* oscillates in synchrony with the resonator:

$$\alpha_{\text{RWA}}(t) = \int_0^t e^{i\omega\tau}\frac{\Omega_0}{2}e^{-i(\omega_d\tau+\varphi)}\mathrm{d}\tau = \frac{\Omega_0 e^{-i\varphi}}{2}\frac{e^{i(\omega-\omega_d)t}-1}{i(\omega-\omega_d)}. \tag{5.53}$$

Close to resonance $\omega_d - \omega \to 0$, it creates a linearly growing displacement $|\alpha_{\text{RWA}}| \sim \Omega_0 t$. The *counter rotating term* has a rapidly oscillating integrand:

$$\alpha_{\text{non-RWA}}(t) = \int_0^t e^{i\omega\tau}\frac{\Omega_0}{2}e^{+i(\omega_d\tau+\varphi)}\mathrm{d}\tau = \frac{\Omega_0 e^{i\varphi}}{2}\frac{e^{i(\omega+\omega_d)t}-1}{i(\omega+\omega_d)}, \tag{5.54}$$

which averages to a smaller displacement $|\alpha_{\text{non-RWA}}| \sim |\Omega_0/(\omega+\omega_d)| \ll 1$. The *rotating wave approximation* (RWA) consists in neglecting this displacement, assuming that the Rabi frequency is smaller than the characteristic frequencies $|\Omega_0| \ll \omega + \omega_d$ and that we are interested in the long-time dynamics $t \gg 1/(\omega+\omega_d)$.

Once we know the outcome of the Heisenberg equations, we can develop a qualitative argument to apply the RWA directly onto the Hamiltonian. This argument goes as follows: Since the cavity field oscillates as $\hat{a}(t) \sim e^{-i\omega t}$, we can neglect any term of the form $e^{-i\omega_d t}\hat{a}(t)$ because such terms will lead to small corrections $\mathcal{O}((\omega+\omega_d)^{-1})$. Applied to the Hamiltonian (5.8), this produces an RWA model:

$$H_{\text{RWA}} = \omega a^\dagger a + i\frac{\Omega_0}{2}\left(e^{-i(\omega_d t+\varphi)}\hat{a}^\dagger - e^{i(\omega_d t+\varphi)}\hat{a}\right), \tag{5.55}$$

which produces the displacement (5.52) with $\alpha_{\text{non-RWA}} = 0$.

Since our ideal resonator has no losses, a resonant driving $\omega_d = \omega$ injects an unlimited amount of energy, which grows quadratically in time $\langle a^\dagger a\rangle \propto t^2$. In Section 5.5.3, we correct this absurdity, collecting the experimental losses from actual resonators – both intrinsic, as well as due to a coupling to the voltage source and the transmission line – in a decay rate κ. Qualitatively, the losses are equivalent to a complex shift of the resonator frequency $\omega \to \omega - i\kappa/2$ $(0 < \kappa \ll \omega)$. With this change, the resonator occupation saturates at around $\sim |\Omega_0|^2/\kappa^2$, a number that can be still very large. For instance, state-of-the-art experiments use this method to engineer displacements with hundreds of photons (Vlastakis et al., 2013) as the starting point for creating macroscopic Schrödinger cats.

5.5.2 Coupling to an Environment

Actual resonators are connected to the outer world. We have already considered the possibility of driving the resonator with microwaves. These signals must be brought into the circuit and into the resonator by long coaxial cables (Figure 5.1d), coplanar waveguides (Figure 5.1c) or some other type of microwave guide. These "cables" introduce classical and quantum microwaves from far away sources ($\hat{b}^{\text{in}}$) to control the circuit and collect signals from the circuit ($\hat{b}^{\text{out}}$) into measurement devices. However, they also connect the resonator to an infinite number of quantum microwave modes that have their own dynamics.

Figure 5.6 sketches two models of an LC circuit connected to a semi-infinite transmission line. Following Sections 4.4 and 4.5, the Hamiltonian of the capacitively coupled LC circuit is

$$\hat{H} = \hat{H}_{\text{t-line}} + \frac{1}{2C_\Sigma}(\hat{Q} - C_g\hat{V}) + \frac{1}{2L}\hat{\Phi}^2. \tag{5.56}$$

Here, $\{\hat{Q}, \hat{\Phi}\}$ are the resonator's charge and flux, $\hat{H}_{\text{t-line}}$ is the Hamiltonian for a semi-infinite transmission line (4.42), and $\hat{V} = \partial_t\hat{\phi}(x_0)$ is the electric potential

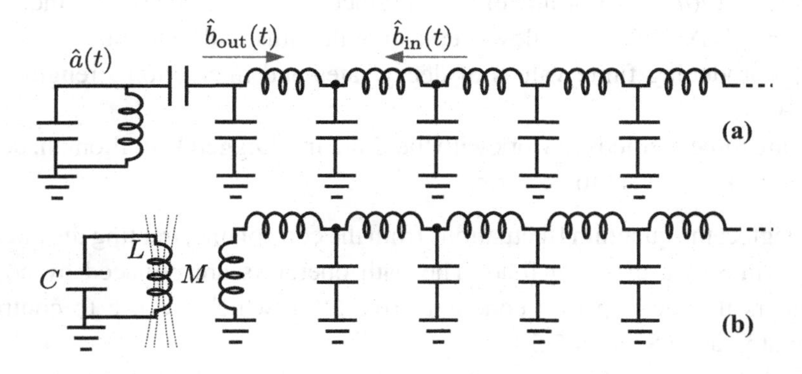

Figure 5.6 A superconducting circuit – in this case, an LC resonator – can be capacitively (a) or inductively (b) coupled to a semi-infinite transmission line, which acts as an environment. Input–output theory (5.65) predicts that field reflected by the resonator $\hat{b}_{\text{out}}(t)$ carries part of the incoming field from the line $\hat{b}_{\text{in}}(t)$ together with a signal that leaks from the coupled circuit.

created by the transmission line at the interface with the LC circuit. If the coupling is inductive, we would instead have

$$\hat{H} = \hat{H}_{\text{t-line}} + \frac{1}{2C}\hat{Q}^2 + \frac{1}{2L}(\hat{\Phi} - ML\hat{I})^2, \tag{5.57}$$

where M is the mutual inductive coupling between the LC and the line (cf. Section 4.7.1), and $\hat{I} = \partial_x\hat{\phi}(x_0)/l_0$ is the transmission line's current close to the place of mutual interaction.

When we quantize either circuit, we obtain a Caldeira–Leggett model for an oscillator $\{\hat{a}, \hat{a}^\dagger\}$ interacting with a *bath* of bosonic modes, corresponding to the microwaves in the transmission line $\{\hat{b}_k, \hat{b}_k^\dagger\}$. If the coupling is so weak that it does not change the eigenmodes of the transmission line, we may write

$$H = \hbar\omega\hat{a}^\dagger\hat{a} + \hbar\sum_k(\hat{a} + \hat{a}^\dagger)\left(g_k\hat{b}_k^\dagger + g_k^*\hat{b}_k\right) + \sum_k\hbar\omega_k\hat{b}_k^\dagger\hat{b}_k. \tag{5.58}$$

Depending on whether the coupling is capacitive or inductive, we obtain slightly different coefficients:

$$\hbar g_k^{(\text{cap})} = \sqrt{\frac{\hbar Z}{2}v\omega_k} \times u_k^*(x_0), \quad \text{or} \quad \hbar g_k^{(\text{ind})} = \sqrt{\frac{\hbar}{2Z}v\omega_k} \times \frac{\partial_x u_k^*(x_0)}{k}. \tag{5.59}$$

These are written in terms of the transmission line's impedance Z, the speed of light on the line $v = 1/\sqrt{cl}$, the dispersion relation $\omega_k \simeq v|k|$, and the eigenmode wavefunction $u_k(x)$. The coupling constants strictly vanish on the zeros of the eigenmodes or their derivatives. For this reason, Figure 5.6 uses different types of transmission lines to engineer a capacitive and an inductive coupling. Moreover, since the wavefunctions are normalized, the constants decrease with the transmission line size d as $g_k \sim \mathcal{O}(d^{-1/2})$. Fortunately, this prefactor is also the spacing in momentum space $g_k \sim g_k^0(\Delta k)^{1/2}$. This allows replacing all sums with integrals $\sum_k |g_k|^2 \to \int \mathrm{d}k|g_k^0|^2$, producing finite values of decay rates and interaction strengths in the limit $d \to \infty$.

There are three methods to work with the Caldeira–Leggett Hamiltonian, depending on the tasks we want to perform:

(1) We neglect the quantum fluctuations from the control line, treating the microwave in the line as a classical field. The bath operators are replaced by complex numbers that add up to a coherent drive $\Omega(t)$, which we use to control the resonator, as in Section 5.5.1.
(2) We study the state of the resonator as a density matrix that results from tracing out the bath modes $\rho_{\text{cav}} = \text{tr}_{\text{bath}}\{|\psi\rangle\langle\psi|\}$. We derive a *master equation* for ρ_{cav} that describes the *leakage of energy* from the resonator into the line, as well as *decoherence* introduced by the fluctuations in the antenna. This description is used in Section 5.5.4.

(3) We study the state of the radiation that leaks from the circuit into the antenna. We derive Heisenberg equations relating the field on the line to the field in the LC circuit, in what is known as *input–output theory*. This approach is used to model the *spectroscopy* of a resonator in Section 5.5.3 and to develop techniques for *measuring the quantum state* of the circuit based on the leaked radiation in Section 5.5.7.

Input–Output Theory

Our goal is to analyze how the quantum fluctuations in the open transmission line affect the resonator. We assume that the exchange of energy between both elements happens at a slow pace, slower than the speed at which the LC circuit evolves, and the speed at which waves propagate through the line. In this limit, the line can control and monitor the resonator, without affecting too much its evolution. This separation of time scales justifies a RWA effective Hamiltonian:

$$H_{\mathrm{RWA}} = \hbar\omega\hat{a}^\dagger\hat{a} + \sum_k \left(\hbar g_k\hat{a}\hat{b}_k^\dagger + \hbar g_k^*\hat{a}^\dagger\hat{b}_k\right) + \sum_k \hbar\omega_k\hat{b}_k^\dagger\hat{b}_k. \tag{5.60}$$

The Heisenberg equations for this model describe a continuous exchange of photons between the oscillator and the bath (cf. Appendix B.3). Integrating out the bath, we get a Langevin equation for the resonator:

$$\frac{\mathrm{d}}{\mathrm{d}t}\hat{a} = -i\omega\hat{a}(t) - i\hat{\xi}(t) - \int_{t_0}^{t} K(t-\tau)\hat{a}(\tau)\mathrm{d}\tau. \tag{5.61}$$

The second term is the input field $\hat{\xi}(t) = \sum_k g_k^* e^{-i\omega_k(t-t_0)}\hat{b}_k(t_0)$ transported by the line. The third term is the field emitted by the resonator into the line in the past $\hat{a}(\tau)$, which may be partially reabsorbed.

The *memory function* $K(\tau)$ is the Fourier transform of the *spectral function*:

$$K(t) = \sum_k g_k g_k^* e^{-i\omega_k(t-\tau)} = \frac{1}{2\pi}\int J^{\mathrm{QO}}(\bar{\omega})e^{-i\bar{\omega}t}\mathrm{d}\bar{\omega}. \tag{5.62}$$

The spectral function of an ideal transmission line grows linearly with the frequency, acting as an *Ohmic bath*:

$$J^{\mathrm{QO}}(\bar{\omega}) = 2\pi\sum_k |g_k|^2\delta(\bar{\omega}-\omega_k) \simeq \pi\alpha\omega^1. \tag{5.63}$$

Whenever the spectral function is a such broad and smooth function, the Fourier transform $K(\tau)$ becomes extremely concentrated around $\tau = 0$. The *Markovian limit* is a regime in which we approximate the memory function with a Dirac delta,

$K(t) \simeq (i\delta_{\text{Lamb}} - \frac{\kappa}{2}) \times \delta(t)$. In the Markovian limit, the bath has no memory of the oscillator's past, and (5.61) becomes local in time:

$$\frac{\mathrm{d}}{\mathrm{d}t}\hat{a}(t) = \left(-i\omega' - \frac{\kappa}{2}\right)\hat{a}(t) - i\sqrt{\kappa}\,\hat{b}^{\text{in}}(t). \tag{5.64}$$

We have to read this equation as follows. First, the environment introduces a *Lamb shift* of the oscillator's resonance, $\omega' = \omega - \delta_{\text{Lamb}}$. This is a slowdown of the resonator caused by "dragging along" the modes of the line as it evolves. However, this slowdown is rarely discussed, because once the resonator is connected to the line we can only measure ω', not ω.

The second term $-(\kappa/2)\hat{a}$ is an exponential attenuation of the resonator field. It is caused by photons leaking into the bath at the *cavity decay rate* $\kappa = J^{\text{QO}}(\omega') \simeq \pi\alpha\omega'$. The ratio κ/ω' is uniform across most of the spectrum: low- and high-frequency resonators decay at the same relative speed. Since κ is the rate at which the bath and the resonator exchange energy, consistency with the RWA in (5.60) imposes the limitation $\kappa/\omega' \ll 1$ for this whole treatment to be justified.

The last term in the equation represents the injection of photons coming from the line into the resonator, $-i\sqrt{\kappa}\hat{b}^{\text{in}}$. When the transmission line contains a coherent microwave drive, such as the $\Omega(t)$ used to control the resonator, it is customary to perform a displacement of the input operators, separating this "classical" contribution $\hat{b}^{\text{in}}(t) \to \Omega(t) + \hat{b}^{\text{in}}(t)$, from a truly quantum noise operator $\hat{b}^{\text{in}}(t)$ that accounts for the quantum fluctuations that the bath injects into the resonator.

The Heisenberg equation for the cavity is accompanied by an *input–output relation* that connects the field reflected by the cavity $\hat{b}^{\text{out}}$ to the input field $\hat{b}^{\text{in}}$ and the field that leaks from the resonator:

$$\hat{b}^{\text{out}}(t) = \hat{b}^{\text{in}}(t) - i\sqrt{\kappa}\hat{a}(t). \tag{5.65}$$

Together with (5.64), this can be used to study cavity spectroscopy and quadrature measurements, as explained in Sections 5.5.3 and 5.5.7.

5.5.3 Cavity Spectroscopy

Spectroscopy is the study of radiation absorbed or emitted by a system, as a function of the frequency and intensity of the "light" with which we illuminate it. Many quantum mechanical systems – atoms, molecules, color centers in diamond, superconducting circuits – have a few experimentally relevant states with well-defined energies. The absorption and emission spectra reveal the transitions that are allowed between those states, the difference between those energies, hidden symmetries of the quantum object, and even sometimes the underlying Hamiltonians.

We have all tools to analyze the spectroscopy of an LC resonator or microwave cavity. Let us begin with a *reflected light* setup, sketched in Figure 5.6a. The input–output theory from Section 5.5.2 predicts the spectroscopic signal $\hat{b}^{\text{out}}$ as a function of the light that illuminates the resonator $\hat{b}^{\text{in}}$ and the instantaneous state of the circuit $\hat{a}(t)$. We can both show how to discriminate the resonator signal $\hat{a}$ from the total reflected field $\hat{b}^{\text{out}}$, as well as give predictions on the combined signal.

To simplify this task, we assume that the probe field is a coherent microwave field accompanied by a thermal microwave background. We split $\hat{b}^{\text{in}}(t) = \Omega(t) + \hat{\varepsilon}(t)$ with a classical displacement $\Omega(t)$ and a noise operator $\hat{\varepsilon}(t)$. The thermal noise averages to zero $\langle\hat{\varepsilon}(t)\rangle = 0$ when we study the first moments of the reflected field $\langle\hat{b}^{\text{out}}\rangle = \Omega(t) - i\sqrt{\kappa}\,\langle\hat{a}(t)\rangle$, giving us direct access to the state of the resonator. Most experiments do spectroscopy with monochromatic beams $\Omega(t) = \Omega_0 e^{-i\omega_d t}$. The Heisenberg equation for the resonator (5.64) predicts a quasistationary displacement of the cavity in the asymptotic limit:

$$\langle\hat{a}(t \to \infty)\rangle = \frac{-i\sqrt{\kappa}}{i(\omega' - \omega_d) + \kappa/2}\Omega_0 e^{-i\omega_d t}. \tag{5.66}$$

As anticipated in Section 5.5.1, the losses stabilize a coherent state with a number of photons around $4|\Omega_0|^2/\kappa^2$. The reemitted signal and the incident microwave interfere, producing the reflected field

$$\langle\hat{b}^{\text{out}}(t \to \infty)\rangle = \frac{i(\omega' - \omega_d) - \kappa/2}{i(\omega' - \omega_d) + \kappa/2}\Omega_0 e^{-i\omega_d t} = r(\omega_d)\Omega(t). \tag{5.67}$$

The cavity reflects the field $\Omega(t)$ with a phase shift $r(\omega_d) = e^{-i2\arctan(2\delta/\kappa)}$. This shift can be experimentally determined (cf. Section 5.5.7) and used to estimate the resonance ω' – not the bare frequency ω! – with very high precision.

Measuring the reflected signal from a circuit is inconvenient, because the signal has to be separated from the original drive using circulators. A common alternative is to work with transmitted microwaves, using the configuration in Figure 5.4a, where we inject the signal into one side of the resonator, and measure the output signal on the other side. We can extend the input–output theory to consider the two decay channels of the $\lambda/2$ resonator: to the left $\hat{a}_L$ and to the right $\hat{a}_R$, with the total decay $\kappa = \kappa_L + \kappa_R$:

$$\frac{\mathrm{d}}{\mathrm{d}t}\hat{a} = (-i\omega' - \kappa/2)\hat{a} - i\sqrt{\kappa_L}\hat{b}_L^{\text{in}} - i\sqrt{\kappa_R}\hat{b}_R^{\text{in}}, \tag{5.68}$$

$$\hat{b}_{L,R}^{\text{out}} = \hat{b}_{L,R}^{\text{in}} - i\sqrt{\kappa_{L,R}}\hat{a}.$$

A signal $\Omega_0 e^{-i\omega_d t}$ coming from the right port $\hat{b}_R^{\text{in}}$ results in a transmitted signal $\hat{b}_L^{\text{out}}$ and a reflected signal $\hat{a}_R^{\text{out}}$. For instance:

$$\langle \hat{b}_L^{\text{out}} \rangle = \frac{\sqrt{\kappa_L \kappa_R}}{i(\omega' - \omega_d) + (\kappa_L + \kappa_R)/2} \Omega_0 e^{-i\omega_d t}, \tag{5.69}$$

$$\langle \hat{b}_R^{\text{out}} \rangle = \frac{i(\omega' - \omega_d) - (\kappa_R - \kappa_L)/2}{i(\omega' - \omega_d) + (\kappa_L + \kappa_R)/2} \Omega_0 e^{-i\omega_d t}. \tag{5.70}$$

The reflected signal is now zero when on resonance $\omega_d = \omega'$, and the transmitted signal only contains the field emitted by the cavity:

$$n(\omega_d) \propto \frac{|\Omega_0|^2}{(\kappa/2)^2 + (\omega_d - \omega)^2}. \tag{5.71}$$

In the transmission spectroscopy setup, the LC circuit is a filter that transmits frequencies on a narrow band around $\omega_d = \omega'$. The *full-half-width* (FHW) of this filter – the separation between the two points at 50% maximum transmitted power – is given by the decay rate κ and by the inverse of the *quality factor* $Q = \omega/\kappa$. A $Q \sim 10^5$ means that the resonator filters out frequencies that are 0.002% outside the central frequency ω'. Such a cavity or resonator is a very good isolator in which to embed other circuits.[11] When those circuits are off-resonant with the cavity, they will be effectively shielded from the environment. We can undo this shielding and allow the circuit talk to the $\hat{b}_k$ modes by shifting its resonances, bringing it closer to the frequencies ω' allowed by the resonator. These ideas will be explored in later chapters about cavity-QED and quantum computing devices.

5.5.4 Losses and Heating

The quantum state of the resonator is described by a reduced density matrix $\rho = \text{tr}_{\text{bath}} |\psi\rangle\langle\psi|$, after tracing out the field. The evolution of $\rho(t)$ can be deduced from the Langevin equation (Gardiner and Zoller, 2004) or directly from the Schrödinger equation (see Appendix B.2). We need to assume that the bath remains in a relatively unperturbed state. More precisely, we separate all coherent microwave fields that propagate through the line, from the quantum fluctuations that are intrinsic to the bath $\hat{b}_k \to \beta_k(t) + \hat{b}'_k$. The complex numbers β_k add up to create $\Omega(t)$. The quantum fluctuations are usually fixed to a Bose–Einstein thermal equilibrium distribution:

$$\bar{n}_\omega = \int \langle \hat{b}'_k \hat{b}'_k \rangle \delta(\omega_k - \omega) \mathrm{d}\omega = \frac{1}{e^{-\hbar\omega_k/k_B T} + 1}. \tag{5.72}$$

[11] Note the superconducting qubit attached to the transmission line resonator in Figure 5.1c. In other montages, the resonator sits in between a transmission line and the quantum register, as seen in Figure 8.2.

The result is a *Lindblad equation* or *master equation*:

$$\partial_t \rho = -i[\hat{H}_{\text{eff}}, \rho] + \frac{\kappa}{2}(\bar{n}_{\omega'} + 1)\left(2\hat{a}\rho\hat{a}^\dagger - \hat{a}^\dagger\hat{a}\rho - \rho\hat{a}^\dagger\hat{a}\right) + \frac{\kappa}{2}\bar{n}_{\omega'}\left(2\hat{a}^\dagger\rho\hat{a} - \hat{a}\hat{a}^\dagger\rho - \rho\hat{a}\hat{a}^\dagger\right). \tag{5.73}$$

The master equation contains an effective Hamiltonian $\hat{H}_{\text{eff}} = \omega'\hat{a}^\dagger\hat{a} + \Omega(t)\hat{a}^\dagger + \Omega(t)^*\hat{a}$, with the same microwave drive Ω and renormalized frequency ω' as before. There are also two Lindblad superoperators cooling and heating the density matrix with rates $\kappa(\bar{n}_{\omega'}+1)$ and $\kappa\bar{n}_{\omega'}$, both of which depend on the temperature, through the thermal occupation of the modes (see Section 4.1.1 and Figure 4.1). For experiments with resonators around 6–10 GHz and temperatures of 10 mK, we typically neglect the thermal occupation $\bar{n}_{\omega'} \simeq 0$ and all the heating terms (cf. Section 5.4.2).

In absence of external driving, the cooling and heating terms reach a balance (Exercise 5.8):

$$\frac{\mathrm{d}}{\mathrm{d}t}\langle\hat{a}^\dagger\hat{a}\rangle = \kappa(\bar{n}_{\omega'} - \langle\hat{a}^\dagger\hat{a}\rangle) \tag{5.74}$$

and the resonator equilibrates with the bath $\langle\hat{a}^\dagger\hat{a}\rangle = \bar{n}_{\omega'}$. We call the timescale for thermal equibration, $1/\kappa$, the T_1 time. During thermalization, the density matrix *decoheres:* The off-diagonal terms $\rho_{n\neq m} = \langle n|\rho|m\rangle$ or *coherences* die off at a slightly lower pace, known as $T_2 \sim 2/\kappa$.

We could slow decoherence and dissipation by insulating the resonator, reducing κ. However, this happens at the expense of our ability to control and measure the resonator, since both $\hat{b}^{\text{out}}, \Omega \propto \sqrt{\kappa}$. For this reason, experiments strike a balance between decoherence and control, engineering κ in the range of kilohertz to megahertz for resonator frequencies of gigahertz. These value lead to *quality factors* $Q = \omega/\kappa$ between 1 000 and 10^6. We may interpret Q as the number of times a photon bounces between the resonator's walls before being lost into the environment. These values are not only competitive with visible range quantum optics, but they are so good that several groups are exploring the use of microwave photons instead of qubits to store and manipulate quantum information.[12]

5.5.5 Beam Splitters and Circulators

There are two devices that are quite common when working with propagating microwaves, both of which implement linear transformations of propagating microwaves. The *beam splitter* is a four-port element dividing the signal of each

[12] In the work by Vlastakis et al. (2013), $\omega \sim 2\pi \times 8.18$ GHz, $\kappa \sim 2\pi \times 7.2$ kHz and $Q \sim 10^6$. This supports extremely long-lived Schrödinger cat states and error correction!

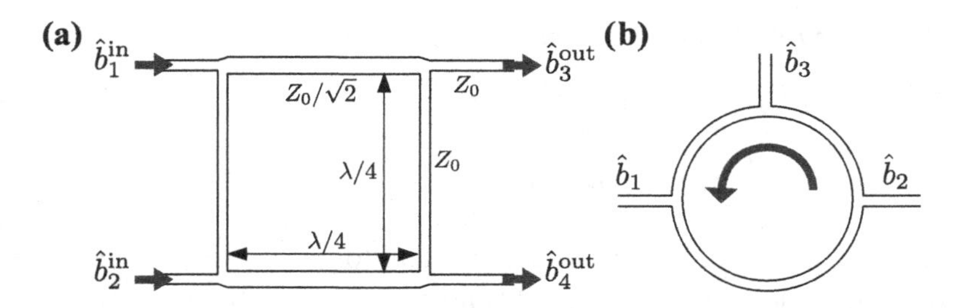

Figure 5.7 (a) Microwave beam-splitter with 90° hybrid design. (b) Scheme of a three-port circulator.

port into two different ports. In visible optics, beam splitters are partially reflective mirrors, but in microwaves we have to use sophisticated designs of waveguides with different lengths and impedances.

Figure 5.7a shows a 90° *hybrid*, a microwave beam splitter with resonance frequency $\lambda/2$. The device is a square with four segments of coplanar waveguides with the same length $\lambda/2$, and different impedances – Z_0 on the vertical arms, $Z_0/\sqrt{2}$ on the horizontal segments. The hybrid connects to four waveguides at its four corners, which we can use to inject and extract microwaves. Right at the frequency resonant with the hybrid's wavelength λ, the input and output modes are related by the simple unitary transformation of a 50–50 beam splitter:

$$\begin{pmatrix} \hat{b}_1^{\text{out}} \\ \hat{b}_2^{\text{out}} \\ \hat{b}_3^{\text{out}} \\ \hat{b}_4^{\text{out}} \end{pmatrix} = \frac{-1}{\sqrt{2}} \begin{pmatrix} 0 & 0 & i & 1 \\ 0 & 0 & 1 & i \\ i & 1 & 0 & 0 \\ 1 & i & 0 & 0 \end{pmatrix} \begin{pmatrix} \hat{b}_1^{\text{in}} \\ \hat{b}_2^{\text{in}} \\ \hat{b}_3^{\text{in}} \\ \hat{b}_4^{\text{in}} \end{pmatrix}. \tag{5.75}$$

If we inject a signal through port 1 or through port 2, it splits into an equal superposition of ports 3 and 4. For instance, $S^\dagger \hat{b}_1^{\text{in}\,\dagger} S = -\frac{i}{\sqrt{2}}\hat{b}_3^{\text{out}\,\dagger} + -\frac{1}{\sqrt{2}}\hat{b}_4^{\text{out}\,\dagger}$. The scattering matrix of the hybrid depends on the frequency of the incoming microwave. Away from resonance, not only is the splitter unbalanced, but some the energy is spread over all four ports – including some undesired back-reflection. However, these effects are negligible within a broad band of 10% the central resonance (Schneider, 2014).

Beam splitters, in this or other designs, are used to divide and to merge signals. For instance, in Section 5.5.7 we divide a signal into two different ports $\hat{b}_{3,4}$ and process those ports separately. This allows us to measure two different quadratures $(\hat{b}_1^\dagger + \hat{b}_1)$ and $i(\hat{b}_1^\dagger - \hat{b}_1)$ of the original input field. Beam splitters can also be used combine light from two different bosonic modes and create entangled microwaves (Menzel et al., 2012).

The *circulator* is a curious device that implements a three-port system without time-reversal symmetry. The scattering matrix of the circulator from Figure 5.7b is

$$\begin{pmatrix} \hat{b}_1^{\text{out}} \\ \hat{b}_2^{\text{out}} \\ \hat{b}_3^{\text{out}} \end{pmatrix} = \begin{pmatrix} 0 & 0 & 1 \\ 1 & 0 & 0 \\ 0 & 1 & 0 \end{pmatrix} \begin{pmatrix} \hat{b}_1^{\text{in}} \\ \hat{b}_2^{\text{in}} \\ \hat{b}_3^{\text{in}} \end{pmatrix}. \tag{5.76}$$

The circulator maps an input signal from one port straight into the following port, with a circular order denoted in circuits by an arrow (see Figure 5.7b). A circulator breaks time reversal symmetry: If we reflect back a signal that exits from port 2, it does not return to the input port 1, but exits from port 3. To break this symmetry, circulators are built with permanent magnets, as bulky devices that cannot be placed on chip – although there is ongoing research in implementing such circulators using 2D electron gases or periodically driven quantum circuits.

Circulators are used to isolate systems or separate signals. In the first type of applications, we connect our circuit to port 1, ground port 3, and connect a measurement device on port 2. This way our circuit is not affected by the noise generated by the measurement device – which sinks into port 2. In the second type of application, the setup is identical but we connect a microwave source to port 3. This way, the measurement device only collects the light reflected by the circuit at port 1 and is not affected by the microwave source.[13]

5.5.6 Amplification

The Gaussian set of operations includes the amplification of one or more quadratures to make them more easily measured. Amplification appears naturally when look at the single-mode (5.38) and two-mode (5.41) squeezing transformations in terms of the gain factor $G = \cosh(r)^2 \geq 1$:

$$\hat{a}(r) = \sqrt{G}\hat{a}(0) - \sqrt{G-1}e^{i2\theta}\hat{a}^\dagger(0), \tag{5.77}$$

$$\hat{b}_1(r) = \sqrt{G}\hat{b}_1(0) - \sqrt{G-1}e^{2i\theta}\hat{b}_2^\dagger(0). \tag{5.78}$$

The first equation is an example of a *degenerate, phase-sensitive linear amplifier*. It is degenerate because it operates using a single mode $\hat{a}$ and just one frequency of photons. It is phase sensitive because amplification only takes place along one direction in phase space, $\hat{p}_\phi(r) = \frac{i}{\sqrt{2}}(\hat{a}^\dagger e^{i\theta} - \hat{a}e^{-i\theta})$, which is the one we should use in the measurements. This type of process is also called *noiseless amplification* because it preserves the relative strength of fluctuations $\Delta\hat{p}/\hat{p}$, without any increase in noise due to the amplification.

[13] See, for instance, the work by Hoi et al. (2011, 2012), where a superconducting qubit operates on microwave light, sorting out photons and creating non-Gaussian states.

The second equation is an example of a *nondegenerate, phase-insensitive linear amplifier*. In this paradigm, we distinguish the *signal* $\hat{b}_1$ from the *noise* mode $\hat{b}_2$. In actual experiments, $\hat{b}_1$ and $\hat{b}_2$ tend to have different frequencies so that they can be filtered separately, hence the name *non-degenerate*. The amplifier is phase insensitive because the signal operator contains two quadratures $\hat{b}_1 = \frac{1}{\sqrt{2}}(\hat{x}_1 + i\hat{p}_1)$ that are simultaneously amplified. In other words, we use the same setup to measure $\hat{x}_1$, $\hat{p}_1$ or any other rotated quadrature and all of them will be amplified with the same gain $\sqrt{G}$. This simultaneous amplification is performed with a cost: The fluctuations of $\hat{b}_2$ add up a contribution to the variance of $\hat{b}_1$, $\hat{x}_1$, and $\hat{p}_1$. Assuming that the amplifier mode is in an uncorrelated thermal state, we write

$$\Delta\hat{b}_1(r)^2 = G\,\Delta\hat{b}_1(0)^2 + (G-1)\left(\bar{n}_{\text{amp}} + \tfrac{1}{2}\right). \tag{5.79}$$

The last term is the number of *photons added* by the amplification process. A *quantum limited amplifier* is one which adds the minimum amount of photons allowed by quantum mechanics $\bar{n}_{\text{amp}} = \langle\hat{b}_2^\dagger\hat{b}_2\rangle = 0$.

Linear Parametric Amplifier

It is possible to build a linear amplifier by periodically modulating the frequency of a tunable resonator (Abdo et al., 2009). In an ideal realization, we modulate the inductance of the resonator with a single frequency $\sim \frac{1}{2}L^{-1}[1+\varepsilon\cos(\Omega t+\theta)]\hat{\phi}^2$. The oscillating perturbation $\propto \varepsilon$ is written down using the Fock operators of the unmodulated Hamiltonian ($\varepsilon = 0$). Expanding the cosine term and applying the RWA leads to the effective model:

$$\hat{H} = \hbar\omega_{\text{amp}}\hat{a}^\dagger\hat{a} + \tfrac{1}{2}\hbar g\left(e^{i(\Omega t+\theta)}\hat{a}^2 + e^{-i(\Omega t+\theta)}\hat{a}^{\dagger 2}\right). \tag{5.80}$$

We will assume that this oscillator is connected to a waveguide that brings the signal and collects the amplified output. Using the input–output theory for a single lead produces two linear equations:

$$\frac{\mathrm{d}}{\mathrm{d}t}\hat{a} = (-i\omega_{\text{amp}} - \kappa/2)\hat{a} - ige^{i(\Omega t+\theta)}\hat{a}^\dagger - i\sqrt{\kappa}\hat{b}^{\text{in}}, \tag{5.81}$$

$$\hat{b}^{\text{out}} = \hat{b}^{\text{in}} - i\sqrt{\kappa}\hat{a}. \tag{5.82}$$

We can solve these equations and establish a linear transformation from input $\hat{b}^{\text{in}}$ to output $\hat{b}^{\text{out}}$ modes in frequency space. Following Roy and Devoret (2016), we express $\hat{a} = (\hat{b}^{\text{in}} - \hat{b}^{\text{out}})/i\sqrt{\kappa}$ and construct an equation that couples annihilation and creation operators at different frequencies:

$$\left(\frac{\mathrm{d}}{\mathrm{d}t}+i\omega_a+\frac{\kappa}{2}\right)\hat{b}^{\text{out}}-ige^{-i(\Omega t+\theta)}\hat{b}^{\text{out}\,\dagger}=\left(\frac{\mathrm{d}}{\mathrm{d}t}+i\omega_a-\frac{\kappa}{2}\right)\hat{b}^{\text{in}}-ige^{-i(\Omega t+\theta)}\hat{b}^{\text{in}\,\dagger}. \tag{5.83}$$

This is more obvious when we Fourier transform the equation, using $\hat{b}(t)=\frac{1}{\sqrt{2\pi}}\int e^{+i\omega t}\hat{b}(\omega)\mathrm{d}\omega$, to derive an algebraic relation between modes at equidistant frequencies from the drive $\Omega=\omega_1+\omega_2$:

$$\frac{1}{\chi(\omega_1)}\hat{b}^{\text{out}}(\omega_1)-i\rho e^{-i\theta}\hat{b}^{\text{out}\,\dagger}(\omega_2)=\frac{-1}{\chi(\omega_1)^*}\hat{b}^{\text{in}}(\omega_1)-i\rho e^{-i\theta}\hat{b}^{\text{out}\,\dagger}(\omega_2). \tag{5.84}$$

We have introduced the single-mode bare susceptibility $\chi(\omega)^{-1}=2i(\omega_a-\omega)/\kappa+1$ and coupling strength $\rho=2g/\kappa$. These equations can be written in scattering form, as a unitary transformation $\det(S)=1$ between fields:

$$\begin{pmatrix}\hat{b}^{\text{out}}(\omega_1)\\ \hat{b}^{\text{out}\,\dagger}(\omega_2)\end{pmatrix}=S(\omega_1,\omega_2,\rho,\kappa)\begin{pmatrix}\hat{b}^{\text{in}}(\omega_1)\\ \hat{b}^{\text{in}\,\dagger}(\omega_2)\end{pmatrix}. \tag{5.85}$$

When the amplifier is operated in non-degenerate mode, we use two different frequencies $\omega_1\neq\omega_2$ and drive close to the parametric resonance $\Omega\simeq 2\omega_a$. The scattering matrix gives us the *direct* $G_D=|S_{11}|^2$ and *indirect gain* $G_I=|S_{12}|^2$ as a function of the rescaled detuning $\delta=(\omega_2-\omega_1)/\kappa$:

$$\begin{aligned}\hat{b}^{\text{out}}(\omega_1)&=\frac{1+\delta^2+\rho^2}{\rho^2+(i+\delta)^2}\hat{b}^{\text{in}}(\omega_1)+\frac{2\rho e^{-i\theta}}{\rho^2+(1+i\delta)^2}\hat{b}^{\text{in}\,\dagger}(\omega_2)\\&=S_{11}\hat{b}^{\text{in}}(\omega_1)+S_{12}\hat{b}^{\text{in}\,\dagger}(\omega_2).\end{aligned} \tag{5.86}$$

When we operate on direct gain, the signal is inserted and recovered at the same frequency, and the amplifier is phase preserving. When we use the indirect gain, the signal is injected in mode $\hat{b}^{\text{in}}(\omega_2)$ and recovered from $\hat{b}^{\text{out}}(\omega_1)\propto\hat{b}^{\text{in}\,\dagger}(\omega_2)$. This is called a *phase-conjugating* amplifier, because it transforms a coherent field $\Omega(t)$ into its complex conjugate $\sqrt{G_I}\Omega(t)^*$. The gains in the nondegenerate parametric amplifier can be quite large: Ideally, both G_D and G_I diverge when the cooperativity approaches $\rho\to 1$. This divergence is associated to instabilities in the driven cavity, which can be cured by shifting Ω slightly away from the parametric resonance, or adding internal losses to the resonator.

We can operate the same device as a phase-sensitive degenerate amplifier, taking the limit $\omega_2=\omega_1=\Omega/2$. The amplifier implements single-mode squeezing on $\hat{b}(\omega_1)$ and $\hat{b}(\omega_1)^\dagger$. The unitary matrix S has two eigenvalues, λ and $1/\lambda$, with the gain factor

$$\lambda=\sqrt{G}=\frac{\sqrt{\rho^2-\Delta^2}+1}{\sqrt{\rho^2-\Delta^2}-1}, \tag{5.87}$$

expressed in terms of the normalized detuning $\Delta=2(\omega_a-\omega_1)/\kappa$.

Josephson Parametric Amplifiers

We have mentioned that the linear parametric amplifier can be implemented using a tunable cavity, such as the one in Figure 5.4c, by periodically modulating the flux through a SQUID (Abdo et al., 2009). This design has two potential problems. The driving mechanism, with an external antenna, can introduce quantum fluctuations in the cavity, causing decoherence. Moreover, once we amplify the signal enough, there will be a large number of photons, and the SQUID will stop acting as a linear inductor.

There exists an alternative setup that embraces this nonlinearity. It is known as a Josephson parametric amplifier (JPA) and consists of a superconducting cavity with an embedded chain of SQUIDs (Castellanos-Beltran et al., 2009). This chain provides a knob to adjust the frequency of the cavity and creates a large nonlinear contribution to the Hamiltonian. The JPA's long $\lambda/2$ also supports two ports: One side of the resonator, strongly coupled, receives the *signal* that will be amplified; the other port, weakly coupled, is illuminated with a strong coherent microwave or *pump*. The pump strongly displaces the cavity and activates the nonlinearity, causing a periodic shift of the cavity's resonance. We treat this modulation as a c-number contribution to the cavity operator $\hat{a}(t) \to \hat{a}(t) + \alpha(t)$, keeping terms of order $|\alpha|$ or larger in the Hamiltonian. The result is a linear quadratic Hamiltonian (5.80).

The JPA is modeled as a single-mode nonlinear oscillator with a Kerr-type interaction between photons:

$$H_0 = \hbar\omega_a \hat{a}^\dagger \hat{a} - \frac{K}{2}\hat{a}^{\dagger 2}\hat{a}^2. \tag{5.88}$$

The Kerr nonlinearity K can be traced back to the Josephson junction potential, $\cos(\hat{\phi}/\varphi_0) = 1 - \frac{1}{2}(\hat{\phi}/\varphi_0)^2 + \frac{1}{4}(\hat{\phi}/\varphi_0)^4 + \cdots$. The input–output theory for the JPA with a signal $\hat{b}_S^{\text{in}}$ and a pump $\hat{b}_{\text{pump}}^{\text{in}}$ is

$$\frac{\mathrm{d}}{\mathrm{d}t}\hat{a} = \left(-i\omega_a - \frac{\kappa_1 + \kappa_2}{2}\right)\hat{a} + iK\hat{a}^\dagger\hat{a}^2 - i\sqrt{\kappa_1}\hat{b}_S^{\text{in}} - i\sqrt{\kappa_1}\hat{b}_{\text{pump}}^{\text{in}}. \tag{5.89}$$

The pump field is replaced with a complex classical drive $\hat{b}_{\text{pump}}^{\text{in}} \sim \beta e^{-i(\omega_p t+\xi)}$ that displaces the cavity by $\sim \alpha_0 e^{-i(\omega_p t+\xi)t} \sim \hat{b}_{\text{pump}}^{\text{in}}/\omega_a$ (see Section 5.5.1). A new bosonic operator $\hat{c} = \hat{a} - \alpha_0 e^{-i(\omega_p t+\xi)t}$ collects the quantum fluctuations around this displacement, producing a familiar Langevin equation:

$$\frac{\mathrm{d}}{\mathrm{d}t}\hat{c} \simeq \left(-i\omega_a - \frac{\kappa_1 + \kappa_2}{2}\right)\hat{c} + i2K\alpha_0^2\hat{c} + iK\alpha_0^2 e^{-i(2\omega_p t+2\xi)}\hat{c}^\dagger - i\sqrt{\kappa_1}\hat{b}_S^{\text{in}}. \tag{5.90}$$

This model is equivalent to the linear parametric amplifier in (5.81), with the pump $2\omega_p = \Omega$, phase $\theta = 2\xi$, intensity-dependent resonator frequency $\omega_{\text{amp}} = \omega_a - 2K\alpha_0^2$, and intensity-dependent coupling $g = K\alpha_0$. Using these identities, we can find the regimes of nondegenerate parametric amplification $\omega_p \simeq \omega_a'$, where the JPA acts as a phase-insensitive amplifier, as well as the phase-sensitive degenerate amplifier regime, $\omega_p \simeq \omega_S$.

JPAs are fantastic amplifiers that power many of today's experiments. Unfortunately, the stronger the amplification, the narrower the bandwidth over which the JPA operates. Typical designs of JPA amplify signals of gigahertz with large gains $G_{D,I} \sim 22$ dB, over a relatively narrow frequency band of around megahertz (Castellanos-Beltran et al., 2009).

There is an alternative to the JPA that supports comparable gains over a broader range of frequencies. These devices are known as *traveling wave parametric amplifiers* (TWPAs). TWPAs were introduced by Cullen (1960), but they have been adapted to work with a long, open string of chained Josephson junctions or SQUIDs. The chain provides a 1D model with a Kerr nonlinearity that is capable of amplifying the waves that propagate through it, with comparable gains of $\sim$20 dB over a bandwidth of 3 GHz (Macklin et al., 2015). A single amplifier can work over multiple control signals for a superconducting circuit, and can be used to amplify the output from many different measurement devices, simultaneously, and without interference between signals. This is ideal for large and complex quantum computing setups, where tens and probably hundreds of qubits must be manipulated. Unfortunately, the theory and details of TWPAs fall outside the scope of this book, but you are encouraged to dig into the literature to learn about this incredibly useful technology.

5.5.7 Photon Quadrature Measurements

There are no good microwave photodetectors and photon counters in the market yet. Instead, superconducting quantum circuits are analyzed using classical devices that measure the power or voltage. These classical detectors sit at the end of a chain of amplifiers – cold (inside the fridge) and hot (outside) mixers and filters that boost, shift around, and clean the signal.

Following input–output theory, the signal that leaves the experiment $\hat{b}^{\text{out}}$ contains information about the instantaneous state of our system, in the form of a small photon flux $\sqrt{\kappa}\hat{a}(t)$. Hence, we describe all these stages with a simple equation:

$$\hat{b}^{\text{amp}} = \sqrt{G}(\hat{b}^{\text{out}} + \hat{\varepsilon}^{\text{noise}}) = \sqrt{\bar{G}}\hat{a}(t) + \sqrt{G-1}\hat{h}^\dagger, \tag{5.91}$$

with a prefactor $\bar{G}$ that includes all gains and losses – also the attenuation at the cavity port κ – and an effective field $\hat{h}$ summing all classical and quantum noise. This includes noise from the amplifiers, beam splitters, detectors, and cables themselves.

The added noise is quantified in the number of photons it adds prior to the gain itself (5.79). This noise can be as low as $1/2$ in ideal quantum-limited amplifiers, or even lower in noiseless phase-sensitive amplifiers. However, when further connections and mixers are accounted for,[14] we find more common values of $1-10$ photons in Josephson-junction-based amplifiers, and $30-50$ in cyrogenic high-electron mobility transistor (HEMT) amplifiers.

Power Measurements

We can feed the signal $\hat{b}^{\text{amp}}(t)$ into a device that measures the integrated or *average power.* A broadband power meter uses diodes to rectify the microwave current, creating a dc signal that is calibrated to determine the average, peak, or integrated power in the signal. In practice, we need to very accurately calibrate the average power P_{noise} of the background noise – i.e., noise introduced by amplification stages, input lines, etc. – to separate the power of the actual signal $P_{\text{signal}} = P - P_{\text{noise}}$, from the measured power P.

Since the dynamical range of a detector can span multiple orders of magnitude, the signal is measured using a logarithmic scale. The dBm unit is a log-10 scale referenced to a standard power of 1 milliwatt. A power P in watts is converted to this scale as $P_{\text{dBm}} = 10\log_{10}(P/10^{-3})$. A very good broadband power detector operating in the gigahertz range can have a minimum sensitivity of -70 dBm. This sensitivity can be improved using spectrometers that work on narrow bands around the desired frequency. A good device can have a detection threshold of around -140 dBm/Hz, relative to the bandwidth. If we wish to detect photons generated by a cavity with a bandwidth κ from 1 kHz to 1 MHz, this means the threshold for detection will lay around -110 and -80 dBm, respectively.

An important problem is that even in this scale, quantum microwave signals can take ridiculously small values. When a cavity releases photons of frequency $\omega \sim 2\pi \times 10$ GHz at a rate $\kappa = 1$ MHz, the output power will be $P_{\text{photon}} = \hbar\omega\kappa \sim 6.626 \times 10^{-18}$ W, or $P_{\text{dBm}} \sim -171$ dBm. This is $\sim 10^{-9}$ or ~ -90 dB to below the threshold of commercial power detectors! Hence the need for various stages of amplification that bridge this gap.

Another problem is obtaining a good signal-to-noise ratio for P_{signal}, estimated as the difference of two large quantities, the measurement with signal P, and a calibration of the noise in the empty line P_{noise}. Both quantities have to be measured with

[14] See Exercise 5.10 to discover how these noises add up.

high precision, over long integration times, and fluctuating experimental conditions. This challenge in stability and calibration makes this measurement technique less desirable than quadrature measurement techniques.

Quadrature Measurements

The use of power meters has been superseded by the use of digital techniques. An analog-to-digital converter (ADC) can periodically sample the voltage of an incoming quantum signal, producing a high-resolution stream of data from which a computer – or a field programmable gate array (FPGA) card (Eichler et al., 2012) – estimates moments $\langle\hat{a}\rangle$, $\langle\hat{a}^\dagger\hat{a}\rangle$, etc. The benefit of this technique is that it provides more information than just a power measurement, but actual implementations face two obstacles. First, the ADC will only sample one quadrature, the voltage $\hat{a}(t) + \hat{a}^\dagger(t)$. Second, we must slow down the signal, because the ADC cannot sample fields $\hat{b}^{\text{amp}}(t) \sim \hat{a}(t)$ that oscillate at frequencies of gigahertz.

Theoretically, both problems have easy solutions. The first obstacle is avoided by dividing the signal $\hat{b}^{\text{amp}}$ into two copies – one in phase, and another one with a 90° rotation $e^{i\pi/2}\hat{b}^{\text{amp}}$ – and feeding each copy to a different ADC. This provides us with two streams of values, associated to the two quadratures $\frac{1}{\sqrt{2}}(\hat{b}^{\text{amp}} + \hat{b}^{\text{amp}\dagger})$ and $\frac{i}{\sqrt{2}}(\hat{b}^{\text{amp}} - \hat{b}^{\text{amp}\dagger})$.

The second problem is solved by looking at how the cavity operator evolves. From input–output theory (cf. Section B.3.2), we know that most of the oscillation in the field is due to "trivial" phases, $\hat{a}(t) = \hat{a}_{\text{slow}}(t)e^{-i\omega t}$. The actual information is encoded an operator that evolves at a much slower pace $\hat{a}_{\text{slow}}(t) = e^{i\omega t}\hat{a}(t)$. We just need to fabricate this operator by mixing the original signal with the opposite field $e^{i\omega t}$.

Both solutions are jointly solved by a type of microwave circuit called *mixers*. A simple mixer is a nonlinear circuit that takes two signals $f(t)$ and $i(t)$, and returns their product $f(t)i(t)$, as sketched in Figure 5.8a. If we feed our signal $\hat{a}(t)$ and a classical oscillating field $i(t) = \cos(\omega_m t + \theta)$ into a mixer, it will produce two copies of the signal at two sidebands. The first sideband oscillates with a lower frequency and is close to our slow modulation $e^{i(\omega_m t)}\hat{a}(t) \simeq e^{i(\omega_m - \omega)tx}\hat{a}_{\text{slow}}(t)$. The second sideband oscillates at higher frequency $\omega + \omega_m$ and can be filtered out. In addition to this, note how the sidebands will be phase-shifted an angle θ determined by the input field. We can use this angle to select which quadrature we wish to explore.

The complete setup is shown in Figure 5.8c. We feed the signal to an amplifier and chain the output into an *IQ mixer*. The IQ mixer uses two mixers to combine the signal with two reference microwaves, $\cos(\omega_m t)$ and $\sin(\omega_m t)$, derived from a stable *local oscillator*. The two outputs of the IQ mixer are two quadratures, $\hat{X}$ and $\hat{P}$, that are sampled by two ADCs. This produces two classical traces $X(t_n)$ and

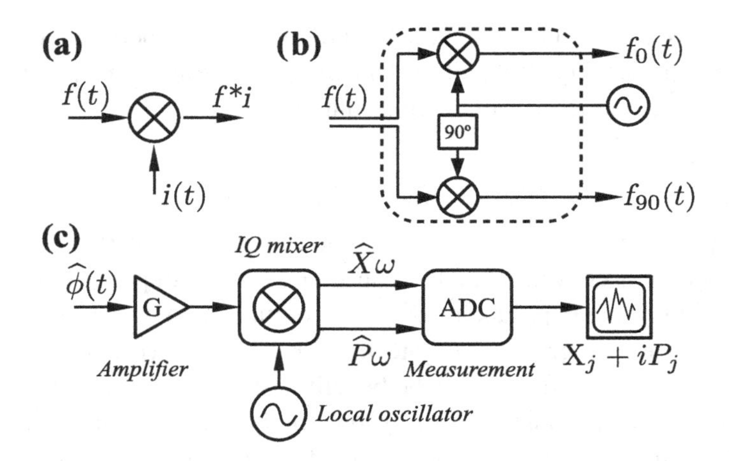

Figure 5.8 (a) A mixer is a nonlinear device that multiplies a signal $f(t)$ with a modulation $i(t)$. (b) An IQ mixer divides the input signal, combining it with two signals in 90° opposition $\cos(\omega t)$ and $\sin(\omega t)$. The output of the mixer contains the two orthogonal quadratures of the signal. (c) The combination of a mixer with a local oscillator, filters, and amplifiers allows us to measure any quadrature $\hat{Q}_{\omega,\theta} = \frac{1}{\sqrt{2}}(a_\omega e^{-i\theta} + a_\omega^\dagger e^{-i\theta})$ in a quantum field $\hat{\phi(t)}$. The classical signal $q(t)$ is a random sequence of voltages associated to one realization. Averages of many realizations of $q(t)$ estimate the expectation value $\langle \hat{Q}_{\omega,\theta}(t)\rangle$.

$P(t_n)$ that combine into a complex number $Z(t_n) = X(t_n) + i\,P(t_n)$. In practical applications, $\omega_m - \omega$ is never zero, but a finite value around MHz. This is slow enough that the ADC can reconstruct the signal, but large enough that the signal is not masked by slow $1/f$ noise in the electronics. If we take into account this residual oscillation and calibrate the gains of our circuit, we can build a sequence $S(t_n) = Z(t_n)e^{-i(\omega_m-\omega)t}/\sqrt{\bar{G}}$, that is a sampled measurement of our slow field with some noise $\hat{s}(t) = \hat{a}_{\text{slow}}(t) + \hat{h}^\dagger(t)$.

The classical values $S(t_n)$ can be used to reconstruct a measurement of resonator quadratures $\hat{a}$. Let us assume that we prepare the LC circuit and stop all external controls. As soon as we do, the cavity field begins to decay exponentially $\hat{a}_{\text{slow}}(t) = \hat{a}(0)e^{-\kappa t/2}$. Using the weight function $w(t) = \kappa e^{-\kappa t/2}$, we recover the cavity operator from traces of our ADC chain:

$$\int w(t)\hat{s}(t)\mathrm{d}t = \hat{a}(0) + \hat{h}_w^\dagger \sim S_w := \sum_n w(t_n)S(t_n)\Delta t. \tag{5.92}$$

In other words, in each experimental realization, the weighted average S_w implements a measurement of the operator $\hat{a}(0) + \hat{h}_w^\dagger$. If we repeat the experiment many times, we can estimate the ensemble averages $\langle S_w\rangle$, $\langle S_w^* S_w\rangle$, $\langle S_w^2\rangle$, ... These values provide the statistics of the cavity because (i) odd moments of the noise cancel out

$\langle \hat{h}_w^\dagger \rangle = 0$, (ii) the statistics of the noise are independent of the signal $\langle \hat{h}_w^\dagger \hat{a} \rangle \simeq \langle \hat{h}_w \rangle \langle \hat{a} \rangle \simeq 0$, and (iii) noise moments – $\langle \hat{h}_w \hat{h}_w^\dagger \rangle$ or higher – can be calibrated from experiments with an empty cavity. This allows us approximate arbitrary moments of the cavity field as $\langle \hat{a}(0) \rangle \simeq \langle S_w \rangle$, $\langle \hat{a}^\dagger(0)\hat{a}(0) \rangle \simeq \langle S_w^* S_w \rangle - \langle h_w \hat{h}_w^\dagger \rangle$, etc. First, second, third, and higher moments are used to test whether a state is Gaussian, and to do Wigner function tomography as explained in Section 5.4.6.

This analysis of bosonic quadratures is a powerful method that has been used to reconstruct the wavefunctions of squeezed states (Mallet et al., 2011) and of single photons (Eichler et al., 2011). The quadratures themselves allow the reconstruction of other properties, such as the energy of a wavepacket with less than one photon – i.e., values around 10^{-26} Joules! – or the spectral distribution of a photon wavepacket (Menzel et al., 2010).

5.6 Conclusion

We close here a long chapter on microwave photons. The chapter began discussing how linear circuits – and nonlinear circuits operating in a linear regime – store energy in quantized units, which we called photons. These bosonic excitations behave as optical photons in all respects, from the linear spectrum of equispaced eigenenergies, $\hbar\omega, 2\hbar\omega, 3\hbar\omega \ldots$, to the quantum description of modes and wavefunctions.

The noninteracting nature of photonic excitations poses both advantadges as well as problems. Among the advantages, we find a relatively simple mathematical description, access to amplification, and real-time tomography of wavefunctions. Photons can also be confined in high-quality environments, for long times of hundreds of microseconds and support long-lived entangled states and quantum superpositions.

The problems begin when we want to create those interesting states or think about developing a quantum computer. We cannot use linear circuits to generate or detect individual photons, and we are constrained to a family of states (Gaussian states) that can be simulated classically and pose no computational interest. To escape these constraints, we must introduce *nonlinearities*. This is the focus of the next chapter, where we study the design and operation of superconducting qubits.

Exercises

5.1 Compute the expected values and uncertainties of $\hat{q}$, $\hat{\phi}$, $\hat{V}$, and $\hat{I}$ for the vacuum state $|0\rangle$ of an LC resonator.

5.2 Plot the average occupation of a microwave cavity as a function of the cavity resonance ω and the temperature of the cryostat, for temperatures ranging from

5–100 mK. What is the minimum frequency at which the occupation number lays below 10%? And 1%?

5.3 Show that the number of photons in a coherent state $|\alpha\rangle$ is $\langle\hat{n}\rangle = |\alpha|^2$. Compute the uncertainty of all quadratures, $\{\hat{\phi}, \hat{q}\}$, on such a state.

5.4 Show that the ground state of Hamiltonian (5.8), for constant drive $\Omega(t) = \Omega_0$, is a coherent state with $\alpha \propto \Omega$. Hint: Use the definition of the coherent state in terms of displacement operators, and the commutation relations between a, $a^\dagger$, and $D(\alpha)$, such as $\hat{a}D(\alpha) = D(\alpha)(\hat{a} + \alpha)$.

5.5 If we start with the vacuum as initial state $|\Psi(0)\rangle = |0\rangle$, the coherent drive $\Omega(t)$ in (5.8) will produce a coherent state $|\Psi(t)\rangle \propto |\alpha(t)\rangle$ with the displacement from (5.10). Relate the Schrödinger and Heisenberg pictures, showing that $U(t)^\dagger \hat{a}\,|\Psi(t)\rangle = \hat{a}(t)\,|\Psi(0)\rangle$. Use the explicit formula for $\hat{a}(t)$ to demonstrate that $|\Psi(t)\rangle$ is a coherent state. Bonus points if you get the phase of the coherent state right.

5.6 Compute the Wigner function for a single photonic mode in (i) a coherent state, (ii) a Fock state with one photon, and (iii) a Schrödinger cat with generic displacement α. Which of these states is Gaussian? Which of them has a negative Wigner function? Which of them is broader in phase space and which of them saturates the Heisenberg uncertainty relation?

5.7 We want to place a magnetic molecule inside a $\lambda/2$ and a $\lambda/4$ transmission line resonator. The molecule couples to the magnetic field generated by the nearby currents in the line.

(1) Which position optimizes the coupling between the molecule and the fundamental mode, ω_0?

(2) Which position optimizes the coupling to the first harmonic of the cavity, ω_1?

(3) Where should we place the molecule so that it couples equally well to both modes, ω_0 and ω_1?

5.8 Cavity spectrum. In Section 5.5.3, we studied how the driven-dissipative resonator stabilizes to an average occupation number. We want to solve the same problem but using master equations:

(a) Starting from the driven model $H = a^\dagger a + (\Omega_0^* e^{-i\omega_d t} a + \Omega_0 e^{i\omega_d t} a^\dagger)$, and following Section B.2, show that the master equation in the interaction picture becomes a function of the detuning, $\delta = \omega_d - \omega$:

$$\partial_t \rho = -i\left[\delta \hat{a}^\dagger \hat{a} + \Omega_0 a + \text{H.c.}, \rho\right] + \frac{\kappa}{2}\left(2\hat{a}\rho\hat{a}^\dagger - \hat{a}^\dagger\hat{a}\rho - \rho\hat{a}^\dagger\hat{a}\right). \quad (5.93)$$

(b) Write and solve the evolution equations for $\langle\hat{a}\rangle = \mathrm{tr}(\hat{a}\rho)$ and for the number of photons $\langle\hat{a}^\dagger\hat{a}\rangle$. Hint: Use the cyclic property of the trace to convert terms of the form $\mathrm{tr}(A\rho B) = \mathrm{tr}(BA\rho) = \langle BA\rangle$.

(c) Show that the FHW linewidth is $\delta = \kappa$.

5.9 Assume a cavity without driving, coupled to an environment at zero temperature. Use the master equation (5.73) to solve analytically the density matrix of the cavity, using as initial conditions the pure states $\rho(0) = |\psi\rangle\langle\psi|$ in a superposition of one photon and the vacuum, $|\psi\rangle = \cos(\theta)|0\rangle + \sin(\theta)|1\rangle$. Show that the average number of photons $\langle\hat{a}^\dagger\hat{a}\rangle = \mathrm{tr}(\rho(t)\hat{a}^\dagger\hat{a})$ in this state follows (5.74). Study the decay of the coherence $\langle 0|\rho(t)|1\rangle$. How fast is it, and how does it relate to κ?

5.10 A microwave signal passes sequentially through two amplifiers with gains $G_{1,2}$ and added number of photons $\bar{n}_{1,2}$. Each stage is described by an equation of the form (5.91), with noise operators $\hat{h}^\dagger_{1,2}$ that are independent from each other – e.g. $\langle\hat{h}_1^{n\dagger}\hat{h}_1^{m\dagger}\hat{h}_2^{r\dagger}\hat{h}_2^{k}\rangle = \langle\hat{h}_1^{n\dagger}\hat{h}_1^{m\dagger}\rangle\langle\hat{h}_2^{r\dagger}\hat{h}_2^{k}\rangle$ – and from the signal. Use this property to estimate the total gain and total number of added photons of the device.

6
Superconducting Qubits

We have already introduced the qubit as the smallest quantum mechanical system that we can study. In this chapter, we further explore qubits, examining how they behave, both formally – through Schrödinger and master equations – and through their implementation in various superconducting circuits. Each design has unique characteristics, but we can develop a common framework for describing the qubit's evolution, its coupling to the environment and to external controls, the qubit's quality and coherence properties, etc. We also discuss in detail the most used qubits in present and past state-of-the-art experiments – charge qubits, transmons, and flux qubits. This chapter intentionally drops some interesting circuits – fluxoniums, g-mons, and other curious animals out there – but it should be enough to approach any qubit architecture critically, deducing their Hamiltonians and coherence properties, and understanding the pros and cons of each those designs critically.

6.1 What Is a Qubit?

6.1.1 From Logical to Physical Qubits

In classical information theory, the bit is the smallest unit of information and corresponds to a discrete variable with two states, labeled 0 and 1. The *qubit* – or *quantum bit* – generalizes this idea, denoting the smallest useful amount of quantum information. A qubit may adopt any possible state in the Hilbert space created by the superpositions of two orthogonal and physically distinguishable states $|0\rangle$ and $|1\rangle$. The continuum of states for a qubit includes not only the basis states $|0\rangle$ and $|1\rangle$, but any other pure state – a linear combination $\alpha\,|0\rangle + \beta\,|1\rangle$ with normalized complex amplitudes $\alpha, \beta \in \mathbb{C}$, $|\alpha|^2 + |\beta|^2 = 1$ – or mixed state – Hermitian matrices $\rho = \sum_{i,j=0,1} \rho_{ij}\,|i\rangle\langle j|$ with $\text{tr}\rho = 1$, and $\rho^\dagger = \rho$ (see Sections 6.1.5 and 6.1.6).

The word *bit* applies to both the unit of information and to the physical system that encodes it. Similarly, a qubit would be the smallest unit of quantum information, as well as any physical system that embodies those two-dimensional superpositions.

Qubits as such do not exist in Nature, because all known physical systems require more than just two quantum states to be described. A photon has a polarization degree of freedom, but also frequency and momentum; an electron has a spin, but also a spatial wavefunction. Creating a *physical qubit* requires controlling the dynamics of a quantum system, constraining it to a subset of all physically available quantum states.

In addition to this simplified dynamics, there other practical requirements, including preparation, read-out, and measurement. We can group them in a "shopping list," a variation of DiVincenzo's famous requirements for a quantum computer (DiVincenzo, 1995):

(1) The physical qubit has two accessible orthogonal eigenstates $|0\rangle$ and $|1\rangle$.
(2) It supports arbitrary, long-lived quantum superpositions $\alpha\,|0\rangle + \beta\,|1\rangle$.
(3) We can reset the qubit to some state, typically $|0\rangle$.
(4) We can do a projective measurent on the computational basis, typically through the observable $\sigma^z = |1\rangle\langle 1| - |0\rangle\langle 0|$.
(5) We can perform (or approximate) arbitrary unitary rotations in the qubit's Hilbert space.
(6) We can implement at least one type of universal two-qubit operation among pairs of qubits.
(7) We can engineer devices with medium to very large numbers of qubits.
(8) Tolerances in all operations – unitary gates, measurements, qubit reset – are tight enough to implement fault-tolerant quantum computation.

Out of this list, items (1)–(5) suffice for many applications in quantum communication and quantum optics. Combined with some type of qubit–qubit interaction – which unlike in term (6) does not need to be very precise (1) this even allows us to build some types of quantum simulators. Conditions (6) and (7) were recently achieved together in setups with more than 50 qubits (Arute et al., 2019; Wu et al., 2021). This enables near-term intermediate scale quantum computers – also known as noisy intermediate-scale quantum (NISQ) computers thanks to Preskill (2018) – to perform tasks, such as the simulation of random quantum circuits, which are arguably difficult in the classical computing world. It also opens the door to other useful applications of imperfect near-term quantum computers, especially in the study of quantum physical systems.

However, the ultimate goal in the field is to achieve *fault-tolerant quantum computation* as a means to build quantum computers that run arbitrarily long quantum algorithms. Fault-tolerant devices need extremely large numbers of qubits to encode

Table 6.1. *Some qubits and energy scales involved.*

Object	**Degree of freedom**	**Energy scales**
Photon	Polarization	IR to UV
"	Frequency: two frequencies	"
"	Which-way: a photon moves through one of two waveguides	"
Electron	Electronic spin	GHz to THz
"	Position in two quantum wells	GHz
Nuclei	Angular momentum	GHz
Atoms, ions molecules, color centers	Electronic state	400 THz–10^{15} Hz
	Hyperfine state	GHz
	Position in lattice	kHz–MHz
Molecules	Rotational states	100 GHz
"	Vibrational states	THz
Superconductors	Charge/current states	GHz

information redundantly. These qubits are operated with arbitrary numbers of measurements and quantum gates to implement quantum algorithms that are resilient to environmental and operational errors. We will discuss this requirement further in Section 8.5. However, as of the writing of this book this is a regime that is far from being achieved, even if we have promising results and proofs of principle demonstrators Chen et al. (2021).

The quest for physical qubits has ran in parallel to – and often been the motivation for – the search for controllable quantum systems in the lab. Table 6.1 enumerates the most successful qubit systems, together with the degrees of freedom that encode the information and the energy scales involved. Indeed, the energy scales of the qubit degrees of freedom are extremely important, because they underlay the general conditions for doing quantum experiments regarding temperature, isolation requirements, cooling, and preparation times (see Section 4.1.1).

However, our shopping list introduces a new and extremely important consideration: the need of single out two states from an experimental device to encode a qubit. This is not as easy as it sounds. Take, for instance, the photonic superconducting circuits from Section 5.4. We could identify the qubit with the states of zero and one photon, $|0\rangle$ and $|1\rangle$. However, as discussed in Problem 6.2, we cannot rely on microwaves to implement transitions between $|0\rangle$ and $|1\rangle$ without involving other photon number states. Linear LC-type circuits therefore do not satisfy the requirements for qubit preparation, arbitrary rotations in the qubit space, or even projective measurements.

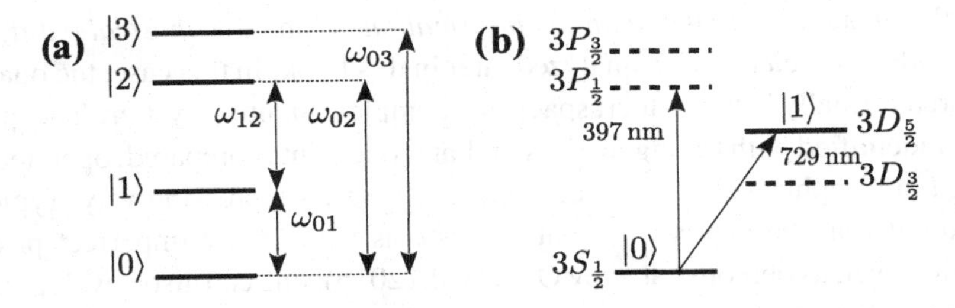

Figure 6.1 (a) A general qubit uses two metastable states $|0\rangle$ and $|1\rangle$, out of a spectrum that is anharmonic – that is, the energy spacing is not uniform. In particular, the qubit ω_{01} is not an integer multiple or fraction of the energy difference to other neighboring states of the system, ω_{0n} or ω_{1n}. (b) Energy levels of the Ca^+ ion, including the two qubit states $|0\rangle = |S_{1/2}\rangle$ and $|1\rangle = |D_{5/2}\rangle$.

One solution is to focus on a degree of freedom that is naturally reduced to a two-dimensional space – e.g., the spin of an electron, the polarization of a propagating photon – while simultaneously "freezing" other properties of the quantum system. In this book, and in most of circuit-QED, we instead look for physical systems with an intrinsically *anharmonic energy spectrum*, such as the one in Figure 6.1a. In this picture, the $|0\rangle$ is mapped to the ground state – facilitating the reset of the qubit – and the $|1\rangle$ is the first excited and long-lived state. Anharmonicity means that the energy gap $\hbar\omega_{01}$ between qubit states is not an integer fraction of the gap to other neigboring states. In other words, ω_{1n}, ω_{0n} cannot be a multiple of ω_{01}. This implies that we can drive transitions between our qubit states $|0\rangle$ and $|1\rangle$ using resonant photons with energy $\hbar\omega_{01}$, without *leakage* – transitions to other states.

The anharmonicity condition is often fulfilled by matter qubits. Take, for instance, a trapped ion quantum computer using Ca^+ atoms (cf. Figure 6.1b). The qubit's zero and one states are associated to the ground state manyfold $S_{1/2}$ and to the metastable state $D_{5/2}$. When we illuminate the ion with a laser at a wavelength 729 nm, we bridge the energy difference between the S and D states, inducing coherent rotations in the qubit space. Since the energy difference between the S and $P_{1/2}$, $P_{3/2}$, or between $D_{3/2}$ and P, is not a multiple of this wavelength, the atom rarely transitions outside the qubit space $\{S_{1/2}, D_{5/2}\}$.

Our goal in this chapter is to recreate these conditions, engineering what we call *superconducting artificial atoms.* These are nonlinear circuits that can store qubit-like degrees of freedom, and where the nonlinearity arises either from the Coulomb repulsion between Cooper pairs or from the nonlinear inductance in a Josephson junction. This includes three large families of qubits: the charge qubit (6.2), the transmon qubit (6.3), and the flux qubits (6.4).

Note that in addition to the *quantum information bit* and the *physical qubit*, there exists the idea of *logical qubit*, analyzed later in this book. In this case, the quantum bit is stored formally in the Hilbert space of a complex quantum system. The qubit is no longer identified with the eigenstates of that model, but is prepared, operated, and measured in a sophisticated, error-correcting, or error-suppressing way. Typically, logical qubits can be synthesized out of systems from many imperfect physical qubits. However, as demonstrated by Ofek et al. (2016), one can also develop logical qubits using complex quantum systems, such as Schödinger cats created by the interaction between superconducting qubit circuits and superconducting cavities. Here the superconducting qubit is just a nonlinear circuit that enables the preparation, operation, measurement, and error correction of the logical qubit, which is no longer an associated to an eigenstate of the cavity but to a very long-lived quantum state.

6.1.2 Qubit Hamiltonian

All artificial atoms that we study in this chapter have a simple Hamiltonian representations:

$$H = \sum_n E_n(\Phi)\,|n\rangle\langle n| + \Omega(t)\hat{d}. \tag{6.1}$$

The levels $|n\rangle$ are the eigenstates of the circuit that makes our atom. The lowest energy states $|0\rangle$ and $|1\rangle$ are identified with our qubit. As sketched in Figure 6.1a, the spectrum is anharmonic and the gap $\hbar\omega_{01} = E_1 - E_0$ is not an integer fraction of the spacing to other levels, $\omega_{02}, \omega_{12} \ldots$

There will be typically two direct controls on any qubit. Tuning an external parameter such as a magnetic flux Φ, we can change the energy levels $E_n(\Phi)$ preserving their population. We can also "illuminate" the qubit with microwaves $\Omega(t)$ that couple to the off-diagonal dipolar moment operator $\langle n|\hat{d}|n\rangle = 0$ and induce transitions between eigenstates.

By using weak or slow controls, and engineering the spacing between levels, we can constrain the dynamics of the artificial atom to just two lowest energy states. This effective *two-level system* lives in a reduced Hilbert space $\mathcal{H}_2$ created by arbitrary linear superpositions of $|0\rangle$ and $|1\rangle$. Following Section 2.2, we model the dynamics in this space using the most general real-valued[1] qubit Hamiltonian:

$$H = \frac{\hbar\Delta(t)}{2}\sigma^z + \frac{\hbar\varepsilon(t)}{2}\sigma^x + E_0\mathbb{1} = E_0 + \mathbf{B}(t)\cdot\boldsymbol{\sigma}. \tag{6.2}$$

[1] It will become evident, when we introduce actual qubit designs, that we can write all models using two observables, one diagonal – the qubit energies – and one off-diagonal – the dipole moment operator. A consistent selection of phases in the definition of states $|n\rangle$ allows us to represent those observables as σ^x and σ^z (see Problem 6.1).

The parameter $\Delta(t)$ is the qubit gap $\hbar\omega_{01} = E_1 - E_0$, associated to $\sigma^z = |1\rangle\langle 1| - |0\rangle\langle 0|$. The transverse driving and the dipole moment strength $\hat{d} \sim \sigma^x = |1\rangle\langle 0| + |0\rangle\langle 1|$ combine into $\varepsilon(t) \propto \Omega(t)$. The energy offset E_0 can be ignored, as it only produces an undetectable global phase shift of the wavefunction.

6.1.3 Interaction Picture

In experiments, qubits spend most of the time parked close to some reference frequency, such as Δ_0. During that time, they evolve freely as

$$U_0(t) = \exp\left(-i\frac{1}{2}\Delta_0\sigma^z t\right) = e^{-i\Delta_0 t/2}\,|1\rangle\langle 1| + e^{i\Delta_0 t/2}\,|0\rangle\langle 0|. \tag{6.3}$$

We can get rid of this "free evolution" and eliminate the dynamical phases by working in a new rotating frame, called the *interaction picture*. This involves redefining the wavefunction and the observables:

$$|\psi_I(t)\rangle = U_0(t)^\dagger\,|\psi(t)\rangle, \quad \hat{O}_I = U_0(t)^\dagger\hat{O}_I U_0(t), \tag{6.4}$$

so that expectation values and predictions remain unaltered $\langle\psi_I|\hat{O}_I|\psi_I\rangle = \langle\psi|\hat{O}|\psi\rangle$. In this frame, diagonal observables such as σ^z remain the same and the evolution of the new wavefunction ψ_I is dictated by an interaction picture Hamiltonian:

$$i\hbar\partial_t\,|\psi_I\rangle = H_I(t)\,|\psi_I\rangle = \left[U_0^\dagger(t)H(t)U_0(t) - i\hbar U_0^\dagger(t)\frac{\mathrm{d}}{\mathrm{d}t}U_0(t)\right]|\psi_I(t)\rangle. \tag{6.5}$$

This transformation helps us in the study of microwave controls:

$$\varepsilon(t) = \varepsilon_0\cos(\omega_0 t + \varphi). \tag{6.6}$$

It is customary to take the microwave field as frame of reference $\Delta_0 = \omega_0$, because it actually provides the experiment with a calibrated time reference. The interaction Hamiltonian for (6.2) becomes

$$H_I(t) = \frac{\hbar(\Delta - \omega_0)}{2}\sigma^z + \frac{\hbar\varepsilon_0}{4}\left(e^{-i\varphi}\sigma^+ + e^{i2\omega_0 t + i\varphi}\sigma^+ + \text{H.c.}\right). \tag{6.7}$$

The diagonal part is proportional to the detuning between the qubit frequency and the drive $\delta = \Delta - \omega_0$. Out of the off-diagonal terms, we can single out the static one $e^{-i\varphi}\sigma^+$, using a rotating wave approximation to neglect the interaction that rotates with twice the frequency $e^{i2\omega_0 t + i\varphi}\sigma^+$.

6.1.4 Single-Qubit Gates

A physical qubit must implement or approximate arbitrary *single-qubit unitary* operations on the qubit Hilbert space $\mathcal{H}_2$. Up to irrelevant global phases, these *single-qubit gates* are rotations in the Bloch sphere:

$$U(\theta,\vec{n}) = \exp(i\theta\vec{n}\cdot\boldsymbol{\sigma}) = \cos(\theta)\mathbb{1} + i\sin(\theta)(\vec{n}\cdot\boldsymbol{\sigma}), \tag{6.8}$$

with angle θ around the rotation axis $\vec{n}\in\mathbb{R}^3$ on the Bloch sphere.

Evolution with the Schrödinger equation (2.16) already implements a unitary operation. If we wish to create a particular gate W, we must investigate the combination of gaps $\Delta(t)$ and external drives $\varepsilon(t)$ that, after a finite time, give rise to the desired operation $U(T) = e^{i\xi}W$, up to an irrelevant global phase ξ. The search for $\Delta(t)$ and $\varepsilon(t)$ is a *quantum control* problem.

Fortunately for us, the interaction picture (6.7) already brought the Hamiltonian into a form $H \propto \vec{n}\cdot\boldsymbol{\sigma}$, where the direction $\vec{n}$ can be fully tuned by changing the qubit frequency Δ, the amplitude ε_0 and the phase φ of the drive. The simplest gates are diagonal in the computational basis, $\vec{n} = \vec{e}_z = (0,0,1)$. They are called *phase gates* because they only impart phases on the qubit states:

$$U_\theta = e^{i\theta\sigma^z} = \begin{pmatrix} e^{i\theta} & 0 \\ 0 & e^{-i\theta}\end{pmatrix} \sim e^{-i\theta}\,|0\rangle\langle 0| + e^{+i\theta}\,|1\rangle\langle 1|. \tag{6.9}$$

This operation is generated by tuning away the qubit from its resting frequency, $\Delta(t) = \delta + \omega_0$. Choosing the time T and detuning δ allows us to adjust the phase to some standard values $\theta = -T\Delta/2 = \pi/2,\ \pi/4$ or $\pi/8 \bmod 2\pi$, which correspond to the Z, S, and T gates:

$$Z = \begin{pmatrix} 1 & 0 \\ 0 & -1\end{pmatrix} = e^{i\frac{\pi}{2}\sigma^z},\ S = \begin{pmatrix} 1 & 0 \\ 0 & i\end{pmatrix} \sim e^{i\frac{\pi}{4}\sigma^z},\ T = \begin{pmatrix} 1 & 0 \\ 0 & e^{i\pi/4}\end{pmatrix} \sim e^{i\frac{\pi}{8}\sigma^z}. \tag{6.10}$$

Alternatively, we can apply a coherent drive (6.6) to the qubit, implementing a rotation around the axis $\vec{n} \propto (\varepsilon_0\cos(\varphi), \varepsilon_0\sin(\varphi), \delta)$, at a frequency $\omega = \frac{1}{2}\sqrt{\delta^2+\varepsilon_0^2}$. In particular, working on resonance $\delta = 0$ with $\varphi = 0$ or $\pi/2$ produces rotations around σ^x or σ^y, respectively. The combination of a rotation around σ^y with a phase creates the ubiquitous Hadamard gate:

$$H = e^{i\pi\sigma^z/2}e^{i\pi\sigma^y/2} = \frac{1}{\sqrt{2}}\begin{pmatrix} 1 & 1 \\ 1 & -1\end{pmatrix}. \tag{6.11}$$

6.1.5 Decoherence and Dephasing

Experiments never create pure states, because of the unavoidable imperfections and limited precisions – both in the creation of the initial states and in the control of the experiments – as much as because of the interaction between our setup and the environment. Instead, experiments must be described using mixed states: averages of different pure states prepared and evolve with random or fluctuating conditions, as sketched in (2.23). This averaging causes *decoherence*,[2] the progressive loss of the quantum fluctuations, due to the destruction of quantum superpositions and entangled states.

The simplest type of decoherence is *dephasing*, a random scrambling of the relative phase between the eigenstates of our artificial atom – or any other quantum system. Dephasing is normally caused by fluctuations in the eigenenergies of our system, which deviate from the ideal values E_n by small amounts that randomly fluctuate in time δE_n.

Obvious sources of dephasing are the control lines used to stabilize and measure qubits and cavities. Those cables and antennas may carry fluctuating electromagnetic fields, white noise that affects the energies of the superconducting circuit. A similar scrambling is caused by electric charges trapped on the substrate of the superconducting chip. These charges act as two-level systems or *quantum fluctuators* that couple to the qubit, shifting their energy levels, typically over longer time scales and sometimes in a quasistatic fashion – i.e., the electric field induced by the charges is static over the duration of an experiment, but may change from realization to realization.

Pure dephasing can be modeled as a Brownian motion of the phase, which accumulates random displacements due to the fluctuations in the qubit's energy levels, $H' \sim \frac{1}{2}(\hbar\Delta + \delta E)\sigma^z$. In the absence of other perturbations, a single realization of the experiment is accounted for by a stochastic equation for the phase:

$$|n\rangle \to e^{-i\varphi_n(t)}\,|n\rangle\,, \quad \hbar\frac{\mathrm{d}}{\mathrm{d}t}\varphi_n = E_n + \delta E_n(t). \tag{6.12}$$

This equation by itself is somewhat useless, because we do not know the noise realization $\delta E_n(t)$ of each experiment. However, we may incorporate our ignorance of the noise, modeled as some probability distribution $p(\delta E_n, t)$, in an ensemble average description of the experiment using mixed states.

Theoreticians like to work with white (or slightly colored) Gaussian noise models, which are uncorrelated in time $\langle \delta E_n(t)\delta E_n(t')\rangle = \hbar^2\sigma^2\delta(t-t')$. In this limit equation (6.12) describes a Brownian particle, with a Gaussian probability distribution of the phase centered on $E_n t/\hbar$, with linearly growing variance $\sigma^2 \propto t$. The evolution of a quantum state $\rho(0)$ is computed as a linear transformation:

$$\rho(t) = \varepsilon_t(\rho(0)) \simeq \int e^{-i\frac{1}{2}\varphi\sigma^z}\rho_0 e^{i\frac{1}{2}\varphi\sigma^z}\frac{e^{-(\varphi-\Delta t)^2/(2\sigma^2 t)}}{\sqrt{2\pi\sigma^2 t}}\mathrm{d}\varphi, \tag{6.13}$$

[2] In recent years, a resource theory has been put forward to quantify *coherence* (Baumgratz et al., 2014) as a function of the off-diagonal elements in the density matrix. From this theory, it follows that *decoherence* is the destruction of such a resource. We are not so rigorous in our description.

that belongs to the family of a *completely positive maps* or *quantum channels* (see Section 8.4.2). For a white noise model, the integral has an analytical expression

$$\rho(t) = \begin{pmatrix} \rho_{11} & e^{(-i\Delta-\gamma_\phi)t}\rho_{10} \\ \rho_{01}e^{(+i\Delta-\gamma_\phi)t} & \rho_{00} \end{pmatrix}. \tag{6.14}$$

The map predicts an exponential decay of the *coherences* – the off-diagonal elements of the density matrix – with a *dephasing rate* $\gamma_\phi = \sigma^2/2$, and a *dephasing time* $T_2 \sim 1/\gamma_\phi$.

The dephasing channel (6.14) is reproduced by the dephasing master equation:

$$\frac{\mathrm{d}}{\mathrm{d}t}\rho = -\frac{i}{\hbar}[H,\rho] + \frac{\gamma_\phi}{2}(\sigma^z\rho\sigma^z - \rho). \tag{6.15}$$

The Hamiltonian $H = \frac{1}{2}\Delta\sigma^z$ accounts for the noise-free evolution, while the Lindblad operator $\mathcal{L}[\rho] = \sigma^z\rho\sigma^z - \rho$ models the destruction of coherences.

The dephasing master equation can be derived – as we did for the cavity environment coupling – from a microscopic model where the energy perturbations $\delta E_n(t)$ arise from a diagonal coupling with a bath, provided the spectrum of fluctuations in the bath's Markovian limit reproduces the desired white noise model over all frequencies. However, the white noise model of dephasing is a bit simplistic. Electrical circuits – and mesoscopic systems in general – are affected by what is known as $1/f$-noise or *slow noise*. When we study the power spectrum of these fluctuations, the contribution of noise diverges at low frequencies. This implies a slow decay of noise correlations – e.g., $\langle\delta E(t')\delta E(t)\rangle \sim \exp(-(t'-t)/T_c)$ with a large T_c that prevents self-averaging. The limit in which the correlation time diverges $T_c \to \infty$ describes a *quasistatic noise*, which remains more or less constant throughout each experimental realization, but fluctuates between runs. Such models are described with variants of (6.13), with other decay forms that depend on the spectral properties of the noise (Ramos and García-Ripoll, 2018), and do not have a Markovian master equation (see Exercise 6.3).

However, even if slow fluctuations are theoretically inconvenient, they may be beneficial, because we can design controls to suppress them. Assume for instance, that the noise δE_n has long time correlations. There is a technique, called *spin-echo*, that exactly cancels this fluctuation, and which consists in introducing a spin flip halfway through the experiment:

$$\rho(t) = e^{-i\delta Et/\hbar}\sigma_x e^{-i\delta Et/\hbar}\rho(0)e^{i\delta Et/\hbar}\sigma_x e^{i\delta Et/\hbar} = \rho(0). \tag{6.16}$$

If the noise is not quasistatic, but has a finite-time correlation T, we can still apply spin-echo at regular intervals with a spacing shorter than the fluctuations, time scale $\delta t \ll T$. This leads to a total or partial cancelation of noise, extending the lifetime of quantum superpositions well beyond the intrinsic T_2 time of the experiment – see, for instance, the work of Yan et al. (2015) with flux qubits, and Problem 6.3.

6.1.6 Relaxation and Heating

Short of losing the qubit itself, the worst type of decoherence is the one that scrambles the probabilities of the basis states ρ_{ii}. As in the photonic case, we find such processes in the interaction between a qubit and a Markovian environment. In this case, the incoherent evolution is described by the qubit's equivalent of a cavity's master equation in Section 5.5.4

$$\frac{\mathrm{d}}{\mathrm{d}t}\rho = -\frac{i}{\hbar}[H,\rho] + [\bar{n}_\Delta + 1]\frac{\gamma}{2}\left(2\sigma^-\rho\sigma^+ - \sigma^+\sigma^-\rho - \rho\sigma^+\sigma^-\right) \qquad (6.17)$$
$$+\bar{n}_\Delta\frac{\gamma}{2}\left(2\sigma^+\rho\sigma^- - \sigma^-\sigma^+\rho - \rho\sigma^-\sigma^+\right).$$

This master equation can be derived from a microscopic model of a qubit interacting with a memoryless environment, with a continuum spectrum that is populated by a thermal distribution of photons $\bar{n}_\omega$. The master equation contains *relaxation* events $\sigma^- |1\rangle\langle 1| \sigma^+$ that take the qubit from the "excited" state $|1\rangle$ to the lower energy state $|0\rangle$. It also includes the converse process, or *heating* $\sigma^+ |0\rangle\langle 0| \sigma^- \rightarrow |1\rangle\langle 1|$. The ratio between both processes is dictated by the temperature of the environment:

$$\frac{\gamma_{\text{heat}}}{\gamma_{\text{cool}}} = \frac{\bar{n}_\Delta}{\bar{n}_\Delta + 1} \propto e^{-\Delta/k_B T}. \qquad (6.18)$$

When the qubit's gap is larger than the effective temperature, we can neglect heating $\bar{n}_\Delta \simeq 0$ and stay with pure relaxation. The exact solution of the master equation in this limit is

$$\rho(t) = \begin{pmatrix} \rho_{11}e^{-t/T_1} & e^{-i\Delta t - t/T_2^*}\rho_{10} \\ \rho_{01}e^{+i\Delta t - t/T_2^*} & \rho_{00} + (1 - e^{t/T_1})\rho_{11} \end{pmatrix}. \qquad (6.19)$$

Note how the excitation probability decays over a time $T_1 = 1/\gamma$, while coherences disappear over a slightly longer timescale $T_2^* = 2T_1$. If an experiment combines dephasing and relaxation, the first timescale remains the same, $\gamma = 1/T_1$, but the decoherence time is reduced (see Problem 6.5):

$$\frac{1}{T_2^*} = \frac{1}{2T_1} + \frac{1}{T_2}, \qquad (6.20)$$

and denoted by the symbol T_2^*. In this scenario, to avoid confusion, $1/\gamma_\phi$ is also often referred to as T_ϕ instead of T_2.

It is important to remark that the losses that we are discussing are not incompatible with the superconducting nature of the circuit. While the circuit has negligible resistance, it can still dissipate energy to the environment, to the substrate, or to other circuit elements – antennas, qubits, resonators – on the same chip. Some of this radiation can be suppressed by placing the qubits in better environments – e.g., enclosing them in the 3D cavities from Section 5.3 – to get qubit lifetimes $T_1 \sim 100\,\mu\text{s}$. Another way to improve the lifetime of the qubit is to use better fabrication processes with different materials or techniques that minimize the imperfections and probability of trapping charges. Recent experiments by Place et al. (2021) have reported lifetimes of $T_1 \sim 300\,\mu\text{s}$ this way. Assuming qubit operation rates of 100 MHz, this is a very long lifetime that allows for 30 000 operations before losses kick in!

6.2 Charge Qubit

After this formal introduction to the world of qubits, we can study our first artificial atoms. We begin with the charge qubit, a small, micro- or nanometer-size metallic island. The island is so small that the energy grows very rapidly with every new Cooper pair we add or subtract from it. This anharmonic growth, known as a *Coulomb blockade*, allows us to engineer a very good qubit eigenspace that can be easily controlled by electrostatic potentials. However, as we will discuss in this section, this facility of control makes the qubit very fragile to dephasing by interaction with the stray electrostatic field and trapped charges.

6.2.1 Coulomb Blockade

Picture a small piece of metal with a lattice of positively charged ions and a sea of electrons binding them. In equilibrium, the metal aims toward *charge neutrality*, a state in which the density of positive charges balances the negative ones. This state has some reference energy E_0. If we add one more electron to this island, we have to pay a small energetic cost $\Delta = E_1 - E_0$, which can be explained by the Coulomb interaction of the quasiparticle with the negatively charged electron plasma and the positively charge ion lattice. Because the average charge of the metal is zero, and because of screening effects in the electron plasma, this interaction is not too large and causes a moderate gap Δ.

The situation changes if we add a *second* charge. This new particle does not face a balanced material, but interacts repulsively with the charge introduced before. The increase in charging energy $\Delta_{\text{Coulomb}} > \Delta$ due to this repulsion is called the *Coulomb blockade.* Because of this blockade, there exists a potential qubit space form by the states with zero $|0\rangle$ and one extra charge $|1\rangle$, with a small splitting among them $\omega_{01} \sim \Delta$, and separated by a large gap $\omega_{02} = \Delta + \Delta_{\text{Coulomb}}$ from higher-charge excitations.

The Coulomb blockade is an energy scale that depends on properties of the conducting material, such as the density of charges and the geometry of the piece. It predates the notion of qubits and the study of quantum information: Coulomb blockade is observed in quantum dots, in transport experiments with molecules, and, of course, also in the superconducting world.

Fortunately, we do not need to worry about microscopic models to describe a *charge qubit*. Since we work with macroscopic devices, with a humongous number of particles, we can treat the superconducting island as a lumped-element circuit, describing the total energy as a quadratic function of the total charge, without caring for how this charge distributes across the island. The work or energy required to charge the island is given by the *self-capacitance* of this metallic object, as expected:

$$E = \frac{1}{2C_{\text{self}}} Q^2. \tag{6.21}$$

If our excitations are electrons with undivisible charge e, we can define a constant called the *charging energy* $E_C = e^2/2C_{\text{self}}$ and express the total cost in term of the number of carriers $n = Q/e$, as in $E = E_C n^2$.

Note the anharmonicity in this expression: The cost of adding one particle is $E_1 - E_0 = E_C$, but the second charge involves a larger penalty $E_2 - E_1 = 3E_C$. We could define our qubit using the Hilbert space of the neutral state and the state with a single Cooper pair, $|0\rangle$ and $|1\rangle$. The total Hamiltonian then would read

$$H = E_C\, |1\rangle\langle 1| = \frac{E_C}{2}\sigma^z + \text{constants}. \tag{6.22}$$

This situation is inconvenient for several reasons: The splitting E_C is too large for a small island, we have no means to modify this splitting, and we also do not yet know how to add other control terms to the Hamiltonian, such as a dipolar coupling with $\sigma^x = |1\rangle\langle 0| + |0\rangle\langle 1|$ (see (6.2)). All this requires a slightly more sophisticated circuit model.

6.2.2 The Actual Superconducting Charge Qubit

The actual superconducting charge qubit represents an upgrade from the ideal superconducting island to a more sophisticated device, shown in Figure 6.2a, with the equivalent circuit analyzed in Section 4.6. The first new ingredient is the capacitive coupling C_g with a conductor – light blue metallic piece in Figure 6.2a – that can transport a microwave, represented as an electrostatic potential $V(t)$.

The second ingredient is a pair of Josephson junctions that connect the qubit to a ground plane, allowing the injection and extraction of charges. The two junctions form a small dc-SQUID, which is threaded by a small magnetic flux Φ. As it was

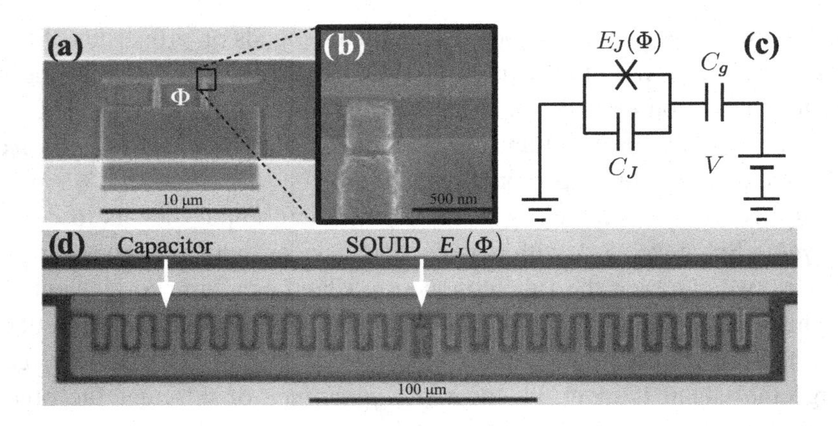

Figure 6.2 (a) Superconducting island coupled to the ground plane through two Josephson junctions. (b) Zoomed-in picture of one junction. Pictures courtesy of Andreas Wallraff, ETH, Zurich. (c) Reminder of the equivalent circuit discussed in Section 4.6, with the superconducting island marked in gray. The two junctions add up to a single effective junction with nonlinear inductance that depends on the flux Φ enclosed by the loop. (d) Image of a transmon qubit, a charge qubit where the capacitance has been dramatically enlarged.

explained in Section 4.7.2, the parallel-junction circuit is equivalent to a Josephson junction $E_J(\Phi)$ that can be tuned with the magnetic flux Φ.

Combined together, the external potential and the SQUID give the lumped-element circuit from Figure 6.2b and the effective nonlinear Hamiltonian:

$$H = \frac{1}{2C_\Sigma}\left(q - q_g\right)^2 - E_J \cos(\phi/\varphi_0). \tag{6.23}$$

The constant C_Σ is the total capacitance of the the charge qubit, including the coupling to the gate voltage, the junction C_J, and the island itself C_{self}. In practice, we cannot tell C_J and C_{self} apart, $C_\sigma \simeq C_J + C_g$. In addition to the dc-SQUID, the external *gate potential* V introduces our second tuning element, the offset charge $q_g = -C_g V$, which determines the equilibrium state of the island.

The charge qubit is best described using the *number state representation* from Section 4.9. The states $\{|n\rangle\}$ are labeled by the excess ($n > 0$) or defect ($n < 0$) of Cooper pairs with respect to charge neutrality ($n = 0$). The charge operator is diagonal in the number basis $\hat{q}\,|n\rangle = -2en\,|n\rangle$, $n \in \mathbb{Z}$, and the electrostatic energy is proportional to the charging energy E_C:

$$E_n = \frac{2e^2}{C_\Sigma}(n - n_g)^2 = 4E_C(n - n_g)^2. \tag{6.24}$$

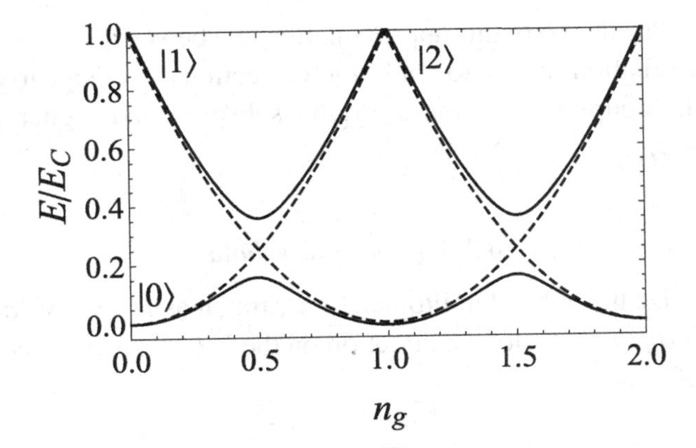

Figure 6.3 Energy levels of the charge qubit without the Josephson junction (dashed) and with the tunneling amplitude $E_J \simeq 0.1 E_C$ (solid). At zero voltage, the lowest eigenstates are $|0\rangle$ and $|1\rangle$. At $n_g = 1/2$, both states become almost degenerate but the tunneling creates new eigenstates that are superpositions of the former $|\pm\rangle = \frac{1}{\sqrt{2}}(|0\rangle \pm |1\rangle)$.

Figure 6.3 shows the energy levels E_n of the charge states in dashed lines. Around the point $n_g = \frac{1}{2}$, states $|0\rangle$ and $|1\rangle$ become degenerate[3] and well separated from all other charge states, $n = 2, 3, \ldots$. This makes them good candidates for qubit states, provided we find ways to reproduce the qubit Hamiltonian (6.2) using the external controls Φ and V.

The first control is given by the external potential. If we perturb the value of the offset charge slightly away from $n_g = \frac{1}{2}$, we obtain an energy splitting between the 0 and 1 charge states that grows as $\hbar\varepsilon = 8E_C(n_g - \frac{1}{2})$. The second control is due to the nonlinear inductance of the SQUID. This effective Josephson junction enables the tunneling of charges in and out of the island (4.63), $\cos(\hat{\phi}/\varphi_0) = \frac{1}{2}\sum_n (|n+1\rangle\langle n| + |n\rangle\langle n+1|)$. Assuming that the capacitive interaction is the dominant energy scale $E_C > E_J$, and that we are close to degeneracy $n_g \simeq 1/2$, the Josephson energy may be truncated to the 0 and 1 charge states, as $-E_J(\Phi)\sigma^x$ with the tunneling operator $\sigma^x = |1\rangle\langle 0| + |0\rangle\langle 1|$. Altogether, we obtain the desired qubit Hamiltonian in a 2×2 matrix form:

$$H = \frac{\hbar\varepsilon}{2}\sigma^z - \frac{\hbar\Delta}{2}\sigma^x, \text{ with } \begin{cases} \hbar\varepsilon = 8E_C\left(n_g - \frac{1}{2}\right), \\ \hbar\Delta = E_J(\Phi). \end{cases} \tag{6.25}$$

[3] There are other points $n_g = \frac{3}{2}, \frac{5}{2}, \ldots$, where different pairs of states become degenerate. However, the physics of those points is similar to that which we discuss later and presents no formal advantage.

As explained in Problem 6.6, this approximation can be verified in a self-consistent way, using perturbation theory to study the corrections that appear when we take into account the coupling between the qubit subspace and higher excited states $|2\rangle, |-1\rangle, |3\rangle, |-2\rangle, \ldots$.

6.2.3 Qubit Hyperbola

We can diagonalize the qubit Hamiltonian by writing it as $H = \frac{1}{2}\Delta E \sigma_{\mathbf{n}}$ with a Pauli matrix operator $\sigma_{\mathbf{n}} = \mathbf{n} \cdot \boldsymbol{\sigma}$ along a direction on the XZ plane, $\mathbf{n} = (\cos\,\theta, 0, \sin\,\theta)$. The overall prefactor

$$\Delta E = \sqrt{\varepsilon^2 + \Delta^2} \tag{6.26}$$

gives the energy $\pm\frac{1}{2}\Delta E$ of the ground and excited states,

$$\begin{aligned} |0_{\mathbf{n}}\rangle &= \cos(\theta/2)\,|0\rangle - \sin(\theta/2)\,|1\rangle\,, \text{ and} \\ |1_{\mathbf{n}}\rangle &= \sin(\theta/2)\,|0\rangle + \cos(\theta/2)\,|1\rangle\,. \end{aligned} \tag{6.27}$$

The energy level structure of the qubit is usually referred to as the *qubit hyperbola*, because E_0 and E_1 satisfy the equation of a hyperbola with asymptotes at $E_{0,1} \simeq \pm\frac{1}{2}|\varepsilon|$ when the applied potential ε is very large.

The qubit hyperbola is symmetric with respect to the *symmetry point* $\varepsilon = 0$, where the separation between energies is minimal $\Delta E \geq \Delta$. In absence of Josephson energy, $\Delta = 0$ and the two charge states $|0\rangle$ and $|1\rangle$ would have the same energy (cf. Figure 6.3 and dashed line at $n_g = 1/2$). However, the introduction of quantum tunneling has activated quantum fluctuations of the charge and it has *broken the degeneracy*, introducing a minimal separation Δ. This separation causes the new eigenstates of the problem to be the symmetric and antisymmetric charge superpositions:

$$|\pm\rangle = \frac{1}{\sqrt{2}}\,|0\rangle \pm \frac{1}{\sqrt{2}}\,|1\rangle\,. \tag{6.28}$$

The *symmetry point* $\varepsilon = 0$ is experimentally relevant, as the configuration in which the qubit experiences the least dephasing. Indeed, if the qubit is surrounded by stray electric fields that introduce random contributions δV, these contributions cancel to first order $\partial\Delta E/\partial V = 0$ at $\varepsilon = 0$. Thus, the energy levels fluctuate the least around this point, and the lifetime of our quantum states is maximized. The notion of a symmetry point for safe operation is a very general concept that appears also in the flux qubit. In Section 6.3 we will push this idea further, ensuring that the energy levels become insensitive to perturbations in the electrostatic field, which gives us the transmon qubit design.

Finally, a remark on the notational discrepancy between our charge qubit Hamiltonian (6.25) and the general qubit model (6.2). This difference does not affect the physics, as both models are related by a unitary transformation that swaps the σ^x and σ^z operators (see also Problem 6.1). This transformation is a relabeling with new qubit states:

$$\left.\begin{array}{l}|\tilde{0}\rangle = |-\rangle \\ |\tilde{1}\rangle = |+\rangle\end{array}\right\} \Rightarrow H = \frac{\hbar\Delta}{2}\tilde{\sigma}^z + \frac{\hbar\varepsilon}{2}\tilde{\sigma}^x. \tag{6.29}$$

The diagonal Pauli matrix is now $\tilde{\sigma}^z = |\tilde{1}\rangle\langle\tilde{1}| - |\tilde{0}\rangle\langle\tilde{1}|$. Some works favor this notation, because only the states $|\tilde{i}\rangle$ – not the $|i\rangle$ – can be experimentally detected, especially when working with qubits inside cavities (cf. Section 7.4.2).

6.2.4 Charge Qubit History

The idea of a small capacitance junction as an implementation of a qubit was put forward by Shnirman et al. (1997), contemporarily with other proposals of universal quantum computers – such as trapped ions by Cirac and Zoller (1995), quantum dots or impurities in solid-state devices by Loss and DiVincenzo (1998), or nulcear magnetic resonance (NMR) experiments by Gershenfeld and Chuang (1997) and Cory et al. (1997). When Shnirman's work was published in 1997, solid-state experiments were still behind AMO[4] experiments in terms of control and coherence, but that was soon to change. Contemporary works by Nakamura et al. (1997) and Bouchiat et al. (1998) validated the two-level model (6.25) for small superconducting islands. These experiments resolved spectroscopically the eigenenergies of the island, confirming the quantization of the charge and the qualitative pictures from Figure 6.3.

Shortly thereafter, Nakamura et al. (1999) provided full evidence of the qubit structure, engineering quantum superpositions of charge states and implementing single-qubit rotations. In this work, the authors prepared a well-defined charge state, in the limit of large external field $|\varepsilon_0| \gg |\Delta|$. They then jumped abruptly to the symmetry point $\varepsilon = 0$, waited for a short time t, and returned to large field limit, to measure the charge state. As explained in Problem 6.7, the final state of the qubit was a quantum superposition $\cos(\Delta t/\hbar)\,|0\rangle + \sin(\Delta t/\hbar)\,|1\rangle$, where the angle $\Delta t/\hbar$ depended both on the time spent by the qubit at the symmetry point as well as on the magnetic flux passing passing through the SQUID, $\Delta = E_J \cos(\Phi/2\varphi_0)$.

The work by Nakamura is an important milestone in the development of solid-state quantum information processors. It is arguably one of the works that changed

[4] Atomic, molecular, and optical.

people's opinion regarding the future of the field, suggesting that it was possible to engineer complex quantum states and quantum registers (Makhlin et al., 2001). Later experiments with similar schemes demonstrated two-qubit Hamiltonians and qubit–qubit interactions (Pashkin et al., 2003; Yamamoto et al., 2003), a crucial ingredient for scalable computations.

However, Nakamura's charge qubit implementation was inadequate for scalable computations. The superconducting islands decohered very quickly $T_1 \sim 2$ ns, because of the surrounding antennas and probes that controlled and measured the qubit. To compensate for this, the qubits were engineered with huge energy gaps, $E_J \simeq 84\,\mu\text{eV} \sim 20$ GHz, and operated very rapidly, and even then only allowed for a few oscillations before losing all quantum properties.

In 2004, the Yale group had the idea of fabricating a charge qubit inside a microwave resonator. As we have seen in Chapter 5, microwave cavitives only absorb excitations at discrete frequencies, $\omega_n = n \times \omega_0$, filtering out other electromagnetic fields. By placing the qubit inside the cavity and ensuring that the qubit's energy gap ΔE was very different from the allowed photons, they could effectively isolate and protect the qubit from most of the electromagnetic fluctuations. The idea of *Purcell filtering* the environment was put forward in the theory work by Blais et al. (2004) and immediately demonstrated by Wallraff et al. (2004) in an experiment that showed times of $\simeq$200 ns (see also Schuster et al., 2005). That was orders of magnitude better than the bare charge qubit demonstrated only a few years before.

The conditions of relative isolation provided by the cavity, together with operating the qubit close to the symmetry point, made it possible that the coherence of the superconducting qubit was no longer limited by relaxation but by environment-induced dephasing. In Yale's charge qubits, dephasing rates of $\gamma_\phi \sim 2\pi \times 750$ kHz or $T_2 \sim 1\,\mu\text{s}$, allowed testing many quantum optics ideas, as well as some primitive quantum information protocols. These were wonderful beginnings with fast-paced developments in the study of the Rabi model, strong coupling, Wigner function tomography, and large entangled states. However, developing scalable quantum computers required improving the quality of the qubits even further. The avenue that has proved most fruitful was to further suppress dephasing, reducing the curvature of the qubit hyperbola even further. This idea led to the transmon qubit in 2007, a significant redesign of the charge qubit's parameter, which we now discuss.

6.3 Transmon Qubit

The charge qubit dephasing around the symmetry point is dominated by second-order fluctuations of the energy levels with respect to the fluctuating fields

$\propto \partial_{g^2}^2 \Delta E$. These corrections can be expressed in terms of perturbations of the charge offset δn_g:

$$\Delta E = \Delta E(n_g = 0) + \frac{1}{2} E_J \left(\frac{E_C}{E_J}\right)^2 \delta n_g + O\left(\delta n_g^2\right). \tag{6.30}$$

A strategy to reduce these corrections is to increase the ratio E_J/E_C. If we move from $E_J/E_C \simeq 1$ for a charge qubit to a new setup with a ratio $E_J/E_C \simeq 50$, our crude estimate predicts a reduction of the dephasing terms by four orders of magnitude 0.0004. The *transmon* is a design of a qubit that achieves this larger ratio by *shunting*[5] the Josephson junction of the charge qubit with a large capacitor, as shown in Figure 6.2d. The increase in the capacitance reduces E_C and leads to an exponential reduction in the dephasing terms. The result is a simpler and larger qubit, easier to fabricate and with greater reproducibility and stability.

6.3.1 Moving Particle Picture and Energy Bands

To understand *why* increasing E_J/E_C decreases the sensitivity to electrostatic fields, we have to reinterpret the charge qubit Hamiltonian (6.23) using the number-phase representation from Section 4.9:

$$H = 4E_C(-i\partial_\varphi - n_g)^2 - E_J \cos(\varphi) \sim \frac{1}{2m} p^2 + V_0 \cos(2\pi x/a). \tag{6.31}$$

This model resembles a particle with position φ moving in a periodic potential $-E_J \cos(\varphi)$, with mass $m \sim \hbar^2/8E_C$. By analogy with the Bloch theory for electrons in a solid, we derive three qubit regimes depending on the relative strength of the potential, E_J/E_C:

(1) When $E_J \simeq 0$, we have an isolated superconducting island. The eigenstates of the problem have a well-defined number of charges n, which we associate to plane waves[6] $\psi_n(\varphi) \propto \exp(-in\varphi)$. The eigenenergies follow a parabolic dispersion $E_n = 4E_C n^2$, reminiscent of the kinetic energy of a free particle (cf. Figure 6.4a, dashed line).

(2) A weak periodic potential $E_J < E_C$ breaks the degeneracy at the symmetry points $n_g = 1/2, 3/2, \ldots$ by an amount proportional to E_J. From Figure 6.3a we see that the original parabolas are transformed into Bloch bands $E_n(n_g)$. These bands match those of a particle moving in a periodic potential with momentum

[5] Shunting means providing an alternative (parallel) path for the charges to flow. The transmon results from *capacitively shunting* a Josephson junction, with one or two capacitors that are connected in parallel to it.

[6] A maximal certainty of the quasimomentum n corresponds to the maximum uncertainty of the position φ.

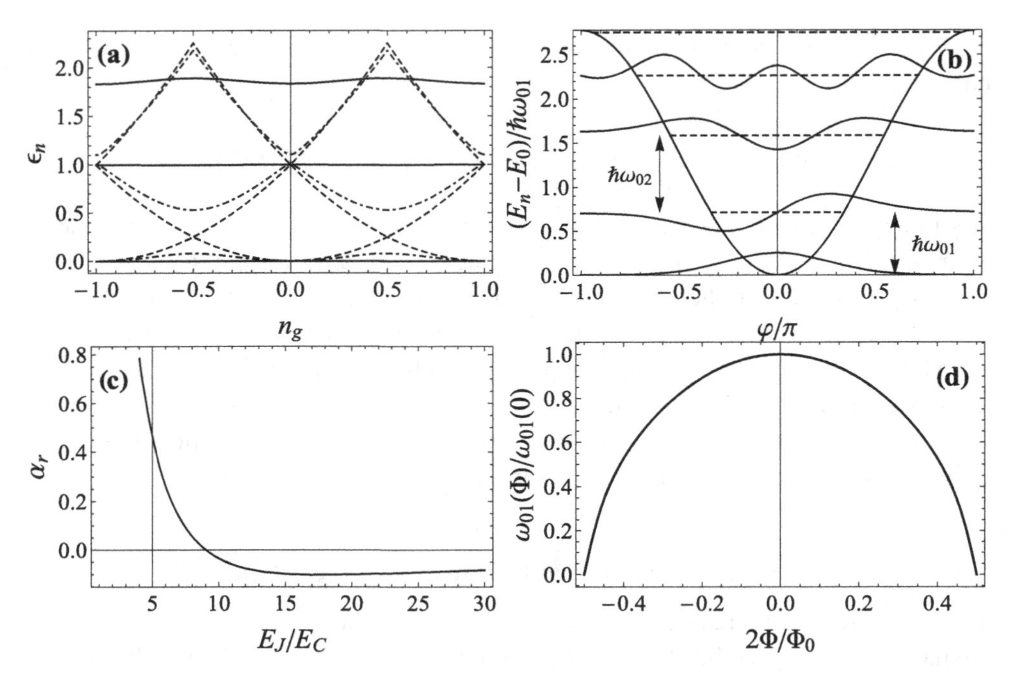

Figure 6.4 (a) Rescaled energy levels of a transmon qubit as a function of the offset charge n_g, for ratios $E_J/E_C = 0.02$ (dashed), 2 (dot-dashed), and 16 (solid). The values plotted lack an irrelevant constant and are scaled to the qubit gap with no offset charge $\varepsilon_n = (E_n(n_g) - E_0(0))/\hbar\omega_{10}$. (b) For a transmon with $E_J/E_C = 16$, we plot the energy levels relative to the ground state $\varepsilon_n(0)$, together with the nonlinear inductive potential $E_J \cos(\varphi)/\hbar\omega_{01}$. On top of each level we plot the eigenfunctions, shifted and scaled arbitrarily. (c) Relative anharmonicity of the transmon qubit $\alpha_r = \omega_{12}/\omega_{01} - 1$. (d) Energy levels of a transmon when we can tune the Josephson energy with a dc-SQUID, as $E_J(\Phi) = E_J \cos(\Phi/\varphi_0)$, for $E_J(0)/E_C = 50$.

$n + n_g$. The value of n is not the charge, but the index of the band, and n_g represents the quasimomentum of the Bloch wavefunction (cf. Figure 6.4a, dot-dash line).

(3) When the confinement is deep, one or more bands fall within the periodic potential $E_J \gg 4E_C n^2$. The splitting between neighboring bands is so large that they become flat. This limit corresponds to a particle hopping among the minima of the inductive potential $\cos(\phi)$. The width of the band $|E_n(n_g = 1/2) - E_n(n_g = 0)|$ is the tunneling amplitude between wells, which decreases exponentially with the barrier height $V_0 \sim E_J$. In other words, when E_J/E_C is very large, the sensitivity of $E_n(n_g)$ to the value of n_g is exponentially small compared to the original charge qubit (cf. Figure 6.4a, solid line). In this limit, we can ignore the bias or set it to $n_g = 0$.

The *transmon qubit* implements the third limit, in which $E_J/E_C \gtrsim 50$ and three or more bands fall inside the Josephson inductive potential (cf. Figure 6.4b). This ratio decreases exponentially the sensitivity to electrostatic fluctuations, which is extremely good for the qubit lifetime, but this advantage comes at a cost: It results in a qubit with a reduced anharmonicity.

6.3.2 Transmon as Anharmonic Oscillator

In the transmon limit, the lowest-energy bands are deep inside the cosine potential. We can approximate the eigenstates by the eigenfunctions of a slightly anharmonic resonator:

$$H \simeq \frac{1}{2C_\Sigma}(\hat{q} - q_g)^2 - E_J + \frac{1}{2}\frac{E_J}{\varphi_0^2}\hat{\phi}^2 - \frac{1}{24}\frac{E_J}{\varphi_0^4}\hat{\phi}^4 + \mathcal{O}(\phi^6). \tag{6.32}$$

We diagonalize the quadratic part for $q_g = 0$ using Fock operators:

$$\hat{q} = -2e\hat{n} = 2e\left(\frac{E_J}{8E_C}\right)^{1/4}\frac{i}{\sqrt{2}}(\hat{a}^\dagger - \hat{a}), \tag{6.33}$$

$$\hat{\phi} = \varphi_0\left(\frac{8E_C}{E_J}\right)^{1/4}\frac{1}{\sqrt{2}}(\hat{a}^\dagger + \hat{a}).$$

The quadratic part becomes an oscilator with equispaced energy levels and a plasma frequency $\hbar\omega_{01} \simeq \sqrt{8E_C E_J}$. We include the external field $q_g\hat{q}/C_\Sigma$ and the Duffing term ϕ^4, working up to second order in perturbation theory. The result is a nonlinear oscillator:

$$H \simeq \hbar\omega_{01}\hat{a}^\dagger\hat{a} - \hbar\alpha\hat{a}^\dagger\hat{a}^\dagger\hat{a}\hat{a} + i\hbar\varepsilon(\hat{a}^\dagger - \hat{a}), \tag{6.34}$$

with a small anharmonicity $\alpha \ll \omega_{01}$ and a driving term $\varepsilon(t)$ that includes q_g with all other transmon parameters.

In this nonlinear oscillator, the first and second excitations $|1\rangle$ and $|2\rangle$ are separated by a gap $\omega_{12} := \omega_{01} - \alpha$ that is slightly *smaller* than the splitting between the ground and first excited state ω_{01}. The spacing between energy levels gets even smaller as we climb up the ladder, because the cosine potential gets softer at higher energies, making it cheaper to continue exciting as we move away from the ground state. Once more, perturbation theory gives an estimate of the relative anharmonicity:

$$\alpha_r = \frac{\alpha}{\omega_{01}} = \frac{\omega_{01} - \omega_{12}}{\omega_{01}} \simeq \sqrt{\frac{8}{E_J/E_C}}. \tag{6.35}$$

For a typical ratio of $E_J/E_C = 50$, the anharmonicity is about 5% the transmon's fundamental frequency. For a qubit with $\omega_{01} \sim 2\pi \times 6$ GHz, this value can be $\sim 2\pi \times 300$ MHz.

The charge operator $\hat{q}$ plays the role of the dipole moment of the transmon, connecting neighboring states that differ in parity – i.e., it allows transitions from 0 to 1, 1 to 2, ... This operator is explicitely coupled in (6.34) to a field $\varepsilon(t)$ that contains the gate voltage n_g or q_g. The source of the field ε can be a classical, coherent pulse, which we treat as a complex, time-dependent pulse with the tools from Section 6.1.3. However, the field may also have a quantum origin and be generated by a resonator, a cavity, or a waveguide, giving rise to the cavity-QED setups in Chapter 7.

It may seem counterintuitive that, since the transmon bands are flat and independent of n_g, an electric field may have any kind of effect on the system. However, notice that the linear term can be approximately absorbed with a unitary redefinition of the Fock operators, $\hat{a} \to \hat{a}+i\varepsilon/\omega_{01}$. This reveals that, up to first order, the energy levels really do not depend on n_g, but this external field may induce a unitary change in the transmon eigenstates $\psi_n(\varphi)$. In other words, starting from our model

$$H = E_1 \left|\psi_1(n_g)\right\rangle\left\langle\psi_1(n_g)\right| + E_0 \left|\psi_0(n_g)\right\rangle\left\langle\psi_0(n_g)\right|, \tag{6.36}$$

changes in the external potential lead to a perturbation of the form

$$H(n_g + \delta n_g) \simeq H + \delta n_g\Big(\big|\partial_{n_g}\psi_1\big\rangle\psi_1 + \big|\partial_{n_g}\psi_0\big\rangle\psi_1 + \text{H.c.}\Big). \tag{6.37}$$

If we project these corrections back onto the basis $\{\left|\psi_0(n_g)\right\rangle, \left|\psi_1(n_g)\right\rangle\}$, we recover the term proportional to $\hat{q}$ or $i(\hat{a}^\dagger - \hat{a})$.

6.3.3 Josephson Junctions and the Mathieu Equation

The properties of the transmon qubit may be obtained analytically, by solving the eigenvalue equation for the Josephson junction Hamiltonian (6.31), with the boundary condition $\psi(\varphi+2\pi) = \psi(\varphi)$. We get rid of n_g changing the boundary condition, with a phase induced by the "gauge field" n_g:

$$\psi(\varphi) = e^{in_g\varphi}\xi(\varphi). \tag{6.38}$$

The new function ξ is not periodic $\xi(\varphi + 2\pi) = e^{-in_g 2\pi}\xi(\varphi)$ and satisfies

$$\left[-4E_C\partial_\varphi^2 - E_J\cos(\varphi)\right]\psi_n(\varphi) = \varepsilon_n\psi_n(\varphi). \tag{6.39}$$

This equation is related to the Mathieu differential equation, which in standard notation (Olver et al., 2010) reads

$$\frac{d^2}{dz^2}u(z;q) + [a - 2q\cos(2z)]u(z;q) = 0. \tag{6.40}$$

Given a value ν, the Mathieu equation admits a complete set of orthonormal solutions labeled by nonnegative integer k such that

$$\int_0^{\pi} u_k(z;q)u_{k'}(-z;q)\mathrm{d}z = \delta_{kk'} \tag{6.41}$$

$$u_k(z+\pi) = e^{i(\nu+2k)\pi}u_k(z),\ n \in \{0,1,2,\ldots\}. \tag{6.42}$$

These solutions are the Floquet functions $u_k(z) = me_{\nu+2k}(z;q)$, where $\nu + 2k$ is the Floquet characteristic exponent of the function $u_k(z)$ after a period π. Note that these functions have the required boundary condition if we identify $2z\varphi$. This suggests that we rescale and shift the phase variable $\varphi = 2z + \pi$, with which $2\partial_\varphi = \partial_z$, so that

$$\left[\partial_z^2 + \frac{\varepsilon}{E_C} - \frac{E_J}{E_C}\cos(2z)\right]\xi(\varphi) = 0. \tag{6.43}$$

From here it follows the identification of physical quantities to Mathieu equation parameters

$$a = \frac{\varepsilon}{E_C},\ q = \frac{E_J}{2E_C},\ \text{and}\ \nu = -2n_g \tag{6.44}$$

and the eigensolutions and eigenenergies

$$\psi_n(\varphi) = \frac{e^{in_g\varphi}}{\sqrt{2}}me_{-2n_g+k(n,n_g)}\left(\frac{\varphi+\pi}{2};\frac{E_J}{2E_C}\right), \tag{6.45}$$

$$E_n = \mathcal{M}_A(r,q), \tag{6.46}$$

where $\mathcal{M}_A(r,q)$ is the Mathieu characteristic function for even functions. It is customary to define the index k as a function of n and n_g that produces ordered energies: $E_n \geq E_m$ for all $n > m$. The function for implementing this ordering is rather convoluted. An expression that works for all values except $2n_g \in \mathbb{Z}$ is[7]

$$k(n,n_g) = -\text{round}\left(n_g\right) + \frac{1}{4}(-1)^{\text{floor}(n_g)}\left[-1 + (-1)^n(1+2n)\right]. \tag{6.47}$$

Here round(x) is the nearest integer to the real number x, and floor(x) is the integer immediately below x. The eigenenergies are given by the Mathieu characteristic function $\mathcal{M}_A(r,q)$ with the preceding identifications.

These formulas allow us to compute the eigenenergies of the transmon qubit for different ratios of E_J/E_C, as shown in Figure 6.4a. Problem 6.11 shows that we also have access to the wavefunctions, a fact that is useful for practical studies of transition matrix elements and coupling strengths. Figure 6.4b illustrates some eigenstates for a ratio $E_J/E_C = 16$, where the two lowest bands are already deep

[7] The formula in Koch et al. (2007) is wrong.

in the noninductive potential. Notice how ω_{12}, the excitation energy from the first excited level, becomes very similar to ω_{01}, the qubit frequency. In this case, the anharmonicity is $\alpha = 8.2\%$.

6.3.4 Transmon as Qubit

The transmon Hamiltonian has two controls: We may tune the plasma frequency ω_{01} using the dc-SQUID, and we may couple the transmon to an external potential. Since the first control is simple to understand, let us instead focus on the influence of a periodically modulated electric potential. We consider a pulse

$$\varepsilon(t) = w(t)\cos(\omega_{01}t + \phi), \tag{6.48}$$

with a carrier frequency ω_{01} and a smooth envelope $w(t)$ that switches the pulse on and off over a time T. The Fourier components of this pulse spread approximately over a frequency band $\omega_{01} \pm 2\pi/T$. Thus, we expect that for long enough pulses, $2\pi/T \ll \alpha$, our external drive will only be able to couple the states $|0\rangle$ and $|1\rangle$, suppressing leakage outside the qubit space $\{|0\rangle, |1\rangle\}$ to unwanted states $|2\rangle$, $|3\rangle$, etc.

In the limit of finite bandwidth controls and no leakage, we can project the nonlinear oscillator model onto the qubit manifold spanned by the ground and first excited state. Up to constants

$$H = \frac{1}{2}\hbar\omega_{01}\sigma^z + \hbar\varepsilon\sigma^y. \tag{6.49}$$

This is essentially our qubit model (6.2) up to an irrelevant change of basis $\sigma^x \to \sigma^y$ (see Problem 6.1). Following Section 6.1.4, the transmon is a fully controllable qubit, in the sense that oscillating electromagnetic fields $\varepsilon(t)$ with tunable amplitude, frequency, and phase can be used to implement any qubit operation – even without tuning the gap ω_{01}.

As in the charge qubit, *frequency tunable transmons* rely on a dc-SQUID with some magnetic flux Φ to control the effective value of the Josephson junction $E_J(\Phi) = E_J(0)\cos(\Phi/\varphi_0)$ and adjust the qubit gap (cf. Figure 6.2d). In particular, the external flux can only *decrease* the plasma frequency from a fixed, maximum value approximately as $\omega_{01}(\Phi) = \omega_{01}(0)\sqrt{\cos(\Phi/\varphi_0)}$.

Frequency tuning finds its use in bringing the transmon qubit on-resonance to other circuits, such as resonators and other qubits, to implement quantum operations and gates (e.g., Section 8.3.3). There are two limitations in this type of control. The first one is that we cannot decrease $E_J(\Phi)$ too much, because then the ratio E_J/E_C becomes so small that decoherence kicks in and spoils the setup. The second problem is that the Fock operators and the basis states implicitely depend on the value of the Josephson energy – see (6.33). Nonadiabatic changes of the Josephson energy must be taken into account by considering the explicit time dependence of the Hamiltonian, via $\partial_t \hat{a}, \partial_t \hat{a}^\dagger$ and induces corrections into the qubit model.

Finally, regarding the detection of the qubit, there are no direct measurement schemes for the transmon. Instead, transmon qubits will be typically connected to one or more resonators, which we will use to indirectly probe the state of the qubit in a nondemolition fashion. Those methods, which are general to all types of qubit, are discussed in Section 7.4.3.

The transmon, in any of its incantations, is *the* most used qubit in all kinds of superconducting experiments, from those working with photons and qubits to reproduce quantum optics, to quantum information experiments with tens of qubits by companies such as IBM, Google, or Rigetti Computing. It is a very popular type of qubit for obvious experimental reasons. First of all, it is a large qubit, with a simple design and easy to fabricate – experimentalists can basically cut and paste masks from one lab to another. Second, the qubit has a very *reproducible* design, because its properties are not extremely sensitive to the underlying parameters. For instance, the plasma frequency goes as $\sqrt{8E_J E_C}$, so that small deviations in the Josephson energy – due to fluctuations in the junction's thickness or size – appear as small corrections of 5% in the final qubit. Finally, transmon qubits have very competitive decoherence times, ranging from $T_2 \sim 20\ \mu\text{s}$ up to 0.3 ms – and maybe better, by the time you read this book.

Interestingly, for a transmon *bigger is better.* We can increase the coherence time, simplify fabrication and reproducibility, and enhance the Josephson energy by making the samples larger. This realization led to the development of the *three-dimensional transmon* by Paik et al. (2011), which is nothing but a very large transmon, with a size of millimeters, fabricated not on a chip, but on a portable sapphire substrate. This device is a tiny object that can be moved around and inserted into different experimental setups, including 3D cavities. Because the transmon frequency can be verified outside the experiment through simple resistive measurements,[8] and because transmons are mobile and can be manually interchanged, the practical need for tunability may disappear from experiments, opening the door to setups with 3D transmon qubits that are extremely long lived, with $T_2^* \simeq 0.092$ ms in state-of-the-art setups (Rigetti et al., 2012).

On the negative side, the operation speed of the transmon is ultimately limited by its anharmonicity. For a qubit gap $\omega_{01} \simeq 6$ GHz, anharmonicity $\alpha \simeq 300$ MHz and $T_2 = 40\ \mu\text{s}$, a rough estimate would set the limit on 10^3 single-qubit rotations within the qubit's lifetime. That is reasonably large for to-date algorithms, but it may fall short for large-scale computations. The situation gets even worse if we consider

[8] The resistance of the junction at room temperature gives an idea of E_J, while E_C can be well estimated from the original design due to geometric considerations.

the interaction speed of the transmon with other circuits, such as cavities and other qubits. In this limit, we can find operation times of $4-40$ MHz, which further lower the size accurate quantum computations. In the following chapter, we will meet the flux qubits, a design that solves both problems, enlarging their anharmonicity and allowing for stronger couplings to other devices.

6.4 Flux Qubit

The charge qubit and the transmon are defined by the distribution of charge accumulated on an effective capacitor, which may be a superconducting island or some large superconducting plates. Flux qubits are instead defined by a superposition of current states, describing the flow of charges in a ring. In this context, the name *flux qubits* alludes to the magnetic flux that is trapped in the ring, once we activate the superconducting currents.

We will study two flux qubits with specific applications in quantum simulation and quantum optics. We begin in Section 6.4.2 with the rf-SQUID, a superconducting loop with a single Josephson junction. This qubit is used in Chapter 9 when discussing the D-Wave "quantum annealer," a simulator for quantum Ising models. We also study the *three-junction flux qubit* or *persistent current qubit* in Section 6.4.3. This is a smaller qubit, with good coherence properties, which has facilitated new regimes of light–matter interaction, discussed in Chapter 7.

6.4.1 Frustration and Current States

Current-based qubits were a very popular device well before the transmon occupied the quantum computing space (Makhlin et al., 2001). Phase qubits, rf-SQUIDs, three- and four-junction loops, or fluxoniums were used to explore quantum superpositions (van der Wal et al., 2000), qubit-cavity interactions (Chiorescu et al., 2004) and quantum state engineering (Hofheinz et al., 2009). However, once the transmon showed its improved coherence and reproducibiliy, it became widely adopted and other qubits lost ground.

Not everything is rosy in the transmon world. The anharmonicity $|\omega_{12}-\omega_{01}|$ sets limits on how fast we can operate the transmon and also how strong it can couple to other devices, such as cavities or other qubits. If we ignore this condition, we compromise the integrity of the qubit space, causing leakage[9] to the excited states $|2\rangle, |3\rangle$, etc.

Flux qubits solve all limitations on coupling strength and speed with a design that, thanks to the use of frustrated supercurrents, produces very anharmonic spectrum.

[9] See Problem 6.10 for some crude estimates.

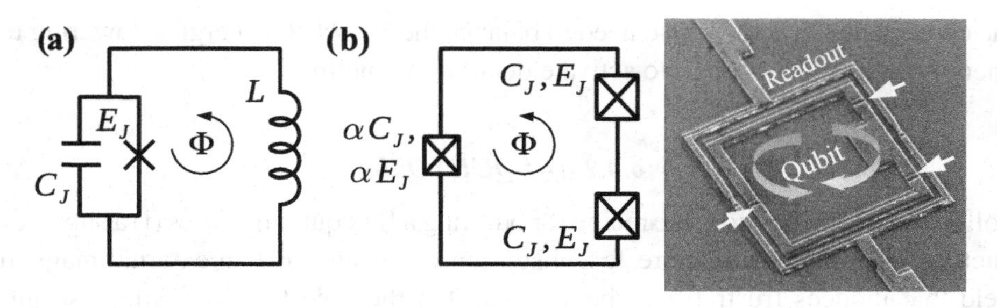

Figure 6.5 Two flux qubit designs. (a) Equivalent circuit for an rf-SQUID qubit, such as the one in Bennett et al. (2007). (b) Three-junction flux qubit circuit (left) and photograph of the same qubit inmersed in a dc-SQUID for readout. Picture courtesy of Clarke and Wilhelm (2008), adapted with permissions.

Figure 6.5 shows different incantations of a flux qubit. The device is essentially a superconducting loop with one or more Josephson junctions. The loop is threaded by a magnetic field, which modifies the inductive energy, causing the appearance of two degenerate ground states with distinguishable current distributions. These will be our qubit states.

The doubly degenerate ground state is explained by the fluxoid quantization. The magnetic flux that passes through the qubit causes a discontinuity in the flux variables along the superconducting circuit $\delta\phi_{\text{loop}}$, which satisfies

$$\oint_C \nabla\phi \cdot d\mathbf{l} = \delta\phi_{\text{loop}} = \Phi_0 \times \mathbb{Z} + \Phi_{\text{ext}}. \tag{6.50}$$

A magnetic flux with a half-integer number of flux quanta $\Phi_{\text{ext}} = \frac{1}{2}\Phi_0$, induces frustration in the circuit: There are two equivalent values of the discontinuity $\delta\phi_{\text{loop}} = \pm\frac{1}{2}\Phi_0$ that fulfill the equation. By virtue of the flux–current relation (3.23), the two values of $\delta\phi_{\text{loop}}$ translate into two distinguishable current states with similar intensity but opposite directions. They are almost our qubit, but not quite yet.

The argument about fluxoid quantization is too general and not tied to any circuit design. A practical qubit has further requirements: two stable, degenerate ground states that can be connected by quantum mechanical means. To satisfy the first requirement, we must stabilize energetically the two values of the flux. For that, we insert Josephson junctions in the superconducting loop, ensuring that their sinusoidal potentials produce two identical minima close to the two frustrated states $\pm\frac{1}{2}\Phi_0$.

The second requirement seems to contradict the previous one: we need to "couple" or enable transitions between the two current states. Part of this requirement is fulfilled by the capacitance of the circuit, which tends to delocalize the ground state wavefunction in flux space, coupling $|+\frac{1}{2}\Phi_0\rangle$ and $|-\frac{1}{2}\Phi_0\rangle$. However, to allow the capacitance to act, we also need to balance the Josephson energies, lowering the energy barriers just enough to activate quantum tunneling.

6.4.2 rf-SQUID Qubit

Following our discussion, the recipe for building a flux qubit includes (i) a superconducting loop, (ii) one or more Josephson junctions, and (iii) an external magnetic field that induces frustration. The rf-SQUID is the oldest and the simplest qubit design that matches all three demands. As shown in Figure 6.5a, the circuit consists of one big loop providing some linear inductive potential, in series with a Josephson junction or a dc-SQUID.

The Hamiltonian of the rf-SQUID was introduced in Section 4.7.1:

$$H \simeq \frac{1}{2C_J}q^2 + \frac{1}{2L}\phi^2 - E_J \cos\left[\frac{1}{\varphi_0}(\phi + \Phi)\right]. \tag{6.51}$$

As with the transmon, we interpret H as the Hamiltonian for a quantum particle of finite mass $m \sim \hbar^2/8E_C$ moving in the inductive potential created by L and E_J. While the inductive terms confine the particle, the capacitive term contributes to its delocalization.

Since the qubit operates in the limit of weak capacitive energies, $E_C/E_J \ll 1$, we develop a good intuition of the qubit's dynamics by studying first the inductive potential. The energy landscape for the rf-SQUID is sketched in Figure 6.6.

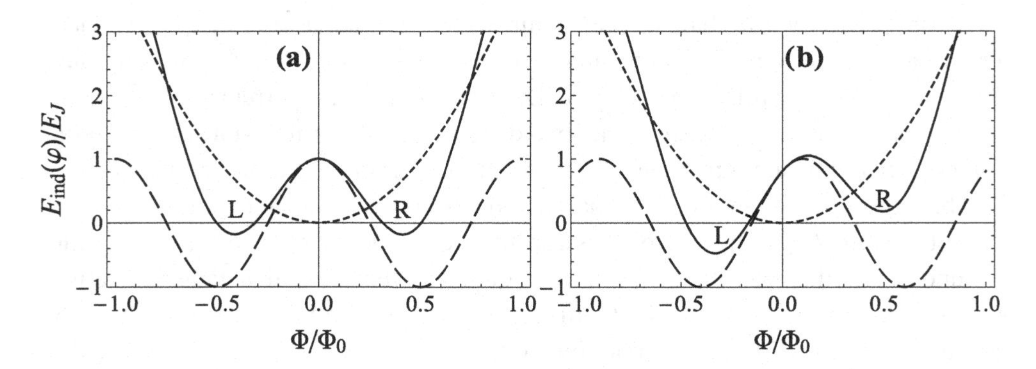

Figure 6.6 Energy levels of the rf-SQUID (a) exactly at the maximal frustration point $\Phi = \Phi_0/2$ and (b) 10% away from this value. Note how the shifted cosine potential (long dashed lines) conspires with the inductor potential (parabola with short dashed lines) to create two local minima, L and R in the frustrated case Note also how a small change in the flux implements a relative energy shift of the two current states $|L\rangle$ and $|R\rangle$. We have exaggerated the plots using a ratio of inductances $L/L_J = 5$ of the loop versus the junction.

It consists of a broad parabolic potential (short dashed lines) on top of which we superimpose a weaker, periodic potential (long dashed lines) with a finite barrier height $E_J \cos((\phi + \Phi)/\varphi_0)$.

When there is no magnetic flux threading the rf-SQUID, the inductor's and the Josephson junction's potential align with each other and define a common ground state around zero flux, $\phi = 0$. On average, this state has no charge and no current, and behaves like a transmon.

We increase the external flux until we reach $\Phi \simeq \frac{1}{2}\Phi_0$. At this point, we achieve *frustration,* because the rf-SQUID cannot minimize the energy of the linear and non-linear terms simultaneously.[10] The two degenerate ground states are the solutions of a transcendental equation:

$$\frac{\partial}{\partial \varphi}\left[\frac{1}{2L}(\varphi_0\varphi)^2 + E_J \cos(\varphi)\right] = 0 \Leftrightarrow \varphi = \frac{E_J L}{\varphi_0^2}\sin(\varphi) = \frac{L}{L_J}\sin(\varphi), \tag{6.52}$$

where $L_J = \varphi_0^2/E_J$ is the effective inductance associated to the Josephson junction (see Section 3.7). Apart from the trivial maximum $\varphi = 0$, this equation reveals two absolute minima whenever $\beta = L_J/L < 1$. As an example, Figure 6.6a shows two absolute minima labeled L and R, both close to the ideal frustrated solution $\phi_{L,R} \simeq \pm\frac{1}{2}\Phi_0$.

Let us complete our model, adding some "quantumness" to the classical picture of well-defined flux states. We first consider each of the local minima separately. If they are deep enough, each of them may be approximated by a harmonic potential. When combined with the capacitive term, the inductive potential gives rise to two effective LC resonator ground states,

$$H_{L,R} \simeq 4E_C\hat{n}^2 + \frac{1}{2}E_J(\varphi - \varphi_{L,R})^2, \tag{6.53}$$

each placed around a different minimum ϕ_L and ϕ_R. The approximate wavefunctions of these two solutions are Gaussians:

$$\psi_{L,R}(\varphi) = \langle\hat{\varphi}|L,R\rangle \propto \exp\left[-\frac{1}{2\sigma_\varphi^2}\left(\varphi - \varphi_{L,R}\right)\right]. \tag{6.54}$$

The wavefunctions of the $|L\rangle$ and $|R\rangle$ states are centered on the left and right potential minima, and correspond to the left- and right-moving currents from Figure 6.5a. The width $\sigma_\varphi \sim \sqrt{8E_C/E_J}$ is both the uncertainty of the phase variable and the spreading of the wavefunction. Note how it depends on the curvature of the inductive potential, which is dominated by the contribution from the Josephson energy.

[10] As explained in Chapter 9, we talk about frustration when we have a functional that is a sum of terms, but we cannot find the minimum of that functional by looking at each term separately.

At low energies, we assume that the qubit remains close to the ground states $|L\rangle$ and $|R\rangle$. We can project the complete Hamiltonian, without approximations, onto the subspace spanned by these two states. Note that $|L\rangle$ and $|R\rangle$ are not orthogonal. The wavefunctions $\psi_{L,R}(\varphi)$ spread outside their respective potential minima, overlapping on the potential barrier. Consequently, the projected Hamiltonian includes terms that enable the tunneling of states in one well $|R\rangle$ to the other one $|L\rangle$, and vice versa. We quantify this probability amplitude with the tunneling matrix element:

$$\hbar t = \langle L|H|R\rangle = \int \psi_L(\varphi)^* H \psi_R(\varphi) \mathrm{d}\varphi \propto E_J e^{-(\varphi_R - \varphi_L)^2/\sigma_\varphi^2} \neq 0. \tag{6.55}$$

The amplitude t is proportional to the matrix element of the inductive energy between flux states and thus grows with E_J. However, the localization of $\psi_{L,R}$ also makes this value exponentially sensitive to the wavepacket size $\sigma_\varphi \propto E_J^{-1/4}$ and to the barrier height.

When performing this projection, we also have to consider the energy of the minima themselves, $\langle L|H|L\rangle$ and $\langle R|H|R\rangle$. Under perfect frustration, shown in Figure 6.6a, both terms would be identical. However, when we change the magnetic flux away from the symmetric point, $\frac{1}{2}\Phi_0 + \delta\Phi$, the flux perturbation $\delta\Phi$ displaces the cosine potential, shifting the energy of the current states in opposite directions (see Figure 6.6b). This energy shift is proportional to the flux variation $\hbar\varepsilon = \mu \times \delta\Phi = \frac{1}{2}(\langle L|H|L\rangle - \langle R|H|R\rangle)$.

When we take both terms into consideration, we arrive at a complete model for our flux qubit, given by

$$H \simeq \mu \times \delta\Phi(|L\rangle\langle L| - |R\rangle\langle R|) + \hbar t\,(|R\rangle\langle L| + |L\rangle\langle R|)\,. \tag{6.56}$$

Since states $|L\rangle$ and $|R\rangle$ are not strictly orthogonal, they do not form a convenient basis. It is much better to work with new qubit states that are superpositions of the original ones:

$$|0\rangle = \frac{1}{\sqrt{2}}(|L\rangle - |R\rangle), \text{ and } |1\rangle = \frac{1}{\sqrt{2}}(|L\rangle + |R\rangle). \tag{6.57}$$

These states produce a qubit Hamiltonian in the usual form:

$$H = \hbar t \sigma^z + \mu\delta\Phi\sigma^x. \tag{6.58}$$

The model contains a fixed, minimum energy gap $\Delta = 2t$ that depends on the tunneling, and an adjustable magnetic field bias $\hbar\varepsilon = \mu\delta\Phi$ that is used to implement single-qubit unitary operations (see Section 6.1.4).

The rf-SQUID has a long history that predates its uses for quantum information processing (Clarke and Braginski, 2004). The circuit itself was conceived as an ultrasensitive magnetometer (Silver and Zimmerman, 1967). It was only much later identified as a qubit candidate due to its two-level subspace and relatively large anharmonicity.

The rf-SQUID is still used in various devices, including the D-Wave quantum annealer or quantum optimizer. Its large size can be an advantage. By stretching the superconducting loops (Harris et al., 2007), a single qubit may physically overlap or come into contact with many other qubits. This is used to create large lattices of thousands of qubits with large connectivity over long distances. The resulting setups can simulate complicated spin Hamiltonians and implement sophisticated quantum optimizations, as explained later in Chapter 9.

The Hamiltonian (6.58) reveals the two main problems of the flux qubit. First, the qubit gap $\Delta \propto 2t$ is exponentially sensitive to the Josephson energy. Small changes in E_J, due to differences in the junction's area or thickness, or due to trapping of charges or impurities in the oxidized layer, can cause large fluctuations in the qubit properties, hampering reproducibility – especially when compared to the transmon.

The second problem is decoherence. The bias field of the qubit is generated by external magnetic fluxes $\delta\Phi$, that are transported close to the qubit by different types of antennas. Unfortunately, as described before when working with cavities, these cables not only transport the static current we need to create the magnetic flux. They also act as bath that couples to the qubit, introducing quantum fluctuations that must be modeled as dephasing and decoherence (see Sections 6.1.5 and 6.1.6). As in the charge qubit (Section 6.2.3), the optimal operation point of the qubit to minimize this environment-induced dephasing lays at the *symmetry point* $\delta\Phi \simeq 0$ of maximum frustration.

However, even close to the symmetry point, the decoherence of the rf-SQUID may be inadequate for quantum information purposes. These qubits are large by design – to have a large coupling strength μ, to overlap with faraway qubits (see Chapter 9) and to have a large linear inductance – and this size comes together with a greater sensitivity to stray fields and decoherence. An obvious solution is to work with smaller qubits, where the frustrated potential is created using only Josephson junctions, as explained in the following section.

6.4.3 Persistent Current Qubit

The flux qubit based on three (or four) junctions was introduced by Mooij et al. (1999) and experimentally demonstrated by van der Wal et al. (2000). The *persistent current qubit* replaces the big inductive loop from the rf-SQUID with a smaller loop where most of the inductance is provided by a ring of Josephson junctions, as shown in Figure 6.5b. When the loop is threaded by half a flux quantum, the qubit develops the same degenerate and frustrated current states as the rf-SQUID. However, the two current states are energetically stabilized in the multiwell inductive landscape generated just by the Josephson junctions, without any linear inductors.

The resulting qubit is smaller than an rf-SQUID – i.e., a few micrometers versus 150 μm structure in Figure 6.5a. This makes it possible to pack more qubits in the same setup, and to embed them in exotic structures, such as a transmission line nanoconstriction, to achieve extreme regimes of interactions with the microwave field.[11] There have also been significant improvements in flux qubit fabrication and experiments by Yan et al. (2015) have demonstrated coherence times that are comparable to those of the best 2D transmons $T_2 \simeq 40\ \mu\text{s}$. All this makes the three-junction qubit the second most used qubit in the field, and usually the one that people refer to when discussing "flux qubits."

The circuit of the persistent current flux qubit was derived in Section 4.8 and it is sketched again in Figure 6.5b. It consists of three junctions. Two junctions are identical and have the same Josephson energy and capacitance E_J and C_J. The third junction's area is smaller by a factor $\alpha < 1$, leading to a reduced Josephson energy $\alpha \times E_J$ and a smaller capacitance $\alpha \times C_J$. The effective Hamiltonian has the form

$$H = \frac{1}{2(1/2+\alpha)C_J} q_+^2 + \frac{1}{2C_J} q_-^2 + E_J V(\phi_-, \phi_+) \tag{6.59}$$

with a nonlinear potential

$$V(\phi_-, \phi_+) = \alpha E_J \cos\left(\frac{\Phi - \phi_+}{\varphi_0}\right) - 2E_J \cos\left(\frac{\phi_+}{2\varphi_0}\right) \cos\left(\frac{\phi_-}{2\varphi_0}\right). \tag{6.60}$$

This Hamiltonian depends on the total flux jump across the two bigger junctions $\phi_+ = \phi_1 + \phi_2$. Note how we have eliminated the flux jump across the smaller junction, using the fluxoid quantization, $\phi_\alpha = \Phi - \phi_+$.

Once more, the flux qubit Hamiltonian joins two competing potentials. A global and fixed inductive potential $\propto E_J \cos(\phi_+/2\varphi_0)$ favors a global minimum at $\phi_+ = \phi_- = 0$. On top of this, we superpose a cosine potential that can be displaced along the ϕ_+ axis using the external flux Φ. When $\Phi = \frac{1}{2}\Phi_0$, this potential favors two minima with opposite signs of the flux.

Figure 6.7 shows two landscapes of the inductive energy: (a) without frustration and (b) right at the maximum frustration point $\Phi = \frac{1}{2}\Phi_0$. The local minima of energy lay always along the line $\phi_- = 0$, where the bigger junctions share the same current $\phi_1 = \phi_2$. However, in the frustrated case we find two minima along the ϕ_+ axis. These minima correspond to two directions of a similar current, i.e., positive

[11] See, for instance, the experiments by Forn-Díaz et al. (2010) and Niemczyk et al. (2010) based on the ideas by Bourassa et al. (2009), or more recent experiments by Forn-Díaz et al. (2016) demonstrating *ultrastrong coupling* between flux qubits and propagating photons.

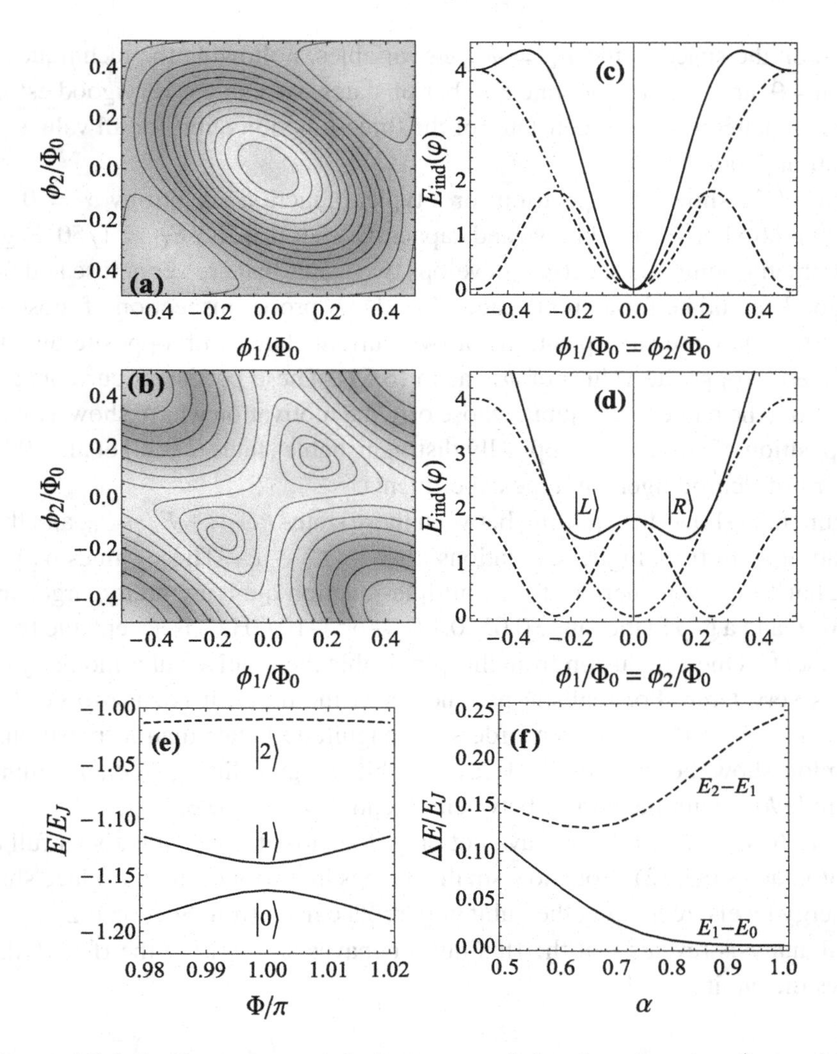

Figure 6.7 Inductive energy and eigenstates of the three-junction flux qubit. Nonlinear inductive potential (6.60) with (a) $\alpha = 0.9$ and no external flux and with (b) $\alpha = 0.9$ and $\Phi/\varphi_0 = \pi$, together with the respective cuts along $\phi_1 = \phi_2$ (c,d). (e) Energy levels of the three-junction flux qubit with anisotropy $\alpha = 0.7$. (f) Energy gap between the "qubit" states $E_1 - E_0$ and between the first and second excited states, $E_2 - E_1$, as a function of the anisotropy, for $\Phi = \Phi_0/2$.

flux jumps and right-moving currents for $\phi_{\alpha,1,2,+} > 0$, and negative jumps or left-moving currents for $\phi_{\alpha,1,2,+} < 0$.

The persistent current qubit is rather complicated and does not bend well to analytical estimates. As explained in Problem 6.13, a numerical method that works well is to diagonalize the qubit Hamiltonian in a basis of number states, defined on the space of the $\varphi_\pm = \phi_\pm/\varphi_0$ variables. Following the techniques from Section 4.9, and with a moderate number of states, we can get very good estimates of the eigenenergies and eigenstates of the three-junction qubit for all values of the frustration parameter Φ.

Figure 6.7e shows the spectrum for a typical junction anisotropy $\alpha = 0.7$, and the usual ratio between inductive and capacitive energies $E_J/E_C = 1/50$. Right on the symmetry point, the spectrum develops two states that are very close and form a very good candidate for a qubit space, $|0\rangle$, $|1\rangle$. A proper inspection of those states reveals that they are superpositions of two current states with opposite directions, associated to opposite values of ϕ_+, as in (6.57). These persistent currents are the ones giving the name to the qubit, whose original motivation was to show a quantum superposition of two macroscopically distinguishable states (Mooij et al., 1999), in the spirit of Schrödinger cat–type experiments.

Figure 6.7e shows the splitting between these states $\hbar\omega_{01} \sim E_1 - E_0$, and the gap separating them from higher excitations $\hbar\omega_{12} \sim E_2 - E_1$. These values may seem small, but Josephson energies for a multiple-junction qubit are pretty large, around 100 GHz, and a gap in the range 0.05–0.1E_J – or 5–10 GHz – is acceptable for most experiments. One conclusion from this plot is that the relative anharmonicity of our qubit is very large. For realistic parameters of the qubit, it is easy to get factors of $\omega_{12}/\omega_{01} \sim 1000\%$ or larger, orders of magnitude better than a transmon. The same plots show the exponential decay of qubit energy splitting – i.e., the tunneling amplitude $\hbar t$ – with the barrier between potential wells $\propto \alpha E_J$.

Figure 6.7e reveals that we have not only good qubit states, but also a full qubit Hamiltonian as in (6.2). Note how small changes in the total flux introduce shifts in the energy levels, recreating the qubit hyperbola curves from Section 6.2.3. A more careful analysis reveals that the flux qubit behaves similarly to the rf-SQUID and follows the qubit model:

$$\Delta = (E_1 - E_0)_{\Phi=\frac{1}{2}\Phi_0}, \quad \text{and} \quad \varepsilon \simeq \mu\left(\Phi - \frac{1}{2}\Phi_0\right). \tag{6.61}$$

The effective magnetic dipole μ can be estimated from the numerical simulations, fitting the energy levels to hyperbola, or semiclassically, studying the energy shifts of the minima as a function of an external magnetic field (see Problem 6.12).

6.4.4 General Operation

Flux qubits can be regarded as large magnetic dipoles with two preferred orientations, associated to the two current states. They couple best to the magnetic fields – generated by nearby antennas, currents in microwave resonators, or other flux qubits – and as a result, this is the usual way to control them. The image of the flux qubit as a pseudospin is supported by their highly anharmonic spectrum. As we have seen, the large nonlinearity of the Josephson potential allows us to engineer qubit states that have a large splitting, $\omega_{01} \simeq 4 - 8$ GHz, protected by an even larger separation from higher excited states, $\omega_{12} > 20$ GHz. The result is a qubit that allows fast operations – i.e., the bandwidth of the pulses that interact with them can be comparable to the gap ω_{01} – and that interact strongly with other superconducting elements – the coupling strength g, as we will soon see, can be of the order of magnitude of $\hbar\omega_{01}$ without yet causing significant leakage to other excited states.

We can implement a projective measurement of the flux qubits in the current basis, detecting the total flux trapped in the loop – i.e., the external flux plus the contribution of the qubit currents. The measurement is performed by a dc-SQUID that is inductively coupled to the qubit. The SQUID, as we have already seen in Section 4.7.2, is a sensitive magnetometer. When placed close to a flux qubit (Figure 6.5a), or completely surrounding it (Figure 6.5b), the SQUID captures some of the qubit's magnetic flux inside its own loop. When we apply a current to the SQUID, it will switch to a voltage state at a critical value of the current. This value depends very sensitively on the trapped flux and can be used to discriminate the qubit's state. Unfortunately, the coupling between the qubit and the SQUID is also a source of decoherence because it involves a permanent interaction between the qubit and the readout device. Nowadays, qubits are more often measured through the cavities they connect to, as in Section 7.4.3, or using some circuit that couples or decouples the readout device, as in the D-Wave architecture (see Section 9.4.1).

There are two intrinsic challenges in the design and operation of flux qubits: reproducibility and decoherence. Regarding the first problem, we have seen how the qubit gap Δ depends exponentially on the junction energy E_J. Small fabrication errors can take us from a situation where the energy gap between qubit states $\omega_{01} \sim \Delta$ is on the order of a few gigahertzs, to a situation where it is so small that the quantum fluctuations are destroyed because of thermal excitations in the chip. This problem is typically solved by tuning the qubit gap: i.e., replacing the $\alpha-$junction with a dc-SQUID that allows us to set the effective value of the anisotropy (Paauw et al., 2009). However, it is quite difficult to inject flux in a dc-SQUID without adding some magnetic flux to the flux qubit's loop itself – unless one creates intricate designs, known as *gradiometric* qubits (see Problem 6.14) – which is another reason why these qubits are not often used.

Regarding coherence times, flux qubits are largely limited by fluctuations in the magnetic flux. These can be induced by the current sources and stray field that are trapped by the qubit – even if it is as small as a persistent current qubit – and cause the dephasing of the qubit. Coherence times of typical persistent current qubits lay around $1-6\,\mu$s, with exceptional values around the tens of microseconds.

Interestingly, we can improve coherence by an order of magnitude, reducing dephasing rates to almost zero. The method is to shunt the flux qubit with a large capacitor. As in the transmon, a large capacitance decreases the sensitivity to external fields, increases coherence times, and enhances reproducibility, all at the expense of anharmonicity. The work by Yan et al. (2016) reports capacitively shunted flux qubits where decoherence is almost all attributed to relaxation $T_2 \simeq T_1/2 \simeq 40\,\mu\text{s}$, with no significant dephasing. The qubit has a smaller but still significant anharmonicity in the range of 500–900 MHz.

Overall, research on flux qubit design and qubit decoherence is a huge and ongoing effort, motivated by applications in quantum simulation and quantum annealing, where we need to preserve quantum superpositions of many interacting qubits for adiabatically long times. However, let us delay those considerations to Chapter 9, ending here our discussion of qubit designs and moving forward to address other important questions, such as understanding how qubits and interact and talk to each other.

6.5 Qubit–Qubit Interactions

Superconducting qubits are usually organized in *quantum registers*,[12] arrays where qubits can be individually controlled and measured, and where they can interact with each other. The natural interactions between qubits are two-body couplings.[13] Two-body interactions can be expanded using the Pauli matrices of the involved qubits. This leads to models often found in textbooks from condensed matter physics and quantum magnetism:

$$H = \sum_i \frac{1}{2}\left(\Delta_i \sigma_i^z + \epsilon_i \sigma^x\right) + \sum_{ij} J_{ij}^x \sigma_i^x \sigma_j^x + \sum_{ij} J_{ij}^z \sigma_i^z \sigma_j^z + \sum_{ij} J_{ij}^y \sigma_i^y \sigma_j^y, \quad (6.62)$$

which combine longitudinal and transverse "magnetic fields" – Δ_i and ϵ_i – with two-qubit interactions $J_{ij}^{x,y,z}$.

Depending on the active parameters, (6.62) receives different names. When Hamiltonian (6.62) only contains one type of Pauli operators – e.g., $J^x, J^y, \varepsilon = 0$ – we have the *classical Ising model*, a model whose eigenstates are all classical states with well-defined values of σ_i^z. If the interaction and the magnetic field appear on different Pauli operators – e.g., $J^x, \Delta \neq 0$ and others zero – we have the *quantum Ising model* or Ising model with a transverse field. Since each direction

[12] See Figure 8.2 in Chapter 8.

[13] As it also occurs in Nature, multibody interactions may appear in some perturbative limits from two-body interactions, but they will not be discussed in this book.

of interaction, X, Y, or Z, is associated with a circuit degree of freedom, it is quite unusual to find two or more types of interactions, but this is possible. For instance, if $J^x = J^y \neq 0$ or $J^x, J^z, \Delta \neq 0$, we arrive at an *XY model*, describing the exchange of quantum excitations between qubits. We also have the well-known *Heisenberg Hamiltonian* $J^x = J^z = J^y \neq 0$, and the general XYZ model $J^x, J^y, J^z \neq 0$.

The design of qubit–qubit interactions has three main applications. First, the implementation of quantum magnetism Hamiltonians (6.62) converts the superconducting circuit into a very accurate *quantum simulator*. We can prepare states and study how they evolve under the effective Ising and XY models to deepen our understanding of quantum many-body physics, quantum phase transitions, many-body localization, termalization, quantum topological order, and many other interesting phenomena. Second, as we will see later in Chapter 9, preparing the ground state of these models is a kind of computation that can be used to solve very hard problems – both classical and quantum. Finally, and most important, activating these qubit–qubit interactions for brief periods of time, we implement universal two-qubit and multiqubit unitary operations, which are the building blocks for scalable *universal quantum computation* (see Chapter 8).

6.5.1 Dipolar Magnetic Interaction

Superconducting qubits can experience a mutual interaction, mediated solely by the electromagnetic field. Consider the two flux qubits in Figure 6.8b. Each qubit's state is a superposition of currents with left- and right-moving orientations, $|L\rangle$

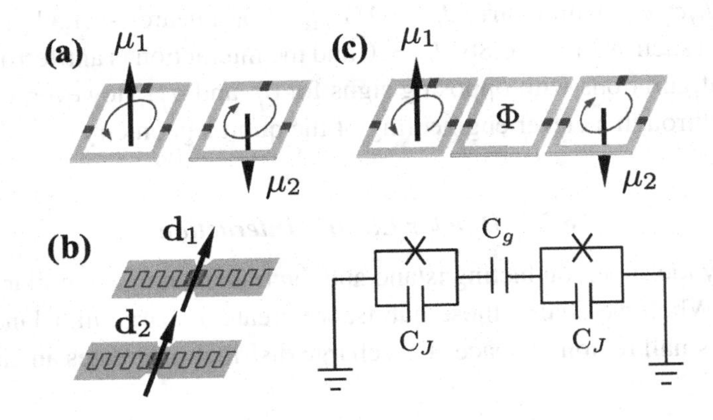

Figure 6.8 Qubit–qubit dipolar interactions. (a) Flux qubits and (b) transmon qubits have an associated magnetic or electric dipole moment. These moments interact with each other through the equations of electromagnetism, $\propto \mathbf{d}_1\mathbf{d}_2$ and $\propto \boldsymbol{\mu}_1 \cdot \boldsymbol{\mu}_2$, giving rise to effective qubit couplings $J\sigma_1^x\sigma_2^x$ or $J\sigma_1^y\sigma_2^y$. Sometimes, as in the magnetic case (c), an auxiliary circuit can be used to mediate and tune the interaction.

and $|R\rangle$. The superconducting current is a σ^x Pauli operator multiplied by an average current intensity[14] $\tilde{I}_c$:

$$\hat{I} = \tilde{I}_c\,|L\rangle\langle L| - \tilde{I}_c\,|R\rangle\langle R| = \tilde{I}_c\sigma^x. \tag{6.63}$$

The current in each qubit creates a magnetic field distribution that is very similar to the one from a magnetic dipole $\boldsymbol{\mu} \propto \hat{I}\mathbf{e}_\perp$ sitting perpendicular to the qubit's plane. From the classical interaction of magnetic dipoles,[15] we expect an *antiferromagnetic interaction* $H_{\text{int}} \sim +M\hat{I}_1\hat{I}_2$ that favors antiparallel orientations of the dipoles – or, equivalently, currents flowing in opposite directions on both qubits.

We can also derive this interaction in the lumped-element circuit model, using geometric properties that can be designed experimentally. In particular, we must introduce, as in the rf-SQUID Hamiltonian (see Section 4.7.1), the *mutual inductance* M between the superconducting qubits' loops. The mutual inductance, which can be estimated from the shape of the circuits, accounts for the change in the magnetic flux of one circuit created by another qubit's currents, and the magnetic energy that results from it. This change is symmetric and is written as a quadratic correction to the energy $E_{\text{int}} = M_{12}I_1I_2$, in terms of the actual currents $I_{1,2}$ flowing in both loops. These currents may have a complicated expression in terms of the circuit's fluxes.[16] When the perturbation is small compared to the inductive potential, we can approximate its contribution to the Hamiltonian with a product $M_{12}\hat{I}_1\hat{I}_2$ of the current operators that we estimated in the qubit basis (6.63). When it is large, on the other hand, M can itself influence the qubit states, leading to renormalizations that must be studied in a self-consistent way.

For a generic array of flux qubits, the inductive coupling adopts the Ising form $J_{ij}\hat{\mu}_i\hat{\mu}_j \sim J_{ij}\sigma_i^x\sigma_j^x$, with matrix $J_{ij} = M\tilde{I}_{ci}\tilde{I}_{cj}$. For a nearest-neighbor and planar configuration such as Figure 6.8b, $J_{ij} > 0$ and the interaction is antiferromagnetic – it favors configurations with opposing signs for σ_i^x and σ_j^x. However, other values are possible through a clever engineering of the qubit's geometry.

6.5.2 Dipolar Electric Interaction

We can study the superconducting island and the transmon in an similar way to the flux qubits. When we charge those qubits, we create a charge imbalance concentrated over a small region of space. This charge distribution creates an electric field

[14] Note that the average value $\tilde{I}_c$ is not exactly the supercurrent of the junctions themselves. For instance, in the rf-SQUID we find $\tilde{I}_c = |\phi_{L,R}|/L$, in terms of the equilibrium value of the flux for either potential well (see Problem 6.15).

[15] For two dipoles $\boldsymbol{\mu}_{1,2}$, their interaction energy goes as $E = -(\mu_0/4\pi|\mathbf{r}|^3)(3(\boldsymbol{\mu}_1\cdot\mathbf{r})(\boldsymbol{\mu}_2\cdot\mathbf{r}) - \boldsymbol{\mu}_1\cdot\boldsymbol{\mu}_2)$ with their separation $\mathbf{r}$. Since both our dipoles sit on the same plane, we expect $E = +\mu_0\mu_1\mu_2/4\pi|\mathbf{r}|^3$.

[16] For instance, in the three-junction flux qubit $I_{1,2} \propto \alpha I_c \sin(\phi_+/\varphi_0)$.

that is very similar to the one from an electric dipole $\mathbf{d}_i \propto \hat{q}_i$. From the classical theory of electrostatic interactions, we expect a contribution to the energy that grows with the product of those dipoles and decreases with their separation. Unlike the magnetic interaction, there is great variability on the sign and strength on the interactions,[17] based on the relative orientations of the qubits and their separation, but we may write something like

$$H_{\text{int}} = \pm \frac{1}{C_{\text{eff}}} \hat{q}_1 \hat{q}_2 \sim \begin{cases} J\sigma_1^x \sigma_2^x, & \text{for charge qubit,} \\ J\sigma_1^y \sigma_2^y, & \text{for transmon.} \end{cases} \tag{6.64}$$

The sign and the effective mutual capacitance C_{eff} can be determined by the microscopic model. The coupling operator may also change depending on our convention for the charge basis: In the superconducting island, $\hat{q} \propto \sigma^x$, while for the transmon with the notation from (6.33), $\hat{q} \propto i(|1\rangle\langle 0| - |0\rangle\langle 1|) \sim \sigma^y$. Sometimes these differences are irrelevant, and may be suppressed with a local unitary transformation, but other times they are not, leading to unequivalent models – e.g., imagine two transmons with SQUIDs interacting also inductively through their loops!

We may be a bit more rigorous in our treatment of the capacitive interactions by considering the effective circuit in Figure 6.8c. The complete model includes both the self-capacitance of the transmon or charge qubit C_J, and a mutual capacitance between both qubits C_g. The whole circuit may be described using the flux degrees of freedom in the two superconducting islands, with a capacitance matrix:

$$C = \begin{pmatrix} C_J + C_g & -C_g \\ -C_g & C_J + C_g \end{pmatrix}, \tag{6.65}$$

where the term $\frac{1}{2} C_g (\dot{\phi}_1 - \dot{\phi}_2)^2$ causes C_g to appear both in the diagonal, renormalizing the effective capacitance of each qubit, and in the off-diagonal, with something that resembles an interaction. This structure is preserved in the effective Hamiltonian, which depends on the inverse capacitance matrix:

$$C^{-1} = \begin{pmatrix} \frac{1+C_g/C_J}{C_J} & \frac{C_g}{C_J^2} \\ \frac{C_g}{C_J^2} & \frac{1+C_g/C_J}{C_J} \end{pmatrix} + \mathcal{O}\left(\left(\frac{C_g}{C_J}\right)^2\right). \tag{6.66}$$

We have a dipolar interaction, but the mutual capacitance $C_{\text{eff}} = C_J \times (C_J/C_g)$ differs from C_g and we can only ignore the renormalization of the qubits when in the limit of weak interactions $C_g/C_J \ll 1$.

[17] For two electric dipoles, the electrostatic energy is $(\mathbf{d}_1 \cdot \mathbf{d}_2 - 3(\mathbf{d}_1 \cdot \mathbf{r})(\mathbf{d}_2 \cdot \mathbf{r})/r^2/(4\pi\epsilon_0 r^3)$ as a function of their separation $\mathbf{r}$.

As in the case of electric dipoles, interactions between qubits decrease with the distance, but are never strictly nearest-neighbor. Take for instance a one-dimensional ring of qubits, coupled with periodic boundary conditions, so that their capacitive energy reads $\sum_i \frac{1}{2}C_J\dot{\phi}_i^2 + \sum_i \frac{1}{2}C_g(\dot{\phi}_i - \dot{\phi}_{i+1})^2$. When we invert the capacitance matrix for this circuit in the limit $C_g/C_J \ll 1$, we obtain an interaction that decays as $C_{ij} \simeq \exp(-\log(C_g/C_J)|j-i-1|)/C_J$, but if we consider two-dimensional arrangements, we may obtain C_{ij} that decay polynomially or even logarithmically up to a critical length (Ortuño et al., 2015). These types of interaction are often neglected, but they contribute to the *residual cross-talk* – all the unavoidable interactions between qubits are not considered in the model, but which appear in the experiment, due to the intervening elements, substrate, cables, qubit separation, etc.

6.5.3 Coupling Tunability

There are scenarios where we only want some qubits to interact with an other. For instance, as explained in Section 8.3, quantum algorithms are decomposed into elementary operations, some of which involve the evolution of one pair of qubits with a given interaction for a brief period of time – e.g., $J\sigma_1^z\sigma_2^z$ to implement a control-phase rotation. During this time, those qubits *must not talk to other qubits*, or otherwise the gate will be spoiled. This motivates our need to activate or deactivate interactions at will.

Looking at qubits that interact with each other, there are two approaches to switch interactions on and off. A hardware-based approach is to change the qubit by modifying its dipole moment. For instance, we could replace principal junction in a flux qubit – e.g., the α loop in the three-junction design – with a tunable dc-SQUID. By threading the SQUID with a magnetic flux, we effectively change the critical current, lowering the value of the interaction. Unfortunately, this also causes us to lose the qubit itself!

A slightly more robust trick is to change the experimental conditions to make interaction less effective. In the dipolar coupling limit, we have two-body interactions along directions orthogonal to the natural Hamiltonian of the qubit. For instance, we find

$$H = \frac{\Delta_1}{2}\sigma_1^z + \frac{\Delta_2}{2}\sigma_2^z + J\sigma_1^x\sigma_2^x, \tag{6.67}$$

where $\Delta_{1,2}$ are the qubits' gaps. The coupling term contains four types of contributions $\sigma_1^x\sigma_2^x = |00\rangle\langle 11| + |01\rangle\langle 10| + |10\rangle\langle 01| + |11\rangle\langle 00|$, which flip or exchange excitations between both qubits. If $|J| \ll |\Delta_{1,2}|$, we may invoke the RWA to neglect the interactions with $|00\rangle$ and $|11\rangle$, because those states have very different energies.

If we further impose a large detuning $\delta = |\Delta_1 - \Delta_2| \gg |J|$, we might even ignore *all interaction terms*, because all the connected states would be off-resonant from each other.

This technique is rather general and often used in the laboratory. It has two shortcomings. The first problem is that we neglect the influence on the dynamics due to the off-resonant interactions, and which to lowest order in perturbation theory are (see Problems 6.16 and 6.17)

$$H_{\text{int}} = \frac{J^2}{2(\Delta_1 + \Delta_2)}(|11\rangle\langle 11| - |00\rangle\langle 00|) + \frac{J^2}{\Delta_2 - \Delta_1}(|01\rangle\langle 01| - |10\rangle\langle 10|) + \cdots . \tag{6.68}$$

These contributions may accumulate over long times, producing large changes in the phases of the qubit, which disturb our computations.

The second problem is that this strategy does not scale up nicely to large setups. We can tune the qubit frequencies only within a finite bandwidth, such as $\delta \sim 1\text{–}2$ GHz. Thus, if we try to park N qubits separated from each other, at least two of them will be at a distance $\mathcal{O}(1/N)$ in the spectrum. This sets a lower bound in the corrections that we neglected, which now grow as $\mathcal{O}(J^2N/\delta)$. This situation, called *frequency crowding*, appears already for setups with a moderate number of qubits, limiting the fidelity of quantum operations (see Problem 6.16). As we will see now, all these obstacles may be strongly suppressed or eliminated when we use other circuits to mediate the interactions between qubits.

6.5.4 Mediated Interactions and Tunable Couplers

In this section, we introduce the notion of *tunable coupler*, a superconducting circuit that acts as a mediator of interactions in a way that can be adjusted by external controls. Mediated interactions are an important element in superconducting circuit technology. Couplers allow us to connect circuits that are far apart – or which at least are not immediate neighbors. We use them to control the strength and sign of interactions, and sometimes they are even used to cancel other residual couplings or cross-talk.

We will now discuss a rather general framework of interactions mediated by "fast" circuits, circuits that never leave their ground state, but which can transport the influence of one circuit to another. We will discuss the particular example of a dc-SQUID, illustrated in Figure 6.8c. In this configuration, the magnetic field from one qubit induces a supercurrent into the mediating loop; this current generates its own magnetic field that influences the far away qubit and vice versa.

When the mediating loop can quickly adapt to the influence of both qubits, it remains in a quasi-instantaneous equilibrium state, and we describe the coupling scheme with the general framework of *adiabatic interactions*. To be precise, the mediating circuit C must satisfy a series of conditions that for our two-qubit setup become the following ones:

(1) The coupler interacts pairwise with the qubits 1 and 2. The total energy of the three systems $E(x_1, x_2, X_C; \Phi_{\text{ext}})$ depends on the states of both qubits $x_{1,2}$, on the state of the coupler X_C and on some control parameter Φ.
(2) The coupler evolves faster than the qubits it talks to. We can apply the Born–Oppenheimer approximation,[18] also known as *adiabatic elimination*, and assume that the coupler is always at the lowest energy state $X_C = X_{\text{eq}}(x_1, x_2; \Phi_{\text{ext}})$.
(3) We can eliminate the coupler variable, which is fixed, obtaining an effective interaction $E_{\text{eff}}(x_1, x_2; \Phi_{\text{ext}}) = E[x_1, x_2, X_{\text{eq}}(x_1, x_2; \Phi_{\text{ext}})]$.
(4) Using the control parameter Φ_{ext}, we can change the form and strength of the interaction energy E_{eff}. We can use the parameter cancel any previous direct interaction or cross-talk between qubits 1 and 2, or to establish a new effective coupling between them.

As shown by van den Brink et al. (2005), this design allows the dc-SQUID from Figure 6.8c to work as tunable inductive coupler. The Hamiltonian for the mediating SQUID, introduced in Section 4.7.2, depends on the total flux Φ that passes through the loop:

$$H = \frac{1}{4C_J}\hat{q}_-^2 + 4E_J \cos\left(\frac{\Phi}{2\varphi_0}\right) \times \cos\left(\frac{\hat{\phi}_-}{\varphi_0}\right). \tag{6.69}$$

We assume that the plasma frequency of the SQUID $\hbar\omega_{dc} \sim \sqrt{8E_C E_J(\Phi)}$ is much faster than the timescales at which Φ will change, $\omega_{dc} \gg \Phi/\dot{\Phi}$. The state of the coupler $X_C \sim \{q_-, \phi_-\}$ adapts to the adiabatic changes in the flux Φ, leaving the SQUID always in the instantaneous ground state defined by the external parameters. The energy of the SQUID can be approximated by $E_{\text{SQUID}}(\Phi) \simeq \hbar\omega_{dc}/2$, as expected from an oscillator in the vacuum state.

We now have to include the effect of the nearby qubits. The flux enclosed by the loop Φ is a sum of a classical control field Φ_{ext} and the flux injected by nearby qubits through their mutual inductance M_i. The combined Hamiltonian for the two qubits with the coupler reads

$$\hat{H} = \hat{H}_1 + \hat{H}_2 + E_{\text{SQUID}}\left(\Phi + M_1\sigma_1^x + M_2\sigma_2^x\right). \tag{6.70}$$

[18] This approximation or way of thinking is quite common in physics. It describes how electrons mediate an interaction between atoms in a molecule, or how phonons bind electrons to form Cooper pairs.

In this Hamiltonian, because of the adiabatic approximation, we replaced the Hamiltonian for the SQUID with its instantaneous ground state energy – a classical function of quantum operators $\sigma^x_{1,2}$. We can rewrite the energy functional as a product of the dipole moments of both qubits:[19]

$$\hat{H} = \hat{H}_1 + \hat{H}_2 + \frac{1}{2} J_{12}(\Phi)\sigma^x_1 \sigma^x_2, \tag{6.71}$$

where $\hat{H}_{1,2}$ are the Hamiltonians for the coupled qubits, and the mutual inductance J_{12} is the second-order contribution:[20]

$$J_{12}(\Phi) \simeq M_1 M_2 \frac{\partial^2}{\partial \Phi^2} E_{\text{SQUID}}(\Phi). \tag{6.72}$$

Many experiments have demonstrated tunable coupling using dc-SQUIDs (Hime et al., 2006), more complicated circuit loops (van der Ploeg et al., 2007) or even a qubit (Niskanen et al., 2007) with faster dynamics. A particularly relevant device is the rf-SQUID coupler developed by Harris et al. (2007), and which together with a somewhat outdated flux qubit design constitutes the basis of the D-Wave quantum annealer discussed in Chapter 9.

6.6 Qubit Coherence

Qubit coherence is one of the limiting factors in the design and implementation of quantum computers and quantum devices. Decoherence destroys the quantum information that is stored in the qubits. It not only acts during the time in which the qubits are "resting," but also affects the qubit during the realization of single-qubit operations and measurements. In presence of decoherence, our controls are no longer the idealized models from Section 6.1.4, but include corrections – such as dephasing, relaxation, or cross-talk – that may limit the quality with which those operations can be performed.

It is very important, and a first step in any quantum technology application, to characterize the coherence properties of the system at hand. This is true not only for fundamental works with new qubits designs – e.g., Nakamura et al. (1997), van der Wal et al. (2000), or Steffen et al. (2010). Higher-order devices, such as multiqubit quantum computers (Arute et al., 2019), and quantum computing infrastructures, such as the IBM Quantum Experience, must continuously calibrate their devices,

[19] Higher powers of Pauli matrices are either the identity, or the matrices themselves. Therefore, the function E_{SQUID} can only produce terms that are proportional to $\sigma^x_{1,2}$ or to the product $\sigma^x_1 \sigma^x_2$, as seen by expanding the function to arbitrary powers in a Taylor series.

[20] Ideally, $\partial E_{\text{SQUID}}/\partial \Phi \simeq 0$. If not, this represents a shift of the qubit gaps, which can be included in the theory.

monitoring the coherence times of the qubits, as well their natural frequencies, quality of measurements, fidelity of operations, etc.

An ab initio determination of how one or more qubits evolve can be a rather cumbersome problem. Chapter 8 will introduce methods for a self-consistent tomographic reconstruction of how a density matrix evolves under a general control. Fortunately, in absence of external controls, qubits tend to follow a simple decoherence model that mostly contains relaxation and dephasing (6.19). This model can be fully determined by calibrating the values of the relaxation time T_1 and of the dephasing time, T_2 or T_ϕ, in different experiments.

To calibrate the relaxation time, we study the evolution of the excited state $|1\rangle$. We reset the qubit to $|0\rangle$, flip the state of the qubit with a π pulse, wait for a time t, and measure the excited population $P_1(t) = \frac{1}{2}\langle 1 + \sigma^z\rangle$. As shown experimentally (Steffen et al., 2010), even if we make small errors in the preparation and measurement, the excited population can be fit to an exponential curve $P_1(t) = P_1(0)\exp(-t/T_1)$, which gives the value of T_1.

To obtain the value of the decoherence time, we must study the evolution of a quantum superposition, using a protocol known as *Ramsey interferometry*. We start with the $|0\rangle$ state, but now implement an incomplete rotation around σ^y with the $\pi/2$ angle, preparing the state $\frac{1}{\sqrt{2}}(|0\rangle - |1\rangle)$. We wait for a time t, repeat the rotation, and measure σ^z. The excited state population now follows a damped oscillation $P_1(t) = \frac{1}{2}(1 + \cos(\Delta t)e^{-t/T_2^*})$.

The coherence time is always limited by relaxation, $T_2^* = 2T_1$. In practice, this time is shorter because of dephasing and other sources of decoherence T_ϕ, as shown in (6.20). In those cases, we can further separate and partially cancel the contribution from dephasing, using spin-echo to obtain a slightly larger effective coherence lifetime T_2^{echo} (see Section 6.1.5).

Figure 6.9 illustrates the experimental coherence times T_2^* (or T_2^{echo}, depending on the work) for a large variety of qubits. There is an overall positive trend, where coherence times almost double every year,[21] starting from the charge qubit in Nakamura et al. (1999), until recent experiments with 2D transmon qubits (Place et al., 2021) and photon-encoded qubits (Rosenblum et al., 2018). We see significant jumps due to the placement of qubits in microwave resonators (Wallraff et al., 2004), the introduction of large three-dimensional transmons (Paik et al., 2011), the development of capacitively-shunted flux qubits (Yan et al., 2016), the encoding of qubits in photonic degrees of freedom (Wang et al., 2016; Rosenblum et al., 2018),

[21] This has been deemed Schoelkopf's law, by analogy with Moore's law for Silicon-based integrated chips.

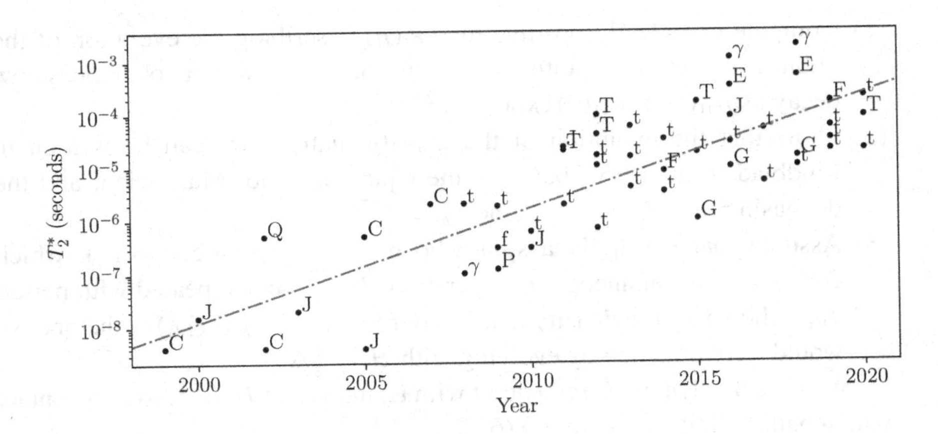

Figure 6.9 Qubit coherence times T_2^* (or T_2 measured after spin-echo) across the years. Letters signify: charge qubit (C), flux qubit (J), fluxonium (f for 2D, F for 3D), phase qubit (P), transmon (t for 2D, T for 3D), photon encoded qubit (γ), error-corrected bosonic qubit (E), and gatemon (G).

and recent work to improve the surface properties of 2D transmons (Place et al., 2021). An important question is whether this trend will saturate, or whether we will discover new and better ways to design, fabricate, and integrate qubits in quantum circuits.

Exercises

6.1 Show that a qubit Hamiltonian of the form $H = \frac{\Delta}{2}\sigma^z + \frac{\varepsilon}{2}\sigma^x$ can be mapped to other forms, such as $H = \frac{\Delta}{2}\sigma^z + \frac{\varepsilon}{2}\sigma^y$ and $H = \frac{\Delta}{2}\sigma^x + \frac{\varepsilon}{2}\sigma^z$, by a choosing a new basis of qubit states $|\tilde{0}\rangle = U\,|0\rangle$ and $|\tilde{1}\rangle = U\,|1\rangle$. Compute the unitary transformation U in those two particular cases.

6.2 A single-mode harmonic oscillator with Gaussian Hamiltonian and linear operations cannot implement a qubit. Assume a general model of the form $H = \omega a^\dagger a + (f(t)a + f(t)^* a^\dagger)$ with a general driving $f(t) \in \mathbb{C}$. Prove that starting from state $|0\rangle$ it is never possible to create a perfect Fock state $|1\rangle$. Discuss how this changes when we add a nonlinearity $Ua^\dagger a^\dagger aa$ to the Hamiltonian, with $U > 0$.

6.3 Let us study the limit of **quasistatic dephasing**. We will use a diagonal perturbation to the dynamics $H' \sim \frac{\hbar}{2}(\Delta + \delta\omega)\sigma^z$, but the random perturbation $\delta\omega$ has no dynamics: It is constant through each experimental realization, but fluctuates from experiment to experiment with Gaussian distribution $P(\delta\omega) = \exp(-\delta\omega^2/2\sigma^2)/\sqrt{2\pi\sigma^2}$.

(1) Compute completely positive map $\varepsilon_t(\rho)$ describing the evolution of the qubit. Compare the resulting map with (6.13). Show that coherences now decay as $\rho_{01}(t) \propto \rho_{01}(0)\exp(-t^2\sigma^2/2)$.

(2) Show that the evolution of the density matrix $\rho(t)$ can be written in Lindblad form (6.15), but now the equation is not Markovian, and the dephasing rate depends on time $\gamma_\phi = t\sigma^2$.

(3) Assume that we apply a spin-echo refocusing protocol (6.16), which consists of instantaneous flip operations $U_{\mathrm{BB}} = \sigma^x$, repeated with period T_{BB}. Show that the density matrix after two pulses $\rho(2T_{\mathrm{BB}})$ is the one we would have obtained by evolving with $H = \frac{1}{2}\Delta\sigma^z$.

6.4 Let us solve the dynamics for a qubit with Hamiltonian $H = \frac{1}{2}\Delta\sigma^z$, in contact with a bath at finite temperature (6.17).

(1) Using the master equation to estimate $\partial_t\rho$, prove that the evolution of an observable for such a qubit is

$$\frac{\mathrm{d}}{\mathrm{d}t}\langle O\rangle = -i\,\langle[O,H]\rangle + \frac{\gamma}{2}\,\langle\sigma^+[O,\sigma^-] + [\sigma^+,O]\sigma^-\rangle\,. \tag{6.73}$$

(2) Next, derive the **Bloch equations** equations for $\langle\sigma^z\rangle$ and $\langle\sigma^\pm\rangle$. Solve those equations in the zero temperature limit $n_\Delta = 0$.

(3) Use those expected values to reconstruct the Bloch vector (2.25). Compute the norm of the vector, and show that it is contracting and the state remains inside the sphere.

(4) Using the Bloch vector $\mathbf{S}(t)$, prove that the density matrix evolves as

$$\rho(t) = \begin{pmatrix} 1 - P_1 + (e^{-t/T_1} - 1)\,[P_1 - \rho_{11}(0)] & \rho_{01}(0)e^{i\Delta t - t/2T_1} \\ \rho_{10}(0)e^{-i\Delta t - t/2T_1} & P_1 + e^{-t/T_2}\,[\rho_{11}(0) - P_1] \end{pmatrix},$$

where $P_1 = n/(1+2n) = \lim_{t\to+\infty}\rho_{11}(t)$ is the asymptotic excitation probability of the qubit and $T_1 = (1-2P_1)/\gamma$ gives the timescale $|T_1|$ at which $\rho(t)$ converges to its asymptotic solution. Relate this solution to the zero temperature limit of pure relaxation.

6.5 Write down a master equation that combines a qubit gap $\hbar\Delta$, dephasing with a rate γ_ϕ and relaxation with a rate γ. Solve the evolution for an initial density matrix ρ_0. Study the evolution of $\rho_{01}(t) = \langle 0|\rho(t)|1\rangle$ and prove (6.20).

6.6 **Charge qubit truncation.** Let us consider the effect of small potential perturbations $n_g = \frac{1}{2} + x$ onto a charge qubit with E_J/E_C, $x \ll 1$. We will rely on the degenerate perturbation theory from Section A.3.2, identifying our target subspace as the qubit space, $P_{0,\alpha=0} = |0\rangle\langle 0| + |1\rangle\langle 1|$. Show that, up to second order in E_J/E_C, it only the coupling to the degenerate subspace $\{|-1\rangle, |2\rangle\}$. Using the series expansion (A.36), you should obtain a diagonal effective Hamiltonian:

$$H_{\text{eff}} = -E_J\sigma^x + \left[4E_C\left(\tfrac{1}{2}+x\right)^2 - \frac{E_J^2}{8E_C(1+2x)}\right]|0\rangle\langle 0|$$
$$+\left[4E_C\left(\tfrac{1}{2}-x\right)^2 - \frac{E_J^2}{8E_C(1-2x)}\right]|1\rangle\langle 1|.$$

Does this model differ from (6.25)? What happens when $x = 0$? And when x is so small that the diagonal terms can be linearized? Can we distinguish the effective theory from the truncated one?

6.7 Following the experiment by Nakamura et al. (1999), we prepare a charge qubit at a point $n_g \simeq 0$, cooled down to charge neutrality $|0\rangle$. We then abruptly move the experiment to the symmetry point $n_g = 1/2$ for a time T, after which we return to $n_g = 0$ and measure the charge in the island. What is the Hamiltonian of the qubit at $n_g = 1/2$ in the charge basis? How does the quantum state evolve, and what is the final state of the qubit at time T? What is the probability that we find one excess Cooper pair at the end of the experiment?

6.8 Let us take the charge qubit by Nakamura et al. (1999), where the energy gap $\Delta \simeq 80$ μeV. We are going to analyze its sensitivity to the environment assuming that there can be trapped charges around the qubit at a distance of $d \sim 4$ μm. As a toy model, we assume that that all of the qubit on average experiences the potential induced by this electron $V = e/4\pi\varepsilon_0 d$. What is the extra energy acquired by the $|1\rangle$ charge state relative to the $|0\rangle$? How do you write this correction into the Hamiltonian? How much does n_g change by this external influence? What about the energy levels ΔE? How much do they change?

6.9 Diagonalize the oscillator part of the transmon Hamiltonian in (6.32), disregarding the contribution of the nonlinear terms, so as to estimate the transmon properties.
(1) Estimate transition frequency $\hbar\omega_{01}$.
(2) Express the charge operator and flux operator in the oscillator basis. What are the matrix elements of those operators between the ground and excited state? Compare with Section 6.3.4.
(3) Using nondegenerate perturbation theory, estimate the negative anharmonicity due to the quartic term $\propto E_J\phi^4$.
(4) Connecting (6.33) to the one for an LC resonator, what is the corresponding impedance of the transmon?

6.10 Let us write down the transmon Hamiltonian for the lowest three levels with an external, time-dependent perturbation:

$$H = (\omega_{12} + \omega_{01})\,|2\rangle\langle 2| + \omega_{01}\,|1\rangle\langle 1| + (\varepsilon_{12}(t)\,|2\rangle\langle 1| + \varepsilon_{01}(t)\,|1\rangle\langle 0| + \text{H.c.}).$$

Assume that $\omega_{12} = \Delta + \alpha$ and $\omega_{01} = \Delta$ are related by a small anharmonicity α, and that $\varepsilon_{12} \simeq \varepsilon_{01} = \varepsilon_0 \cos(\omega t)$. We are going to study the dynamics and a possible leakage outside the qubit space:

(1) Perform a unitary transformation with $U = \exp(-iH_0t/\hbar)$, where $H_0 = \omega(|2\rangle\langle 2| + \omega\,|1\rangle\langle 1|)$, to arrive at an equation similar to (6.7). In that equation, neglect the terms that oscillate with frequencies $\omega_{01} + \omega$, $\omega_{12} + \omega$, etc. Show that the outcome reads

$$H_{\text{eff}} = (\delta + \alpha)\,|2\rangle\langle 2| + \delta\,|1\rangle\langle 1| + \varepsilon_0\,(|2\rangle\langle 1| + |1\rangle\langle 0| + \text{H.c.})\,.$$

(2) Assume a large anharmonicity $|\delta + \alpha| \gg |\delta|$, and ignore the third state. Solve the dynamics of an initial state that is in the ground state, $|\psi(0)\rangle = |0\rangle$, as a function of time, detuning and coupling. What conditions do you need to reach the perfect state $|1\rangle$, and how does the time T_{flip} required depend on ε_0?

(3) Consider leakage into the third state perturbatively, while on resonance, $\delta = 0$. For that, realize that this matrix will have two eigenstates that will be close to $|0\rangle$ and $|1\rangle$ but will be mixed with a small population of $|2\rangle$. Show that the excitation probability grows as $|\varepsilon_0|^2/|\alpha|^2$. If the anharmonicity is 5% of the gap Δ, what does this tell us about the maximum coupling strength $|\varepsilon_0|$ of the transmon to other systems?

(4) For a transmon with an anharmonicity of $2\pi \times 400$ GHz, what would be the leakage error for implementing a gate in $T_{\text{flip}} = 10$ ns? What is the shortest pulse we can afford to reduce leakage below 1%?

6.11 Using the relations from Section 6.3.3 and the software package *Mathematica*, we can easily compute the transmon eigenenergies and eigenfunctions.[22] The eigenenergies are given by the Mathieu characteristic function, which Mathematica labels as $\mathcal{M}_A(r,q) =$ `MathieuCharacteristicA[r,q]`. The eigenfunctions can be reconstructed from the Mathieu sine and cosine functions, $se_\nu(z)$ and $ce_\nu(z)$, as $me_{\nu+2k}(z,q) = ce_{\nu+2k}(z,q) + i\,se_{\nu+2k}(z,q)$, which Mathematica expresses as $ce_r(z,q) =$ `MathieuC[r,q,z]` and $se_r(z,q) =$ `MathieuS[r,q,z]`.

(1) Plot the function $\mathcal{M}_A(r,q)$ for different ratios of $q = E_J/2E_C = 1, 10$, 20, and 50 over a range $r \in [-6, 6]$. Can you locate the different bands?

(2) Using the index function $k(n, n_g)$ and the characteristic function, plot the energy levels of the transmon for $E_J/E_C = 1, 10, 20, 50$. Also reconstruct the anharmonicity plot.

[22] As a curious detail, our definition of z and of q makes q positive, and later versions of Mathematica seem to behave more stably when $q \geq 0$.

(3) From the wavefunctions, by means of Mathematica's numerical integration, compute the expectation values of the charge operator among the ground state and first excited state:

$$q_{01} = \int_0^{2\pi} \psi_1(\varphi)^*(-i2e\partial_\varphi)\psi_0(\varphi)\mathrm{d}\varphi. \tag{6.74}$$

Plot its behavior as a function of the E_J/E_C ratio both for $n_g = 0$ and $n_g = 1/2$. Hint: Mathematica also has the derivatives $ce'_\nu(z)$ and $se'_\nu(z)$.

6.12 Let us look at the **three-junction persistent current qubit**, from Section 6.4.3. In the seminal experiment by van der Wal et al. (2000), the two biggest junctions had critical currents and capacitances $I_p \simeq 570\,\text{nA}$ and $C \simeq 2.6\,\text{fF}$; the smallest junction was so by a factor $\alpha = 0.82$, and the total qubit had an area of about $10\,\mu\text{m}^2$. Using the circuit model from Section 4.8 and the preceding theory, answer these questions:

(1) How large is the energy barrier between the two potential wells along $\phi_- = 0$? How does it compare to the total depth of the potential?

(2) Estimate the gauge-invariant phase φ across the big junctions and relate it to the current around the loop.

(3) Write down the inductive energy E_{ind} of the three Josephson junctions operating at the flux qubit point, $\Phi = 0.5\varphi_0 + \Phi_B$, where Φ_B is the flux induced by a uniform magnetic field perpendicular to the loop. Estimate the qubit's effective dipole as the change in energy with respect to B, $\mu \propto \partial E/\partial B$.

(4) How does the effective magnetic moment (in Bohr magnetons) compare to the permanent magnetic moment of atoms such as chromium? Discuss the reason for the difference.

6.13 We can diagonalize the flux qubit Hamiltonian (6.59) using the number-phase representation from Section 4.9. Note that we have to use two integer numbers $|n_+, n_-\rangle$, one for each flux degree of freedom.

(1) Adimensionalize Hamiltonian H expressing all terms as a function of ratio E_J/E_C.

(2) Write down the Hamiltonian H in the two-dimensional number basis $|n_+, n_-\rangle$. Your result should produce diagonal expressions for $q_\pm$ and introduce charge increments or decrements associated to $\cos(\phi_\pm/2\varphi_0)$ and similar terms.

(3) Translate the previous expression into a small Mathematica or Matlab program that works with a truncated space, $n_\pm \in \{-N, \dots, N\}$. Typically, $N \simeq 10$ will suffice.

(4) Diagonalize the Hamiltonian to compute the eigenenergies $E_{0,1,2}$ around the frustrated point, $\Phi \simeq \frac{1}{2}\Phi_0$. Try to reproduce the curves from Figure 6.7e,f, for both the qubit gap and the separation from the second excited state.

(5) Demonstrate that the Hamiltonian at any other flux point $H(\frac{1}{2}\Phi_0 + \delta\Phi)$ can be decomposed according to the formula (6.2). For that, compute the two eigenfunctions $|0\rangle$ and $|1\rangle$ at the symmetry point $\frac{1}{2}\Phi_0$ and project the Hamiltonian at any other value onto those states, computing the effective matrix:

$$H_{\text{eff}} = \begin{pmatrix} \langle 0|H_{\delta\Phi}|0\rangle & \langle 0|H_{\delta\Phi}|1\rangle \\ \langle 1|H_{\delta\Phi}|0\rangle & \langle 1|H_{\delta\Phi}|1\rangle \end{pmatrix},$$

and expand in the basis of Pauli matrices.

6.14 Take the qubit design by Paauw et al. (2009). In this *gradiometric qubit*, the flux is spread over two symmetrically placed loops. Show that the qubit operation point depends only on the flux difference $\Phi_{\text{up}} - \Phi_{\text{down}}$. The qubit is therefore insensitive to longer-scale fluctuations of the magnetic field, which may be favorable for reducing sensitivity to environmental noise.

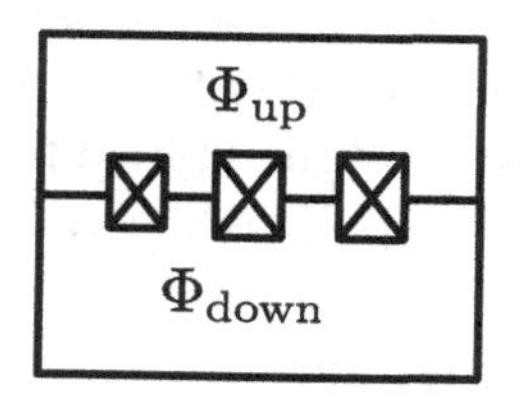

6.15 **Flux qubit current operator.** Solve approximately the equations for the minima of the inductive potential $\phi_{L,R}$, in the limit of large L/L_J ratio. Show that in this limit, the phase φ can be written as $\pm(\pi - \varepsilon)$, with a small correction $\varepsilon \propto L/L_J$. Use the derivation in Section 4.7.1 to write down the current operator for the rf-SQUID in terms of the flux jump along the inductor $\hat{\phi}$. Estimate the matrix elements of the current operator on the $|L\rangle$ and $|R\rangle$ states. Show that it is approximately diagonal in these states (6.63). Using the change of basis (6.57), prove that the current operator has approximately the form $\tilde{I}_c\sigma^x$, finding the expression for $\tilde{I}_c = |\phi_{L,R}|/L$.

6.16 The **rotating wave approximation** also applies to the dynamics of qubits. Let us consider the interacting qubit model (6.67), in the limit of resonant qubits $\Delta_1 = \Delta_2$.

(1) Use the decomposition $\sigma^x = \sigma^+ + \sigma^-$ to find what would be the RWA Hamiltonian H_{RWA} associated to this model.

(2) Show that H can be decomposed into two 2×2 matrices acting in the even and odd subspaces, $\{|00\rangle, |11\rangle\}$, and odd $\{|01\rangle, |10\rangle\}$. Use this to compute all eigenvectors and eigenenergies.

(3) Analyze the eigenvectors and show that in the limit $|g| \ll |\Delta|$, this matrix has approximately the same eigenstates as the Hamiltonian H_{RWA} that results from neglecting the counterrotating terms. Estimate the correction to the states introduced by not neglecting the counterrotating terms and show that the associated excitation probability scales as $|J/\Delta|^2$.

6.17 Return to the interacting qubit model (6.67), but now assume that the qubits are far apart in frequency space, $\delta = \Delta_1 - \Delta_2 \gg |J|$.

(1) Show that the Hamiltonian H still can be decomposed into even and odd subspaces, as in Problem 6.16.

(2) Show that, in the limit $\Delta_1 \neq \Delta_2$ the spectrum is formed by four different eigenenergies. Analyze the same problem using second-order nondegenerate perturbation theory (see Section A.3.1). Find out the effective model for this Hamiltonian in the strongly off-resonant limit, considering corrections up to $\mathcal{O}(J^2/\delta)$. Based on your exact diagionalizations, at which limit of the interaction does perturbation theory fail?

(3) Solve analytically the Schrödinger equation for two qubits that start in the state $|01\rangle$ and evolve with the full Hamiltonian (6.67) and with an effective Hamiltonian where you neglect all interactions. Call those solutions $|\psi_{01}^{\mathrm{eff}}(t)\rangle$ and $|\psi_{01}^{\mathrm{exact}}(t)\rangle$.

(4) Estimate the fidelity $F = |\langle\psi^{\mathrm{eff}}|\psi^{\mathrm{exact}}\rangle|^2$ between the states evolved with the idealized model, where effective interactions have been disconnected, and the real one. What reasonable values can J, Δ_1, Δ_2 and t take to achieve an infidelity $1 - F \leq 10^{-1}, 10^{-2}$ and 10^{-4}, respectively?

(5) Repeat the previous calculation using the qubit parameters from the work by Barends et al. (2014).

7

Qubit–Photon Interaction

It can be argued that the work by Wallraff et al. (2004) included the two important breakthroughs that superconducting circuits needed to become a mainstream quantum technology. The first breakthrough was the jump in a qubit's coherence time that appeared by placing those qubits in the gap of a microwave resonator. The improvement was so impressive that superconducting qubits regained momentum – and strongly surpassed other solid state platforms – in the quest for scalable quantum computers.

The second breakthrough in this work was to show how qubits could interact coherently with the photons trapped in the resonator that protected the qubit. The combined qubit–resonator sample evidenced an exchange of photons that was compatible with the canonical Jaynes–Cummings or Rabi models from quantum optics. Thus emerged the field of quantum optics with superconducting circuits, *circuit quantum electrodynamics* or simply *circuit-QED*. This field brings to the lab old ideas from quantum optics, such as quantum superpositions of photons (Hofheinz et al., 2009), Schrödinger cat states (Vlastakis et al., 2013), and collective effects in superradiance (Mlynek et al., 2014; Nataf and Ciuti, 2010). A large body of fundamental results was accompanied with crucial developments in the preparation, control, and measurement of superconducting circuis. Microwave resonators became *the* tool to control and measure superconducting qubits (Schuster et al., 2005), to create entanglement (Majer et al., 2007) and to assist quantum gates between qubits (Rigetti and Devoret, 2010). Understanding how this happened is the goal of this chapter.

We begin this important segment of the book introducing a model for light–matter interaction between superconducting circuits and microwave photons. Roughly, we quantize a large circuit with one or more qubits in a microwave transmission line, we then identify linear and nonlinear excitations – photons and qubits – and establish parallelisms between the resulting Hamiltonian and atoms in optical environments. We first apply this model to study the dynamics of a qubit in an open environment, explaining how an atom can emit photons, one at a time, and interact with a propagating microwave field. We then make a drastic change, placing our qubits inside microwave resonators. We show that photons in a resonator can interact more strongly and for longer periods of time with our artificial atom. This leads to simpler, radically more useful models, such as the Jaynes–Cummings Hamiltonian. We explore applications of these circuit-QED setups to the engineering and tomography of qubit and photon states, introducing the dispersive coupling, single- and two-tone spectroscopy and quantum nondemolition (QND) measurements of qubits.

7.1 Qubit–Line Interaction Models

The central idea in circuit-QED is to engineer matter-like and photon-like excitations using different plasmonic excitations of a superconducting quantum circuit. The fact that such artificial photons and atoms are made of the same material means we can design them with a good *impedance match,* favoring an easy, fast, and lossless exchange of energy and information. In circuit-QED setups, the interaction between the microwave photons and the superconducting qubits is comparatively stronger than that of optical photons with atoms and molecules. This allows individual qubits to fully reflect photons in transmission lines (Astafiev et al., 2010) and to reach the ultrastrong coupling regime and break down the RWA (Forn-Díaz et al., 2010; Niemczyk et al., 2010).

A typical setup, sketched in Figure 7.1, consists of one or more qubits coupled to waveguides or other microwave resonators. In 2D waveguide setups, qubits sit between the central conductor (bottom of the figure) and the ground plane (top), while in 3D waveguides the qubits are inserted into the hollow of the copper or aluminum tubes. As usual, the qubits exhibit localize excitations with an anharmonic spectrum. These atom-like states couple to the electric and magnetic fields that propagate along the waveguide, and that are bound to the photon-like plasmonic excitations of the superconducting microwave guide.

7.1.1 Dipolar Interaction

Let us begin with a transmon inside the coplanar coaxial waveguide from Figure 7.2. As suggested by the sketch, the charge accummulated on the transmon interacts with the capacitor formed by the central conductor and the ground plane. We can describe this system using the mutual capacitance between the transmon and the small segment of waveguide around the qubit. Alternatively, and consistent with Section 6.5, we can model the transmon as an electric dipole sitting on the electric field confined inside the waveguide. Both images are equivalent and lead to the same mathematical description.

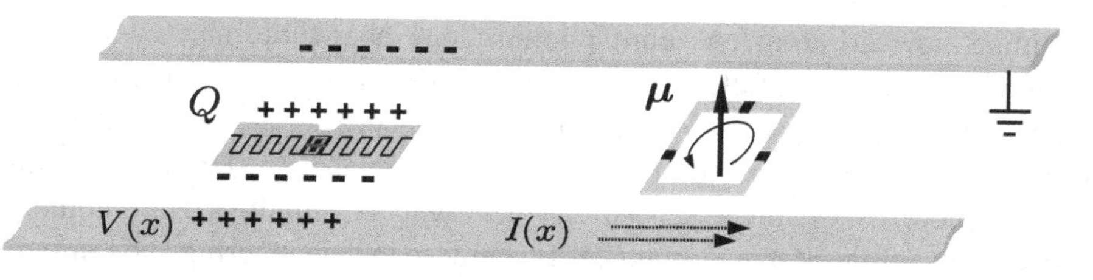

Figure 7.1 Transmon and flux qubit sitting inside a one-dimensional microwave transmission line – central conductor at the bottom, ground plate on the top. The qubits interact preferably with either the voltage or the current transported by the waveguide, through their electric and magnetic dipoles.

In the lumped-element circuit model, the waveguide enters as a potential biasing the qubit's equilibrium point. The Hamiltonian is the one in Section 4.6, but the gate voltage is an operator $\hat{V}(x)$ that contains the waveguide's field around the qubit. The effective model is

$$H = \frac{1}{2C_\Sigma}\left(\hat{q} - C_g\hat{V}(x)\right)^2 - E_J \cos\left(\hat{\phi}/\varphi_0\right) + H_{\text{photons}}. \tag{7.1}$$

We split this Hamiltonian into three contributions, $H_{\text{qubit}} + H'_{\text{photons}} + H_{\text{int}}$, two slightly renormalized Hamiltonians for the qubit and the waveguide, and an interaction term. The interaction between charge qubits and photons, when projected onto the qubit eigenbasis, adopts the dipolar form[1] from Section 6.5

$$H_{\text{int}} = -\frac{C_g}{C_\Sigma}\hat{q}\hat{V}(x) \simeq \varepsilon\sigma^x\hat{V}(x). \tag{7.2}$$

As before, the voltage stimulates transitions between both qubit states.

In addition to the coupling term, the qubit-line interaction has slightly modified the qubit and transmission line properties. It adds a correction to the transmon's capacitance, which is now larger $C_\Sigma \simeq C_J + C_{\text{qubit-line}}$. It also adds a local term to the line $\hat{V}(x)^2$, causing a *renormalization* of the transmission line's properties. Potentially, this effect can be so large as to expel all plasmons away from the qubit's position x_{qb}, zeroing $\hat{V}(x_{\text{qb}})$ and decoupling the qubit from the line (García-Ripoll et al., 2015). In most experiments, however, the qubit-line coupling is small, and both effects can be neglected – or maybe even undistinguished from a change in the qubit's properties during fabrication.

We can build a similar argument for the flux qubit. The qubit's loop hosts a superposition of circulating currents and behaves as a small magnetic dipole $\boldsymbol{\mu}$. This dipole interacts with the magnetic field generated by the curents on the line, as

[1] If we follow the same convention as before, the transmon would couple using the σ^y operator. However, for all applications discussed in this chapter, we can use a local unitary transformation to transform $\sigma^y \to \sigma^x$.

in $g\boldsymbol{\mu} \cdot \mathbf{B}(x)$. Another way to reach the same conclusion is to see the currents in the line as generators of a magnetic field that modifies the flux inside the qubit Φ. As in Section 6.4, we find this perturbation induces opposite energy shifts of the $|L\rangle$ and $|R\rangle$ current states. When we define the qubit states at the symmetry point, the shift adopts the same dipolar coupling form as for the charge qubit:

$$H_{\text{int}} = M\hat{I}_{\text{qubit}}\hat{I}(x) \sim \varepsilon\sigma^x \hat{I}(x). \tag{7.3}$$

The coupling constant ε now accounts for the qubit–line mutual inductance M and the qubit's supercurrent intensity $\tilde{I}_c$.

7.1.2 Spin-Boson Hamiltonian

To complete our model, we must include the quantum nature of the transmission line, replacing voltage and current operators with their expressions in terms of photons. We assume a homogeneous waveguide, with uniform inductance and capacitance per unit length l and c. The current is defined as $\hat{I}(x) = \partial_x\hat{\phi}(x)/l$, while the voltage can be derived from the Heisenberg equation $\hat{V}(x) = \partial_t\hat{\phi}(x,t) = -i[\hat{\phi}, \hat{H}_{\text{photons}}]/\hbar$. Using waveguide modes (5.19),

$$\hat{V}(x_m) = \sum_k \sqrt{\frac{\hbar\omega_k}{2c}} \left(\frac{e^{-ikx_m}}{\sqrt{d}}\hat{b}_k^\dagger - \frac{e^{ikx_m}}{\sqrt{d}}\hat{b}_k \right), \tag{7.4}$$

$$\hat{I}(x_m) = \sum_k \sqrt{\frac{\hbar k^2}{2c\omega_k}} \frac{k}{l} i \left(\frac{e^{ikx_m}}{\sqrt{d}}\hat{b}_k - \frac{e^{-ikx_m}}{\sqrt{d}}\hat{b}_k^\dagger \right).$$

Inserting these in the qubit–line Hamiltonian, we get the *spin-boson model*:

$$\hat{H} = \sum_k \hbar\omega_k a_k^\dagger a_k + \frac{\hbar\Delta}{2}\sigma^z + \sum_k \sigma^x \left(g_k a_k + g_k^* a_k^\dagger \right). \tag{7.5}$$

The Hamiltonian is the same for capacitively and for inductively coupled qubits, with minor changes in the coupling constants:

$$g_k^{\text{cap}} = -iG^{\text{cap}}\sqrt{\frac{\hbar\omega_k}{2cd}}, \quad \text{and} \quad g_k^{\text{ind}} = iG^{\text{ind}}\frac{k}{l}\sqrt{\frac{\hbar}{2c\omega_k d}}. \tag{7.6}$$

The prefactor G encapsulates the microscopic details of the qubit–line coupling. For the transmon, it is proportional to the qubit–line capacitance $G^{\text{cap}} = -2eC_g/C_\Sigma$. For a flux qubit, it combines the qubit–line mutual inductance M with the qubit's critical supercurrent $G^{\text{ind}} = M\tilde{I}_c$.

Many 2D waveguides exhibit a close to linear dispersion relation $\omega_k = v|k|$, with group velocity $v^2 = 1/cl$. We can then show that both couplings scale identically with frequency $\sim\mathcal{O}(\sqrt{\omega_k})$ and have a simple expression in terms v and the waveguide's impedance $Z = \sqrt{l/c}$, as

$$g_k^{\text{cap}} = -iG^{\text{cap}}\sqrt{\frac{\hbar Z}{2} \times \frac{v\omega_k}{d}}, \qquad g_k^{\text{ind}} = i\,\text{sgn}(k)G^{\text{ind}}\sqrt{\frac{\hbar}{2Z} \times \frac{v\omega_k}{d}}. \tag{7.7}$$

The spin-boson Hamiltonian (Leggett et al., 1987) is one of the canonical models for studying dissipation in quantum mechanics.[2] The spin-boson model is the simplest mathematical realization of a small and discrete quantum system – e.g., an electron in an atom or a molecule – coupled to a big bosonic environment. In experiments, the environment is so large and evolves so quickly that it is immune to the qubit's tiny perturbations. As we did with the resonator, we can trace the environment out and develop a master equation just for the qubit. The resulting equation, already introduced in Section 6.1.6, models the relaxation and decoherence of the qubit accurately, with few parameters to calibrate experimentally.

In this chapter, we plan to dig deeper, using the spin-boson model also as the fundamental tool in our study of circuit-QED and qubit–photon interactions. Instead of ignoring or tracing out the environment, we will treat both the qubit and the waveguide on equal footing. By studying the influence of the qubit on the environment, we will see that decoherence and losses are a consequence of the qubit's capacity to absorb and emit quantua of excitations or photons. These photons can be used for spectroscopy, monitoring the qubit, transporting information, and mediating interactions between quantum devices.

7.1.3 Spectral Function and Spin-Boson Regimes

The spin-boson model (7.5) has three sets of parameters: the qubit's gap Δ, the spectrum of bosons ω_k, and the distribution of coupling constants g_k. We can combine the last two sets of parameters into the *spectral function* $J^{\text{QO}}(\omega)$ from (5.63). This was a mathematical object that we used to analyze the memory function of a bosonic bath, when coupled to a microwave resonator. We then argued that the spectral function evaluated at the cavity's resonance frequency ω provided a good estimate of the decay rate $\kappa = J^{\text{QO}}(\omega)$ at which the cavity equilibrates to the bath's temperature. We can extend this intuition to the study of a qubit in a photonic environment, but we have to consider additional subtleties.

[2] The other one, the Caldeira–Leggett model, was used in Chapter 5 to describe losses in a resonator.

The first subtlety is the distinction between *continuous* and *discrete* spectral functions. The former are associated to extremely long waveguides,[3] supporting all frequencies and wavelengths. Discrete functions, on the other hand, appear when we place the a superconducting qubit inside a $\lambda/2$ or $\lambda/4$ resonator, and we can resolve the individual eigenfrequencies – with the qubit typically interacting with just one or few of those modes.

Even in the continuous case, we can also find *gapped* spectral functions, which are zero over an extended range of frequencies. A common example is a 3D microwave guide made from a long tube of aluminum or copper. The waveguide only supports modes propagating with a frequency that lays above a minimum energy $\omega_k \geq \omega_{\text{cutoff}}$, which is needed to excite the transverse electromagnetic profile of the waveguide's modes. Below this frequency, the waveguide has no excitations that can absorb energy from the qubit. When the qubit is placed inside the *bandgap*, $\Delta \in [0, \omega_{\text{cutoff}}]$, the waveguide acts as a filter that blocks all excitations, protecting the qubit from decoherence.[4]

Conventional waveguides on 2D chips are approximately *gapless*, offering excitations from very low frequencies up to some intrinsic cutoff.[5] The spectral function of such waveguides grows linearly with the frequency of the excitations, describing an *Ohmic spin-boson model*:

$$J^{\text{QO}}(\omega) = \pi\alpha\omega^1. \tag{7.8}$$

This regime is to be contrasted with other polynomial spectral functions $J(\omega) \simeq \omega^s$ with smaller and bigger exponents, called the *sub-Ohmic* $s < 1$ and *super-Ohmic* $s > 1$ regimes.

We can verify (7.8), rewriting the coupling constants from (7.7) as $|g_k| = \hbar\sqrt{\pi\alpha v\omega_k/2d}$, and inserting them into the definition of the spectral function (5.63):

$$J^{\text{QO}}(\omega) = 2 \times \sum_{k>0} \frac{2\pi v}{d} \frac{\pi\alpha\omega_k}{2} \delta(\omega - \omega_k) \simeq \int d\omega_k \pi\alpha\omega_k \delta(\omega - \omega_k) \sim \pi\alpha\omega^1. \tag{7.9}$$

This assumed a linear dispersion relation, a uniform density of states $d\omega_k \sim v dk \simeq 2\pi v/d$, and two directions of propagation for each frequency.

The Ohmic spin-boson model exhibits a rich phenomenology, depending on the dimensionless ratio α, describing the qubit–photon interaction strength and the competition between incoherent processes (decay) and coherent processes

[3] Experiments don't allow "infinite" lengths, but a waveguide shorted by an impedance-matched resistor behaves as an infinitely long medium, too (Devoret, 1995).

[4] The qubit can still radiate to free space and into its substrate. These are other environments that are rarely modeled, but grouped into *intrinsic or nonradiative losses*.

[5] The superconducting gap is at least one possible cutoff.

Table 7.1. *Regimes of the Ohmic spin-boson model.*

$0 < \alpha < 1/2$	Markovian	Exponentially damped oscillations of the qubit to a state with a small, nonzero excitation probability.
$\alpha = 1/2$	Tolouse point	Exactly solvable point where exponential decay is exact.
$1/2 < \alpha < 1$	Kondo regime	Mapping to the Kondo problem. Non-exponential decay. Problem cannot be approximated by a master equation. Strong correlation between the qubit and the photonic environment.
$1 < \alpha$	Localization	Quantum tunneling between qubit states is suppresed. Dynamics amounts to dephasing due to the bosonic environment.

(qubit rotation with the Hamiltonian). As we summarize in Table 7.1, there are four distinct phases and behaviors depending on α.

The first region $\alpha \leq 1/2$ is where coherent dynamics overcomes decoherence at short times: The qubit evolves between the excited and ground states, with oscillations that dampen at a rate γ_Δ. This dynamics is very well approximated by a memoryless or Markovian master equation (6.17), with a decay rate that, for small values $\alpha \ll 1/2$, is provided by the spectral function $\gamma_\Delta = J^{\mathrm{QO}}(\Delta) \simeq \pi\alpha\Delta$. Note how $\gamma/\Delta = \pi\alpha$ so that α can be regarded as the relative speed of spontaneous emission, as compared to the qubit's intrinsic dynamics Δ.

As α approaches $1/2$, the dynamics of the qubit follows a slightly corrected Markovian dynamics, due to the entanglement between the qubit and its electromagnetic environment. We regard the qubit as dressed by a cloud of photons that slows it down, shifting its frequency Δ to lower values, and changing the decay rate γ_Δ, which is no longer proportional to the spectral function. These corrections were estimated analytically by Shi et al. (2018) and observed experimentally by Forn-Díaz et al. (2016).

When the coupling strength increases above $\alpha = 1/2$, correlations between the qubit and its surrounding photon cloud prevent any Markovian description. The dynamics of this strongly correlated system is faster than exponential and requires very different (mostly numerical) descriptions.[6] The theory around this point describes a qubit decay that is accompanied by the emission of odd numbers of photons – one, three, five, etc. – with varying combinations of frequencies.

[6] Path integral calculations, the Kondo renormalization group, or matrix product states (MPS) (Peropadre et al., 2013; Shi et al., 2018) provide insight into this region.

Finally, if we push the coupling strength above $\alpha = 1$ we cross a quantum phase transition into the *localization* phase. In this region, the qubit is effectively frozen and quantum tunneling is suppressed. Qualitatively, this may be derived from the spin-boson model in the limit, $\Delta \simeq 0$. In this limit, the coupling $\sigma_x(a_k + a_k^\dagger)$ produces two eigenstates with opposite orientations of the qubit, $|\pm\rangle \propto |0\rangle \pm |1\rangle$. These states are accompanied by displacements of the bosonic field along opposite directions $|\pm g_k/\omega_k\rangle$. When g is very large, both ground states are so different that tunneling between them is virtually impossible and the qubit is frozen, suffering at most a dephasing in the $|\pm\rangle$ basis due to the evolution of the different photonic environments.

Approaching or exceeding the Tolousse point $\alpha = 1/2$ is extremely hard for atomic and molecular systems. However, experiments by Forn-Díaz et al. (2016) have showed a very good impedance match between flux qubits and photons that approaches $\alpha = 1/2$ and even overflows into the $\alpha > 1$ region. These experiments – and also others that take place in cavities under similar conditions (Niemczyk et al., 2010; Peropadre et al., 2010; Yoshihara et al., 2016) – are generically referred to as achieving the *ultrastrong* or even *deep-strong coupling regime.* They offer interesting possibilities for studying non-Markovian open quantum systems, strong light–matter and photon–photon correlation effects, and applications to the design of quantum simulators (Kurcz et al., 2014; Pino and García-Ripoll, 2018; Shi et al., 2018). However, before discussing these interesting regimes, we will first focus on the equally useful regime of $\alpha \ll 1/2$, where most quantum computing and circuit-QED experiments take place.

7.1.4 Rotating Wave Approximation

When the qubit–bath coupling is the slowest timescale, $\alpha \ll 1/2$ and we can make the *rotating wave approximation* (see Section 5.5). We neglect counter-rotating terms of the form $\sigma^+ a_k^\dagger + \sigma^- a_k$, and keep the coherent exchange between the qubit and the bosonic field $\sigma^+ a_k + \sigma^- a_k^\dagger$. The resulting model is the *Jaynes–Cummings* Hamiltonian:

$$\hat{H} = \sum_k \hbar\omega_k \hat{a}_k^\dagger \hat{a}_k + \frac{\hbar\Delta}{2}\sigma^z + \sum_k \left(g_k \sigma^+ \hat{a}_k + g_k^* \hat{a}_k^\dagger \sigma^-\right). \tag{7.10}$$

In the RWA limit, every excitation of the qubit is accompanied by the subtraction of a photon from the environment. Such dynamics conserves the total number of excitations given by

$$\hat{N} = \sum_k \hat{a}_k^\dagger \hat{a}_k + \frac{\sigma^z + 1}{2}. \tag{7.11}$$

This operator commutes with the Hamiltonian $[\hat{N}, H] = 0$, is a constant of motion $\frac{\mathrm{d}}{\mathrm{d}t}\langle\hat{N}\rangle = 0$, and can be used to classify the eigenstates of the problem. The Hilbert space splits into subspaces $\mathcal{H} = \mathcal{H}_0 \oplus \mathcal{H}_1 \oplus \mathcal{H}_2 \oplus \cdots$, labeled by the integer number of excitations $\hat{N} = 0, 1, 2, \ldots$, which are not coupled by the dynamics with the Hamiltonian.

The ground state manifold contains a single state $\mathcal{H}_0 = \{|0, \mathrm{vac}\rangle\}$ with the qubit in the unexcited state and an empty waveguide without photons – the *vacuum state*. The single excitation manifold $\mathcal{H}_1$ contains an infinite family of wavefunctions with a simple parameterization, describing quantum superpositions of the excited qubit $|1, \mathrm{vac}\rangle$ and a propagating photon $|0, \phi(x)\rangle$ with any possible spatial profile $\phi(x)$. This set of states will be used in Section 7.2 to study the spontaneous decay of a qubit and the interaction of propagating photons with a two-level system. Finally, states with two or more photons $\mathcal{H}_{N\geq 2}$ require more sophisticated techniques – e.g., input–output theory, scattering theory, path integral formulations, or numerics – and will be mostly ignored in this book.

It is important to note that we can derive RWA models that go beyond the two-level approximation. These models are relevant for studying transmons in open transmission lines. In this case, the interaction term will be slightly more complex and involves transitions between more than two states, i.e.,

$$\hat{H} = \sum_i \Delta_i\, |i\rangle\langle i| + \sum_{i,j} g_k^{i,j}\, |i, j\rangle\langle\hat{a}|_k + \mathrm{H.c.} + \cdots. \tag{7.12}$$

This is still a relatively simple Hamiltonian, with a ground state that consists of the vacuum in the bosonic environment and an unexcited transmon, $|0\rangle \otimes |\mathrm{vac}\rangle$. Interestingly, if we allow for cyclic transitions, such as $g^{01}, g^{12}, g^{02} \neq 0$, the number of excitations will no longer be conserved. This prevents using the theoretical methods in this chapter, but enables interesting side-effects and processes such as splitting a photon into two other frequencies (Sánchez-Burillo et al., 2016) or facilitating interactions between photons at different frequencies (Hoi et al., 2013).

7.2 Waveguide-QED

Waveguide-QED is a relatively new subfield of quantum optics that studies the interaction between propagating photons confined in 1D systems and few-level systems. The term includes broad families of experiments. The photonic medium could be nanofibers, plasmons, or photonic crystals that interact with ultracold atoms, quantum dots, or color centers. Of course, waveguide-QED also describes the superconducting setups from Figure 7.1.

In this section, we introduce various tools that are developed in the context of waveguide-QED, focusing on the RWA limit with excitations and states $|\psi\rangle \in \mathcal{H}_1$

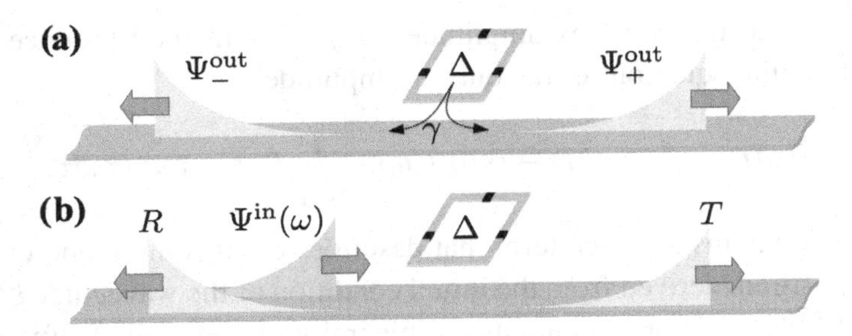

Figure 7.2 (a) An excited qubit relaxes by emitting one photon in a superposition of directions in an empty line. (b) Conversely, if we send a photon against a qubit, it will be partially reflected and partially transmitted, with amplitudes R and T.

that satisfy $\hat{N}\,|\psi\rangle = 1 \times |\psi\rangle$. As sketched in Figure 7.2 in this regime we expect only two relevant physical processes: the *spontaneous emission* of a photon by an excited qubit; and the *scattering* of a propagating photon by a relaxed qubit. Both processes can be modeled using a quantum superposition of qubit and photon excitations, known as the Wigner–Weisskopf ansatz. Despite its simplicity, the wavefunction model captures important concepts, including spectroscopy, Markovianity, and decoherence.

7.2.1 Wigner–Weisskopf Approximation

The *Wigner–Weisskopf model* is a mathematical ansatz or trial state, built from a coherent superposition of an excited qubit and a propagating boson:

$$|\psi(t)\rangle = \left[c_1(t)\sigma^+ + \sum_k \psi_k a_k^\dagger\right]|0, \text{vac}\rangle\,. \tag{7.13}$$

This ansatz represents the most generic state within the single-excitation sector of the RWA Hamiltonian (7.10). We therefore make no mistake when we project the Schrödinger equation for this Hamiltonian onto this subspace, and obtain a differential equation for the amplitudes of the qubit's $c_1(t)$ and one for the boson's excitations $\psi_k(t)$ (see Problem 7.1):

$$i\partial_t c_1 = \Delta c_1 + \sum_k g_k \psi_k, \quad \text{and} \quad i\partial_t \psi_k = \omega_k \psi_k + g_k^* c_1. \tag{7.14}$$

As in our study of the lossy cavity, we integrate the equation for the propagating photon $\psi_k(t)$, beginning from a time far away in the past $t_0 \to -\infty$. The mathematical procedure is identical[7] to the one used in Section B.3.2, replacing the

[7] The solutions are formally identical because we are dealing with linear differential equations!

operator $a(t)$ with the complex amplitude $c_1(t)$. This method produces an *exact* integro-differential equation for the qubit's amplitude:

$$\partial_t c_1(t) = -i\Delta c_1(t) - i\xi(0,t,t_0) - \int_{t_0}^{t} K(t-\tau)c_1(\tau)\mathrm{d}\tau. \tag{7.15}$$

The equation contains a source term that describes the incoming photonic field at the qubit's position, derived from the initial condition of the waveguide $\xi(x,t,t_0) = \sum_k g_k e^{ikx-i\omega_k t}\psi_k(t_0)$: It also contains an integral with the memory function $K(t)$ of the bath (5.62). In the Markovian limit, where the bath loses all memory of the qubit's past, it is replaced with a complex number $K(t-\tau) \simeq (i\delta_{\text{Lamb}} - \frac{1}{2}\gamma)\delta(t-\tau)$:

$$i\partial_t c_1 = \left(\Delta' - i\frac{1}{2}\gamma\right)c_1 + \xi(x,t,t_0). \tag{7.16}$$

As in the cavity, the Lamb shift $\delta_{\text{Lamb}} = \Delta - \Delta'$ describes the slowdown of the qubit due to the surrounding photons, while the spontaneous emission or qubit decay rate $\gamma \simeq J^{\text{QO}}(\Delta)$ models the relaxation of the qubit to the ground state.

7.2.2 Input–Output Relations

We can establish a relation between the field that illuminates the qubit $\psi_k(t \to -\infty)$, the photons absorbed and emitted by the qubit, and the total field that departs from the qubit position $\psi_k(t \to +\infty)$. The result is analogous to the input–output relations from Section 5.5.2.

To discriminate incoming from scattered photons, we work in position space, defining left- and right-moving fields that group positive and negative momenta $\Psi_\pm(x,t) = d^{-1/2}\sum_{k\in\mathbb{R}^+} e^{\pm ikx-i\omega_k t}\psi_k(t)$. As usual, d represents the length of the waveguide, and the open transmission line is recovered in the limit $d \to \infty$. Using the formal solution for $\psi_k(t)$, we can express these fields:

$$\Psi_\pm(x,t) = \Psi_\pm(x,t,t_0) - i\sum_{k\in\mathbb{R}^+}\int_{t_0}^{t} g_k^* \frac{e^{ikx-i\omega_k(t-\tau)}}{\sqrt{d}} c_1(\tau)\mathrm{d}\tau. \tag{7.17}$$

This expressions combines a term $\Psi_\pm(x,t,t_0)$ that depends on the initial condition $\psi_k(t_0)$, with a photon emitted by the qubit $\propto c_1(t)$. We distinguish two solutions depending on whether our boundary condition is set back in the past – an *input field* $\Psi^{\text{in}}(x,t) = \Psi_\pm(x,t,-\infty)$ – or in the far future – the *output* field $\Psi^{\text{out}}(x,t) = \Psi_\pm(x,t,+\infty)$. To move further, we use our recurrent assumptions:

(1) Without loss of generality, we place the qubit at $x = 0$.
(2) We assume a uniform spacing of momenta, $\sim 2\pi/d$.

(3) The coupling strength g_k will change very little around the qubit frequency, scaling as $g_k = g/\sqrt{d}$.
(4) The density of states around Δ is approximately uniform, and we can linearize the dispersion relation $\omega_k \simeq v\,|k|$ with group velocity v.

We can relate the coupling strength g to the spontaneous emision rate:

$$J^{\text{QO}}(\omega) = 2\pi \sum_k |g_k|^2 \delta(\omega - \omega_k) = 2\frac{|g|^2}{v} \int \delta(\omega - \omega_k) \mathrm{d}\omega_k, \tag{7.18}$$

which implies that $\gamma = 2\,|g|^2/v$ and $g_k = \sqrt{\gamma v/2d}$. With this parameterization and the Markovian assumption from earlier sections, we compute the emission of the qubit:

$$\begin{aligned}
\Psi_\pm(x,t) - \Psi_\pm(x,t,\pm\infty) &= -i \sum_{k\in\mathbb{R}^+} \int_{t_0}^{t} \sqrt{\frac{\gamma v}{2d}} \frac{e^{\pm ikx - i\omega_k(t-\tau)}}{\sqrt{d}} c_1(\tau)\mathrm{d}\tau \\
&\simeq -ic_1(t)\sqrt{\frac{\gamma v}{2}} \int_0^{+\infty} \frac{\mathrm{d}\omega_k}{2\pi v} \int_{t_0}^{t} \mathrm{d}\tau\, e^{\pm i\omega_k x/v - i\omega_k(t-\tau)} \\
&= -i\ \mathrm{sign}(t_0)\frac{1}{2}\sqrt{\frac{\gamma}{2v}} c_1(t \mp x/v).
\end{aligned} \tag{7.19}$$

The photon field that reaches the qubit from the left or from the right $\Psi_\pm(x,t)$ cannot depend on whether we used the input or output boundary condition to derive it. Equating both alternatives gives an *input–output relation* between incoming and outgoing fields on both directions:

$$\Psi_\pm^{\text{out}}(0,t) = \Psi_\pm^{\text{in}}(0,t) - i\sqrt{\frac{\gamma}{2v}} c_1(t). \tag{7.20}$$

We must read this as a relation between the field just before and just after interacting with the qubit, Ψ^{in} and Ψ^{out}, and the photons absorbed and emitted by the qubit. When we look at this equation, it may seem as if right-moving ($\Psi_+^{\text{in,out}}$) and left-moving ($\Psi_-^{\text{in,out}}$) fields are decoupled. This is not true, because the qubit field $c_1(t)$ is fed by both sets of photons and present in both equations. This means the qubit can absorb a left-moving photon and reemit it as a right-moving photon, and vice versa.

7.2.3 Spontaneous Emission Spectrum

Consider an excited qubit $c_1(t_0) = 1$ sitting on an empty waveguide $\psi_k(t_0) = 0$. The Wigner–Weisskop theory predicts that the qubit will exponentially decay to its

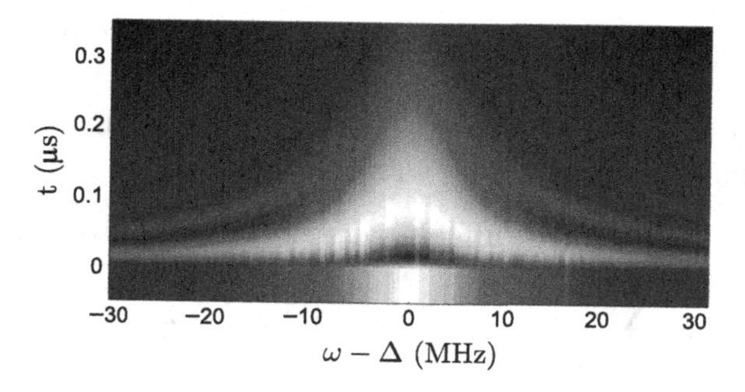

Figure 7.3 Spontaneous emission of a photon by a transmon qubit, as observed in the experiment by Sharafiev et al. (2021). At times $t < 0$, the transmon is being constantly driven. The drive stops at $t = 0$. The transmon then relaxes to a ground state, emitting a photon. Despite some experimental artifacts caused by amplification – e.g., dark fringe close around $t = 0$, one may appreciate the narrowing of the photon spectrum over a time $t \simeq 1/\gamma \sim 0.3\ \mu$s, until it converges to the Lorentzian profile (7.22).

ground state $P_1 = |c_1|^2 = \exp(-\gamma t)$, emitting a photon into the waveguide. This phenomenon is called *spontaneous emission*, and the constant γ is the *spontaneous emission rate* of the qubit.

The input–output relations (7.20) allow us to compute the wavepacket of the emitted photon in terms of the qubit's dynamics. This photon is a truncated exponential,[8] with a maximum amplitude at the wavefront and an exponentially decaying tail:

$$\Psi_{\pm}^{\text{out}}(x,t) = -i\Theta(t)\sqrt{\frac{\gamma}{2v}}e^{(-i\Delta'-\gamma/2)(t\mp x/v)}. \tag{7.21}$$

Once the photon has been fully emitted, the Fourier transform of the wavepacket reveals a normalized Lorentzian spectrum,[9] centered on the qubit's frequency Δ:

$$|\psi_k|^2 = \frac{1}{\pi}\frac{\gamma/2}{\gamma^2/4 + (\Delta - \omega_k)^2}. \tag{7.22}$$

The *full-half-width* (FHW) of this profile – i.e., the bandwidth of the photon or linewidth of the two-level system – is just the emission rate γ.

The spontaneous emission of a photon has been experimentally observed in frequency and in time by Sharafiev et al. (2021). The spectrum of the emitted photon changes in time, as shown in Figure 7.3. The Lorentzian prediction is only strictly true at long times $t \gtrsim 1/\gamma$, once the photon has been fully emitted. At very short times $1/t \simeq \Delta \gg \gamma$, the Markovian approximation still does not apply and the qubit deposits energy in all frequency modes, with a rather homogeneous spectrum

[8] With the Heaviside function $\Theta(t) = 0$ for $t < 0$, $\Theta(t) = 1$ for $t > 0$.
[9] Compare this with the cavity transmission spectrum from Section 5.5.3.

$|\psi_k|^2 \propto |g_k|^2$. At intermediate times, the photon looks like a truncated version of the exponential wavepacket, shortened to a length t from its wavefront. During those times, the width of the photon goes as $\sim 1/t \gg \gamma$, fulfilling the so-called time-energy Heisenberg uncertainty relation – i.e., the shorter time we measure, the longer the uncertainty in the energy we obtain.

During spontaneous emission, the qubit and the waveguide entangle with each other. For this reason, if we trace out the waveguide we obtain a quantum state with less information – i.e., a state that has decohered. Consider a qubit that is initially prepared in the quantum superposition, in an empty waveguide:

$$|\psi(0)\rangle = (\alpha\,|0\rangle + \beta\,|1\rangle) \otimes |\text{vac}\rangle\,. \tag{7.23}$$

This state will decay in time, producing a photon only when the qubit is in the $|1\rangle$ state. We can derive the dynamics of the qubit-waveguide system using the Wigner–Weisskopf ansatz, including the state that does change $|0, \text{vac}\rangle$. When we trace out the photonic environment, the resulting density matrix $\rho_{\text{qb}} = \text{tr}_{\text{photons}}(\rho)$ is exactly

$$\rho_{\text{qb}} = \begin{pmatrix} |\alpha|^2 + (1 - |\beta|^2\, e^{-\gamma t}) & \alpha\beta^* e^{-i\Delta t - \gamma t/2} \\ \alpha^* \beta e^{+i\Delta t - \gamma t/2} & |\beta|^2\, e^{-\gamma t} \end{pmatrix}. \tag{7.24}$$

The qubit begins in a pure state[10] in a quantum superposition with coherences $\rho_{01} \propto \alpha^*\beta$. The superposition deteriorates, as evidenced by the decay of the coherences and of the excited state's probability. Asymptotically, as t grows to infinity, the qubit returns to a pure ground state. At this point, all the quantum information from the qubit has been transferred to a quantum superposition of a propagating photon and a vacuum state.

The time evolution of the qubit's density matrix is analogous to the one predicted for a qubit in a zero-temperature bath (6.19). It exhibits a decay time $T_1 = 1/\gamma$ and a decoherence time $T_2 = 2/\gamma$, given by the qubit's spontaneous emission rate. As shown in Problem 7.4, this analysis can be generalized, verifying that both the Wigner–Weisskopf ansatz and the master equation (6.17) provide equivalent descriptions of the emission process.

7.2.4 Single-Photon Scattering

Imagine that, as depicted in Figure 7.2c, we send one photon against a relaxed qubit in a transmission line. Since the number of excitations is conserved and the qubit always relaxes to the ground state, the outcome of this experiment must be a transmitted and/or a reflected photon. The analysis of the transmitted and reflected

10 Pure states are projectors, which satisfy $\rho_{\text{qb}}(0)^2 = \rho_{\text{qb}}(0)$.

fractions is called *single-photon spectroscopy*. Like the coherent spectroscopy from the cavity in Section 5.5.3, it informs us about the qubit renormalized frequency Δ' in the open environment, and the combination of the qubit's spontaneous emission rate γ with other nonradiative decay and dephasing channels.

We can use the input–output relations to analyze this experiment. We rewrite (7.16), including the input photons that approach and drive the qubit:

$$i\partial_t c_1(t) = \sqrt{\frac{\gamma}{2v}}\left[\Psi_+^{\text{in}}(0,t) + \Psi_-^{\text{in}}(0,t)\right] + (\Delta' - i\gamma)c_1(t). \tag{7.25}$$

This equation is solved as a convolution of the input field with the exponentially decaying wavefront of a reemitted photon:

$$c_1(t) = -i\sqrt{\frac{\gamma}{2v}}\int_{t_{\text{in}}}^{t}\left[\Psi_+^{\text{in}}(0,\tau) + \Psi_-^{\text{in}}(0,\tau)\right]e^{(-i\Delta'-\gamma/2)(t-\tau)}\mathrm{d}\tau. \tag{7.26}$$

As usual, the output field is derived from the qubit's state by (7.20).

Since the scattering relation is linear, we can decompose the incoming wavepackets into simpler states and analyze their scattering. We choose the basis of quasi-monochromatic fields that are adiabatically switched on:[11]

$$\Psi_\pm^{\text{in}}(0,t) \propto \Theta(-t)a_\pm e^{(-i\omega+0^+)t} = \begin{cases} a_\pm e^{(-i\omega+0^+)t}, & t < 0 \\ 0, & \text{otherwise.} \end{cases} \tag{7.27}$$

Such wavepackets are infinitesimally narrow in frequency space. When integrated, they give the following output fields for $t \in (-\infty, 0]$:

$$\Psi_\pm^{\text{out}}(0,t) = b_\pm e^{(i\omega+0^+)t} = e^{(i\omega+0^+)t}\left[a_\pm - \frac{\gamma}{2}\frac{a_+ + a_-}{i(\Delta'-\omega)+\gamma/2}\right]. \tag{7.28}$$

The way we read this equation is that an input field coming from the left ($a_+ = 1$) or from the right ($a_- = 1$) is reemitted by the qubit in all directions, producing nonzero amplitudes b_+, b_-.

The input amplitudes and output amplitudes, $a_\pm$ and $b_\pm$, are related by a unitary transformation, called the scattering matrix, which contains the transmission and reflection coefficients, T_ω and R_ω:

$$\begin{pmatrix} b_+ \\ b_- \end{pmatrix} = \begin{pmatrix} T_\omega & R_\omega \\ R_\omega & T_\omega \end{pmatrix}\begin{pmatrix} a_+ \\ a_- \end{pmatrix}, \quad \text{with} \begin{cases} T_\omega & = 1 + R_\omega, \\ R_\omega & = -\frac{\gamma}{2}\frac{1}{i(\Delta'-\omega)+\gamma/2}. \end{cases} \tag{7.29}$$

Figure 7.4 illustrates the transmission and reflection probabilities, $|T_\omega|^2$ and $|R_\omega|^2$, and the frequency-dependent phase shifts experienced by the photon. Note how

[11] The term e^{0^+t} is a normalized exponential $\frac{1}{\epsilon}e^{\epsilon t/2}$ with $\epsilon > 0$, so that the wave vanishes in the distant past $t \to -\infty$. At the end of the calculations, we take the limit in which ϵ approaches 0 from above to recover monocromatic fields.

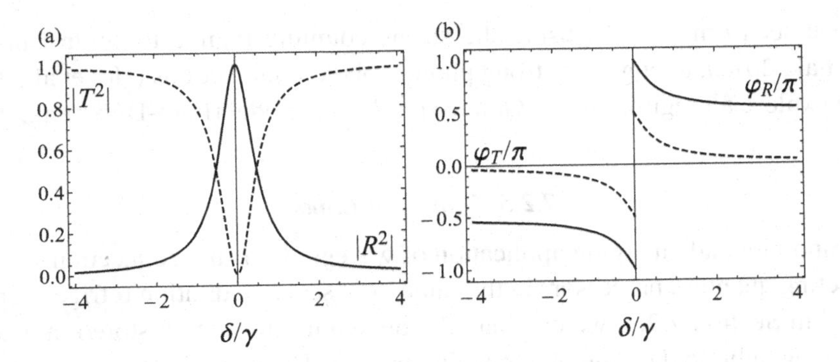

Figure 7.4 (a) Tranmitted and reflected probabilities for an incoming photon, as a function of the photon–qubit detuning $\delta = \omega - \Delta'$. (b) Phase slip (in units of π) of the transmitted and reflected photon, φ_T and φ_R, respectively.

$|R_\omega|^2$ displayw a Lorentzian profile, centered on the qubit's resonance. This is to be expected, since the reflected field only contains the energy produced by the qubit's reemission of the input field. This Lorentzian peaks exactly on resonance $\Delta' = \omega$, a point at which $R_\omega = -1$, $T_\omega = 0$, and the qubit acts as a perfect mirror for a single photon.

The scattering profiles from a qubit, and the fact that it acts as a perfect mirror for individual photons, were first demonstrated by Astafiev et al. (2010), using flux qubits. In this experiment, the qubit had some intrinsic dephasing γ_ϕ and experienced some intrinsic nonradiative losses γ', meaning that it could dissipate energy to other environments, different from the waveguide. The transmission spectrum under nonideal conditions can be studied with a small modification of the theory, which treats the qubit using a more general master equation while preserving the input–output relations. This gives

$$T_\omega = 1 - \frac{\gamma}{2} \frac{1}{i(\Delta' - \omega) + \gamma_\phi + (\gamma + \gamma')/2}. \tag{7.30}$$

The single-photon spectroscopy reveals the qubit's renormalized frequency Δ', but the linewidth is a function of the combined decoherence rate $\gamma_\phi + (\gamma + \gamma')/2$. The qubit is only a good mirror when it is in *strong coupling* regime, that is, $\gamma \gg \gamma', \gamma_\phi$.

In the waveguide-QED literature, the strong-coupling regime is characterized by the ratio $\beta = \gamma/(\gamma + \gamma') \simeq 1$. Astafiev et al. (2010) were the first researchers to simultaneously demonstrate strong coupling between a qubit and an open waveguide, and to implement single-photon scattering for that setup. In contrast to early attempts with atoms or quantum dots, the superconducting setup benefited from the good impedance match between the qubit and the line, and from the good isolation provided by the coplanar waveguide. After the work by Astafiev et al. (2010), we have seen other works using the strong-coupling regime to control photons (Hoi et al., 2012), to engineer strong photon–photon interactions (Hoi et al., 2013), and to explore all regimes in the Ohmic spin-boson model (Forn-Díaz et al., 2016).

7.2.5 Quantum Links

One important and emerging application of waveguides is the connection of superconducting quantum processors in the same or in separate dilution refrigerators. As we saw in Section 7.2.3, we can map the quantum information stored in a superconducting qubit to the state of a traveling photon. The goal of these quantum links is to do this mapping in a transitive way, mapping the state of a qubit into a photon that is then absorbed by another qubit at the other side of the waveguide.

In order to achieve perfect state transfer, we need three ingredients. First, we need to control the coupling between the qubit and the waveguide, achieving a time-dependent control of $\gamma(t)$. In particular, we need to switch off the coupling at the end of the transfer, to ensure that the qubit that absorbed the incoming photon does not decay again!

Second, we need to control the direction along which the photons propagate, to ensure that all the information goes from the first to the second qubit. In absence of chiral waveguides, where photons can only move along one direction, this is achieved by placing both qubits on the opposite sides of the guide.

Third, we need to control the shape of the photon wavepacket and the control $\gamma(t)$ to ensure that the second qubit experiences the time-reversed process of the first qubit's emission. As pointed out by Cirac et al. (1999), we need to engineer the decay rate of the first qubit $\gamma_1(t)$ to produce a photon that has a time-symmetric wavepacket. We then design the spontaneous emission rate of the second qubit as the time-reversed profile $\gamma_2(t_{\text{prop}} - t)$, where $t_{\text{prop}} = d/v$ is the time for the photon to travel along the whole length of the waveguide.

This problem can be modeled as a Wigner–Weisskopf problem with two qubits instead of the one we have used so far. Doing things properly, we will be able to transfer a arbitrary quantum state (7.23) from one qubit to another, over relatively short times $T \simeq t_{\text{prop}} + \mathcal{O}(1/\gamma)$, of a few tens of nanoseconds for medium-size waveguides. Interestingly, one may show that it is possible to implement perfect state transfer protocol for photons whose duration $\mathcal{O}(1/\gamma)$ exceeds the propagation time between qubits. This is a regime where the input–output formalism does not really apply, because the spacing between frequency modes in the waveguide – the free spectral range – is too large for a continuum limit description. However, Wigner–Weisskopf theory still produces a valid protocol, provided that we tune the qubit close to one of the waveguide's mode $\Delta' \simeq \omega_k$.

As mentioned previously, Magnard et al. (2020) have demonstrated quantum state transfer between qubits hosted at separate refrigerators, over distances of 5 m. The experiment used a long waveguide cooled at temperatures of tens of millikelvin. It also used a sophisticated method to achieve the tunable qubit–photon interactions where a microwave drive controls the transfer of the excitation $|1\rangle$ from each transmon to a cavity, from which a photon rapidly leaks into the waveguide that connects both fridges. Despite this complication, the physics still fits into a Wigner–Weisskopf model, where the amplitude and phase of the microwave control a time-dependent qubit–photon coupling, which is used to create and absorb the symmetric photons required for state transfer. We expect that similar designs will enable future local networks of quantum computers, supporting both quantum communication and distributed quantum computation.

7.3 Cavity-QED

An artificial atom interacts coherently with the photons in a transmission line, but these photons quickly fly away, leaving us with a decohered qubit. We can increase the interaction time between the qubit and the photon by cutting the waveguide to a shorter length, as shown in Figure 7.5. In this setup, the waveguide becomes a microwave resonator or a cavity, and we are exploring the field of *cavity quantum electrodynamics*, or cavity-QED.

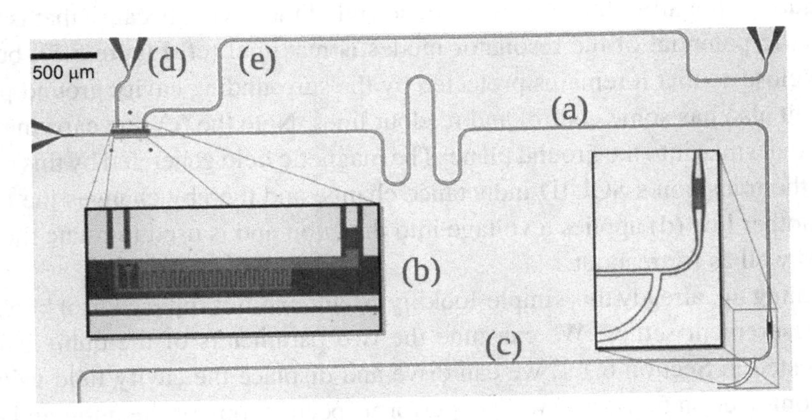

Figure 7.5 Photography of a transmon qubit standing near a microwave resonator. The picture shows the transmission line that forms the cavity (a), a zoom of the qubit (b), the lines that control the cavity (c) and the transmon's gap (d), and a line that interacts with the transmon to control its state (e). Picture courtesy of A. Wallraff, with minor edits.

As explained in Chapter 5, the cavity is a *quantum filter* that only allows certain microwave photons in. If the resonator is short, those frequencies are broadly spaced (see Section 5.2.2), opening large gaps in the spectral function. If we park the frequency of the qubit to lay deep in those gaps, the qubit is prevented from emitting photons, and we preserve its quantum state for long times, only limited by the qubit's other nonradiative decoherence mechanisms. If instead we tune the qubit close to a cavity's resonance, the qubit will be able to emit photons that will bounce back and forth among the resonator's walls. The photons now interact with the qubit for longer periods of time, only limited by the cavity's decay rate. The *Purcell enhancement* of the qubit–photon interaction allows the qubit to endow the resonator with highly nonlinear properties, creating multiphoton states and engineering sophisticated non-Gaussian states of light. Similarly, when the resonator is connected to the outer world through a transmission line, we can use the cavity to efficiently *control and measure* the state of the qubit.

Figure 7.5 shows a state-of-the-art circuit-QED experimental setup with the most common ingredients. You should recognize (a) a transmission line resonator and (b) a transmon qubit. The $\lambda/2$-resonator has an approximate length $d = \lambda/2 \sim 1$ cm and supports an infinite number of harmonic resonances, with regularly spaced frequencies $\omega_n = n\omega_1$, for $n = 1, 2, \ldots$ (cf. Section 5.2.2 and Figure 5.3). The resonator interacts capacitively with a semi-infinite transmission line, used to read the state of the cavity and to inject energy into it.

The transmon qubit (b) has a size of about 300 μm and sits inside the resonator, between the core of the transmission line and one of the ground planes. The position of the qubit is not arbitrary: It is close to the end of the cavity, because that is where the electric potential of the resonator modes is maximal (cf. Figure 5.3), but it is not too close so that it remains protected by the surrounding cavity ground planes. The qubit also has some control and readout lines. Note the (c) line carrying some current that sinks into the ground plane: The magnetic field generated by this current affects the transmon's SQUID inductance change and thereby changes the qubit's gap. Another line (d) applies a voltage into the qubit and is used to rotate the qubit basis, as well as for readout.

Summing up, already this simple-looking circuit exhibits three control knobs and two measurement setups. We can tune the two parameters of the qubit Δ and ε as requested in Section 6.1.1, we can drive and displace the cavity field using the ideas from Section 5.5.1, and we can perform spectroscopy of the qubit and of the cavity by sending photons and studing the reflected signals. As we explain now, these are extremely powerful controls that can be used as much for quantum optics experiments as to engineer scalable quantum computers.

7.3.1 Quantum Rabi and Jaynes–Cummings Models

Consider a qubit inside a zero-mode microwave resonator, which is itself connected to a transmission line. This is a special case of the spin-boson model studied in Section 7.1, where the qubit sees a big Lorentzian resonance:[12]

$$J^{\mathrm{QO}}(\omega) \simeq |g|^2 \frac{4\kappa\omega_0\omega}{\left(\omega_0^2-\omega^2\right)^2+(\kappa\omega)^2}, \tag{7.31}$$

centered on the microwave cavity's frequency ω_0, with a linewidth proportional to cavity decay rate κ, and with a parameter g that describes the speed at which the cavity and the qubit interact.

Typically, we would like to bring the qubit close to resonance with the cavity $\Delta \simeq \omega_0$. We also want the qubit–photon coupling to be in some type of *strong-coupling regime*, in which the qubit and the cavity exchange photons faster than the speed at which photons escape from the cavity, $|g| \gg \kappa$. This way, the qubit can perform meaningful operations in the photonic field before it decays.

Under these conditions, we begin to see that we cannot treat all photonic degrees of freedom as a unit. The coherent exchange of excitations between the cavity and the qubit is a highly non-Markovian situation. This is confirmed by the dramatic change of the spectral function $J^{\mathrm{QO}}(\omega)$ across the range of frequencies involved in the dynamics $\sim(\omega_0-g,\omega_0+g)$. Since we cannot approximate J^{QO} as a constant, the bosonic kernel $K(t)$ becomes a function that oscillates rapidly, with frequency $\sim g$; only at long timescales $1/\kappa$ the bosonic environment loses all memory about the interaction with the qubit.

The solution is to consider the qubit *with* the cavity as a combined quantum system, deriving a quantum Hamiltonian for both interacting objects, and only including *afterward* the waveguide as a truly Markovian environment. When we quantize the qubit in contact with the resonator, we obtain the single-mode limit of (7.5), which is known as the *quantum Rabi model*:

$$\hat{H} = \frac{\hbar\Delta}{2}\sigma^z + \hbar\omega_0\hat{a}^\dagger\hat{a} + \hbar\sigma^x\left(g\hat{a}+g^*\hat{a}^\dagger\right). \tag{7.32}$$

The first and second terms are the qubit and cavity Hamiltonians, which are coupled by a capacitive or inductive dipolar interaction g.

The quantum Rabi model is named after Austrian physicist I. I. Rabi (1898–1988), who received the Nobel Prize for his studies on nuclear magnetic resonance.

12 See Zueco and García-Ripoll (2019) for a complete derivation of the cavity-QED model for all regimes of losses in the cavity and the qubit.

NMR is the problem of one or more magnetic moments or spin driven by an oscillating electromagnetic field. We addressed a reduced version of this problem in Section 6.1.4, when studying the control of a qubit by a classical field. Equation (7.32) is a more complicated version where the driving field is quantum mechanical in nature.

In most experiments, the Rabi coupling is slower than the qubit's and cavity's frequencies, $|g| \ll \omega, \Delta$. By the RWA, we can eliminate the counter-rotating terms $\sigma^+ a + a^\dagger \sigma^-$, to produce the *Jaynes–Cummings Hamiltonian*:[13]

$$\hat{H}_{\text{JC}} = \hbar\omega\hat{a}^\dagger\hat{a} + \hbar\left(g\sigma^+\hat{a} + g^*\hat{a}^\dagger\sigma^-\right) + \frac{\hbar\Delta}{2}\sigma^z, \tag{7.33}$$

a model that is simpler and analytically solvable.

Using the theory of open quantum systems and the methods from Chapters 5 and 6, we can derive a master equation that includes the decay of the excitations from the cavity into the waveguide with rate κ, coherent microwave drives for the qubit $\varepsilon(t)$ and the cavity $\Omega(t)$, and the qubit's intrinsic losses γ and dephasing γ_ϕ:

$$\begin{aligned}\partial_t\rho = & -i\left[\hat{H}_{\text{Rabi/JC}} + \hbar\varepsilon(t)\sigma^x + \hbar\Omega_{\text{cav}}(\hat{a} + \hat{a}^\dagger), \rho\right] \\ & + \frac{\gamma}{2}\left(2\sigma^-\rho\sigma^+ - \sigma^+\sigma^-\rho - \rho\sigma^+\sigma^-\right) + \gamma_\phi\left(\sigma^z\rho\sigma^z - \rho\right) \\ & + \frac{\kappa}{2}\left(2\hat{a}\rho\hat{a}^\dagger - \hat{a}^\dagger\hat{a}\rho - \rho\hat{a}^\dagger\hat{a}\right).\end{aligned} \tag{7.34}$$

This master equation is accompanied by an input–output relation between the cavity and the waveguide, which relates the field leaked into the line with the instantaneous state of the cavity (5.65).

While it is clear that we need γ and γ_ϕ as small as possible – ideally below the megahertz regime – the cavity decay rate κ requires a delicate compromise between coherence and controllability. On the one hand, a small value of κ enables the qubit and the photons to interact for longer times. On the other hand, a small value of κ reduces our capacity to control and measure the experiment – the signals that leak from the cavity are weaker (5.65), and it becomes harder to inject energy into the microwave mode. Finally, for the coherent coupling g, we aim to be in the *strong-coupling* regime, which in this case translates into $|g| \gg \gamma, \gamma_\phi, \kappa$.

As an example, the original setup by Wallraff et al. (2004) used qubits and cavities with frequencies around 6 GHz, with $\kappa \sim 0.8 \times 2\pi$ MHz, $\gamma \sim 0.7 \times 2\pi$ MHz, and $g \sim 5.5 \times 2\pi$ MHz. The more modern three-qubit mini-quantum-computing setup by DiCarlo et al. (2010) explores a similar range of qubit and cavity frequencies, but

[13] Interestingly, models (7.5), (7.32), and (7.33) were all posed by physicists Jaynes and Cummings to describe an idealized atom in a quantum electromagnetic field. However, the Jaynes–Cummings or JC Hamiltonian nowadays refers to RWA version (7.33).

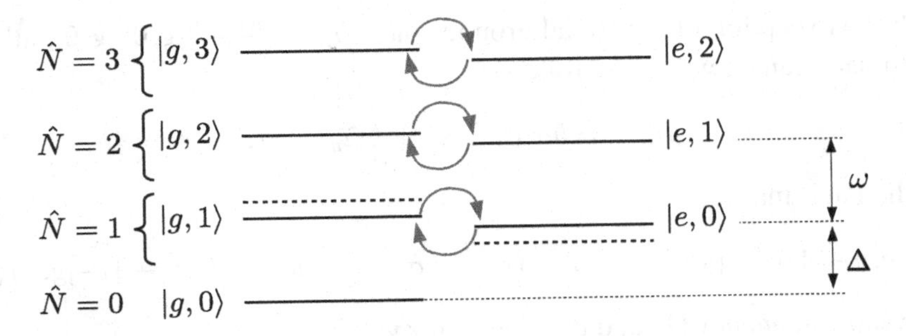

Figure 7.6 Energy-level structure of a qubit and a cavity without coupling (solid), as described by the Jaynes–Cummings model (7.33). The interaction couples neighboring states $|g,n\rangle$ and $|e,n-1\rangle$, creating new eigenstates that are superpositions of the original ones, and whose energies split even further (dashed).

now the cavity–qubit couplings were around $2\pi \times 220$ MHz, two orders of magnitude larger than the cavity losses, $\kappa = 2\pi \times 2.4$ MHz, and $\gamma \sim 0.8-1.2\times 2\pi$ MHz.

7.3.2 *Jaynes–Cummings Ladder*

For times shorter than the decoherence times, $\sim 1/\gamma,\ 1/\kappa$, we can study the cavity-QED setup using the Jayness–Cummings (JC) model, without the master equation (7.34). Since the JC model is a variant of the RWA spin-boson model, it also conserves the total number of excitations $\hat{N}$. In cavity-QED, it is customary to label the ground and excited qubit states as $|g\rangle = |0_{\text{qubit}}\rangle$ and $|e\rangle = |1_{\text{qubit}}\rangle$, which allows writing

$$\hat{N} = |e\rangle\langle e| + \hat{a}^\dagger\hat{a} = \hat{a}^\dagger\hat{a} + \frac{1}{2}(\sigma^z + 1). \tag{7.35}$$

The Hilbert space splits into a *Jaynes–Cummings ladder* of decoupled subspaces with different values of $\hat{N}$ and increasing energy, as sketched in Figure 7.6. At the bottom of the ladder, we find the zero-dimensional subspace for the ground state $\mathcal{H}_0 = \{|g,0\rangle\}$. All other subspaces are formed by all possible linear combinations of two basis states, $\mathcal{H}_{n\geq 1} = \text{lin}\{|g,n\rangle\,,\ |e,n-1\rangle\}$, one with the unexcited qubit and n photons $|g,n\rangle$, and one where the qubit got excited at the expense of one cavity photon $|e,n-1\rangle$.

Since the number of excitations commutes with the Hamiltonian $[\hat{N},\hat{H}] = 0$, the Hamiltonian also decomposes as a box-diagonal matrix, the direct sum of 2×2 matrices $\hat{H}_{\text{JC}} = \hat{H}_0 \oplus \hat{H}_1 \oplus \cdots$ of 1×1, each coupling one or two states within a rung $\mathcal{H}_n$ of the JC ladder. This subdivision means that the whole dynamics is *integrable* – in other words, we can analytical solve the evolution of the cavity–qubit setup in *all* those subspaces, not just for one photon.

With the exception of the trivial ground state, $H_0 = \frac{\hbar(1-\Delta)}{2}\,|g,0\rangle\langle g,0|$, all other Hamiltonians can be written as follows:

$$H_{n\geq 1} = \hbar\omega\left(n - \tfrac{1}{2}\right) + \hbar\Omega_n \mathbf{v}_n \cdot \tilde{\boldsymbol{\sigma}}, \tag{7.36}$$

using the Pauli matrices

$$\tilde{\sigma}^x := |e,n-1\rangle\langle g,n| + |g,n\rangle\langle e,n-1|, \qquad \tilde{\sigma}^z := |e,n-1\rangle\langle e,n-1| - |g,n\rangle\langle g,n|,$$

and the *Rabi frequency* Ω_n and unitary vector $\mathbf{v}$:

$$\Omega_n = \sqrt{\frac{1}{2}(\Delta-\omega)^2 + |g|^2 n}, \qquad \mathbf{v}_n = \left(\frac{\mathrm{Re}(g)\sqrt{n}}{\Omega_n}, \frac{\mathrm{Im}(g)\sqrt{n}}{\Omega_n}, \frac{\Delta-\omega}{2\Omega_n}\right). \tag{7.37}$$

The two-dimensional Hamiltonians can be diagonalized with the usual tricks from Section 2.2. Each rung of the ladder is associated two eigenenergies:

$$E_{n,\pm} = \hbar\omega\left(n - \tfrac{1}{2}\right) \pm \hbar\Omega_n, \tag{7.38}$$

which split symmetrically around $\hbar\omega\left(n - \frac{1}{2}\right)$. We call these the *lower and upper-polariton branches*, E_{n-} and E_{n+}, because the eigenstates are hybridized states (*polaritons*) of excitations living on the cavity $|g,n\rangle$ and on the qubit $|e,n-1\rangle$. The qubit–photon coupling introduces a level repulsion between the polariton branches, which separate by an amount proportional to the Rabi frequency $2\Omega_n$.

When the cavity and the qubit are straight on resonance $\Delta = \omega$, the splitting is minimal $2\sqrt{n}|g|$ and the two polariton states become symmetric superpositions:

$$|\psi_{n\pm}\rangle = \frac{1}{\sqrt{2}}\left(|g,n\rangle \pm |e,n-1\rangle\right), \quad n \geq 1. \tag{7.39}$$

If, on the other hand, the detuning is very large $|\Delta - \omega| \gg \sqrt{n}|g|$, the coupling strength will be insufficient to hybridize the qubit and the cavity. The two eigenstates will be then approximately close to the bare states $|g,n\rangle$ and $|e,n-1\rangle$, and will have close noninteracting eigenenergies $E_{n\pm} \sim \{\hbar\omega n - \frac{1}{2}\hbar\Delta, \hbar\omega(n-1) + \frac{1}{2}\hbar\omega\}$, with small dispersive corrections to be discussed in Section 7.3.7.

7.3.3 Vacuum Rabi splitting

The single-polariton energy gaps $E_{1-} - E_0$ and $E_{1+} - E_0$ are visible as the smallest quanta of energy that an empty cavity-QED system can absorb. They were first probed spectroscopically by Wallraff et al. (2004), measuring which microwaves were transmitted by a cavity that contained a frequency-tunable charge qubit. Figure 7.7a sketches the results from a similar experiment, as we move the qubit's frequency in and out of resonance with the cavity.

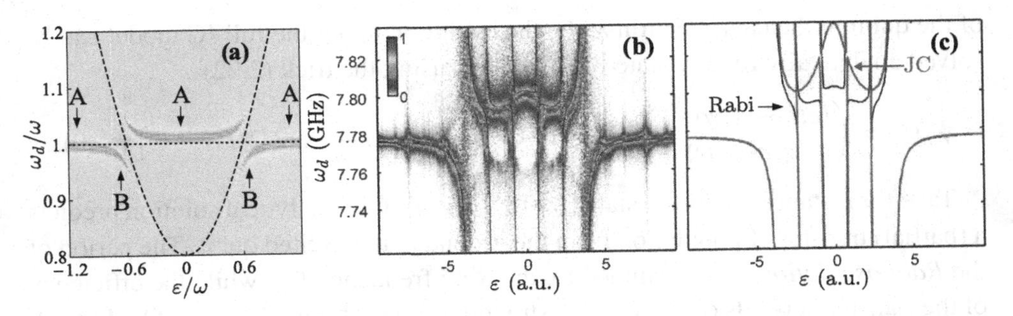

Figure 7.7 (a) Transmission through the cavity as a function of the qubit bias, ε, for $\Delta = 0.7\omega$ and $g = 0.05\omega$. With the dashed line we plot the qubit energy levels for $g = 0$. The linewidths assume $\sqrt{\gamma\kappa} = 0.005\omega$. (b) Transmission spectroscopy for a flux qubit interacting with the third mode of a microwave resonator, in the ultrstrastrong coupling regime $g/\omega = 16\%$. (c) Two different fits with the RWA and Jaynes–Cummings and the full Rabi model.

As explained previously, in the regime of large detuning (points "A" in Figure 7.7a), the qubit and the cavity cannot exchange excitations. In this situation, the setup develops one qubit-like and one cavity-like eigenstates, of which only the latter can be excited by the coherent microwave drive. For instance, if $\Delta \ll \omega$, the incoming microwaves only connect the ground state $|g,0\rangle$ with the the upper polariton branch, which consists mostly of $|g,1\rangle$ state. The result is a single peak in the cavity transmission sitting at $E_{1+} \simeq \hbar\omega$, with a linewidth proportional to the decay rate of the cavity $\sim\kappa$.

If we instead bring the qubit close to resonance with the cavity $\Delta \simeq \omega$ (points "B" in the same plot), the energy that we inject in the cavity will be efficiently exchanged with the qubit. In this case, the two polariton branches will appear as resonances, symmetrically placed around the cavity's frequency with a separation proportional to the qubit–cavity coupling strength $\hbar\omega_\pm \sim \hbar\omega \pm 2\hbar|g|$, and a linewidth proportional to the decoherence rate of the combined qubit–cavity system.

This separation of peaks is known as the *vacuum Rabi splitting,* and it is a smoking gun for coherent qubit–cavity coupling in circuit-QED. Naturally, for the peaks to be distinguishable, their linewidth must be narrower than the splitting, which takes us back to the strong coupling condition $|g| \gg \kappa, \gamma$ introduced before.

7.3.4 Rabi Oscillations: Weak and Strong Coupling

Spectroscopy is not a conclusive evidence of the "quantumness" of a circuit – after all, interacting classical oscillators also exhibit resonances and splittings, even at finite temperature. A more convincing result would be to reconstruct the dynamics of the qubit in the cavity. Fortunately, the evolution under the full JC model can be solved analytically for any state $|\psi(0)\rangle \in \mathcal{H}_n$, using the trick (2.22)

$$\boldsymbol{\psi}_n(t) = \begin{pmatrix} \langle e, n-1|\psi(t)\rangle \\ \langle g, n|\psi(t)\rangle \end{pmatrix} = e^{-i\omega(n-\frac{1}{2})t}\left[\cos(\Omega_n t) - i\sin(\Omega_n t)\mathbf{v}_n \cdot \tilde{\sigma}\right]\boldsymbol{\psi}_n(0).$$

Take for instance the initial state $|\psi(0)\rangle = |g, n\rangle$. Our analytical solution predicts a (partial) transfer of population from the ground to the excited qubit. The period of the *Rabi oscillations* is determined by the Rabi frequency Ω_n, while the efficiency of the transfer depends on the relative strength of the detuning $|\Delta - \omega|/\Omega_n$. Indeed, the transfer is only perfect in the resonant case $\Delta = \omega$,

$$|\psi(t)\rangle = \cos(\Omega_n t)\,|e, n-1\rangle - i\sin(\Omega_n t)\,|g, n\rangle\,, \quad n \geq 1, \tag{7.40}$$

where it happens with a frequency $\Omega_n = g\sqrt{n}$ that is half the Rabi splitting.

Actual Rabi oscillations are always damped due to decoherence in the cavity or the qubit. To analyze this, let us project our density matrix onto the different sectors with a fixed number of excitations $\rho_N(t) = \mathrm{P}_N\rho(t)\mathrm{P}_N$. Following (7.34), these matrices satisfy the master equation:

$$\partial_t\rho_N = -\frac{i}{\hbar}\left(H_{\text{eff}}\rho_N - \rho_N H_{\text{eff}}^{\dagger}\right) + \kappa\hat{a}\rho_{N+1}\hat{a}^{\dagger} + \gamma\sigma^{-}\rho_{N+1}\sigma^{+}, \quad \text{with} \tag{7.41}$$

$$H_{\text{eff}} = H_{\text{JC}} - i\frac{\hbar\gamma}{2}\sigma^{+}\sigma^{-} - i\frac{\hbar\kappa}{2}\hat{a}^{\dagger}\hat{a}.$$

The matrix ρ_N is fed by excitations from higher sections ρ_{N+1} and itself evolves with a non-Hermitian "Hamiltonian" H_{eff}, which is excitation number conserving, and where the frequencies of the qubit and the cavity now contain imaginary corrections. If our initial condition has N or less excitations, $\rho_{N+1}(t) = 0$, we can solve analytically the dynamics of ρ_N as a Schrödinger-like equation:

$$\rho_N(t) = e^{-iH_{\text{eff}}t/\hbar}\rho_N(0)e^{+iH_{\text{eff}}^{\dagger}/\hbar}. \tag{7.42}$$

We then find that the role of γ and κ in H_{eff} is to exponentially reduce in time the probability $P_N = \text{tr}(\rho_N)$ of having N excitations (see Problem 7.9), as one would expect from their interpretation as qubit and cavity loss rates.

Let us apply this idea to study the Rabi oscillations of an excited qubit in an empty cavity $\rho_1(0) = |e,0\rangle\langle e,0|$. The single excitation component of the density matrix is a projector $\rho_1(t) = |\psi(t)\rangle\langle\psi(t)|$, onto an unnormalized state $|\psi(t)\rangle = c_e(t)\,|e,0\rangle + c_g(t)\,|g,1\rangle$, which follows the equation

$$\mathbf{c}(t) = e^{-i\frac{1}{2}(\Delta+\omega)t - \frac{1}{4}(\kappa+\gamma)t} \times \left[\cos(\bar{\Omega}t)\mathbb{1} - i\sin(\bar{\Omega}t)\bar{\mathbf{v}}\cdot\boldsymbol{\sigma}\right]\mathbf{c}(0), \tag{7.43}$$

$$\bar{\Omega} = \sqrt{g^2 + \left(\frac{\Delta-\omega}{2} + i\frac{\kappa-\gamma}{4}\right)^2}, \qquad \bar{\mathbf{v}} = \left(\frac{g}{\bar{\Omega}}, 0, \frac{\Delta-\omega}{2\bar{\Omega}} + i\frac{\kappa-\gamma}{4\bar{\Omega}}\right).$$

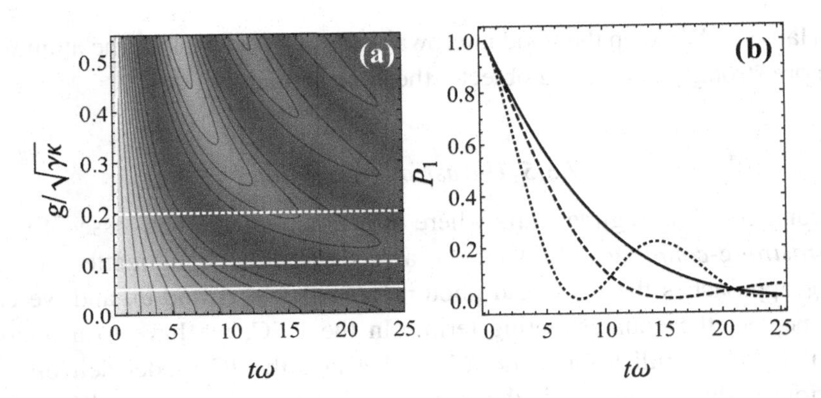

Figure 7.8 Dynamics in the weak and strong coupling regimes. (a) Excited state population for the cavity–qubit system starting in the $|e,0\rangle$ state, as a function of time and coupling strength g. (b) Horizontal cuts for $g = 0.05\gamma, 0.1\gamma$ and 0.2γ. Plots use $\omega = \Delta = 1$, $\gamma = \kappa = 0.1$.

The norm of $|\psi\rangle$ decreases exponentially, as does the probability of having one excitation in the system $\text{tr}(\rho_1) = \|\psi\|^2 \sim \exp(-(\kappa+\gamma)t/4)$. By virtue of (7.41), this probability is transferred to the ground state component of the density matrix $\rho_0(t)$, through processes of cavity decay and the qubit's nonradiative losses.

Our analytic solution predicts Rabi-type oscillations between the $|e,0\rangle$ and $|g,1\rangle$ states, modulated by an exponential envelope. Figure 7.8b explores these oscillations for a broad range of the *cooperativity factor* $g/\sqrt{\gamma\kappa}$, with select cuts plotted in Figure 7.8a. In the *weak-coupling regime*, decoherence is stronger than qubit–photon interactions $g \ll \gamma, \kappa$, and we cannot resolve the oscillations. This is exemplified by the exponentially decaying solid line in Figure 7.8b. The opposite regime, called *strong coupling*, appears when $g \gg \gamma, \kappa$. Thanks to a longer coherence time, we can detect one or several Rabi oscillations in the dashed and dotted lines from Figure 7.8b.

The strong-coupling regime was considered the Holy Grail of quantum optics for a long time. One of the difficulties in reaching strong coupling was the need for a cavity with a large *quality factor*[14] $Q = \omega/\kappa$. The first AMO experiments in the strong-coupling regime used superconducting microwave cavities that could trap photons for a long time. These cavities were made to interact with flying photons in Rydberg states in a series of experiments that culminated with the measurement of Rabi oscillations and the creation of Schrödinger cats (Haroche, 2013). In retrospect, circuit-QED is the natural evolution of the experiments by Nobel Prize winner

14 Q loosely represents the number of times a photon can oscillate in a cavity before it decays.

Serge Haroche. We keep the good microwave cavities, but replace the atom with an even more strongly interacting object – the superconducting qubit.

7.3.5 Ultrastrong Coupling

The strong-coupling regime starts where interaction overcomes losses. Similarly, the *ultrastrong-coupling* (USC) *regime* arises when the light–matter interaction speed g approaches the qubit and photon frequencies, Δ and ω, and we can no longer neglect the counterrotating terms. In the USC, we have to abandon the RWA and use the full Rabi model (7.32) because the JC model delivers wrong predictions in the eigenstates, in the associated spectra (cf. Section 7.4.1) and in the qubit–cavity dynamics.

First of all, in the USC regime, the qubit gets highly entangled with the photonic field, to a point that even the ground state becomes a superposition of vacuum $|g,0\rangle$ with even-excitation states, such as $|e,1\rangle$, $|g,2\rangle$, etc. These changes in the ground and excited states are enabled by the Rabi coupling $\sim\sigma^x(g\hat{a}+g^*\hat{a}^\dagger)$, a term that favors a hybridization of the cavity with the qubit, through a state-dependent displacement $\hat{a}\to\hat{a}-g^*\sigma^x/\omega$. This qubit–photon hybridization also activates transitions between states that do not conserve the number of excitations, making the dynamics richer and more interesting.

The differences are remarkable when we compare the spectra delivered by experiments, such as Figure 7.7b, with the numerical predictions of the JC and Rabi model, shown in Figure 7.7c. The existence of non-excitation-conserving processes modifies the eigenenergies not only around the crossing points $\Delta\simeq\omega$, but also at points where the qubit is highly off-resonant from the cavity. Notice how around those points – $\varepsilon=0$ in Figure 7.7c – the JC model predicts a huge dispersive energy shift of the cavity, which is corrected by the Bloch–Siegert shift from the Rabi model (Forn-Díaz et al., 2010).

The USC regime with a single quantum emitter was first demonstrated by Niemczyk et al. (2010) and Forn-Díaz et al. (2010), using flux qubits in microwave resonators and reaching relative coupling strengths $g/\Delta\sim 12\%$. Later experiments by Yoshihara et al. (2016) have surpassed these values, obtaining ratios g/ω close to and above 100%. In order to achieve these coupling strengths, those experiments required the use of flux qubits, the only qubits with a large enough anharmonicity as not to be "bridged" by the large interactions. The USC regime was possible because, as mentioned before, superconducting microwave photons and superconducting qubits are very similar excitations that can interact very efficiently. It must be remarked that, beyond mere curiosity and a challenging regime, the USC regime is also a useful tool in the development of quantum simulators, because it enables the implementation of tunable qubit–qubit Ising Hamiltonians and exploration of interesting quantum phase transitions (Kurcz et al., 2014). However, we will not have space available to discuss these applications in this book.

7.3.6 Multiple Qubits

Let us now consider an experimental setup where one resonator hosts multiple qubits, such as DiCarlo et al. (2010). The RWA model for those setups is the *Tavis–Cummings* Hamiltonian, an extension of the JC model:

$$H_{\text{TC}} = \sum_{i=1}^{N} \frac{\hbar\Delta_i}{2}\sigma_i^z + \hbar\omega\hat{a}^\dagger\hat{a} + \sum_{i=1}^{N}\left(\hbar g_i\sigma_i^+\hat{a} + \text{H.c.}\right), \tag{7.44}$$

which can also be upgraded to a full master equation (7.34). In this model, the qubits talk "collectively" with the resonator. The cavity exchanges excitations with all of them and mediates interactions among the qubits.

Spectroscopically, the features of this model are very similar to the JC Hamiltonian. For simplicity, we will discuss the case of identical emitters, with $\Delta_i = \Delta$ and $g_i = g$. In this limit, the TC becomes an RWA version of the Dicke model:

$$H = \frac{1}{2}\Delta\hat{S}^z + \hbar\omega\hat{a}^\dagger\hat{a} + g\hat{S}^+\hat{a} + g^*\hat{S}^-\hat{a}^\dagger, \tag{7.45}$$

with the collective operators $\hat{S}^\pm = \sum_{i=1}^N \sigma^\pm$ and $\hat{S}^z = \sum_{i=1}^N \sigma^z$.

The cavity is coupled to a symmetric superposition of excitations in the qubit space. This collective excitation acts as a new quasiparticle $\hat{b}^\dagger \sim S^+/\sqrt{N}$ with bosonic statistics, which interacts more strongly than the single-qubit case. In the limit of many qubits,

$$H \simeq \hbar\Delta\hat{b}^\dagger\hat{b} + \hbar\omega\hat{a}^\dagger\hat{a} + \sqrt{N}|g|(\hat{b}^\dagger\hat{a} + \text{H.c.}). \tag{7.46}$$

This physics is analogous to many other collective phenomena, from lasing to Bose–Einstein condensation, where a bunch of identical quantum systems act coherently in a synchronized way, and this way they become "better" at doing something – in this case, at producing a photon.

The ground state of this model is the trivial vacuum state $|\tilde{0}\rangle := |g, g \ldots, g, 0\rangle$. The single-excitation space contains $N+1$ states with one excitation on any qubit $\{\sigma_i^+ |\tilde{0}\rangle\}$ or on the cavity $\hat{a}^\dagger |\tilde{0}\rangle$. These states form one upper and one lower polariton branch, with eigenfrequencies

$$E_{1\pm} = \hbar\omega \pm \sqrt{\frac{1}{2}(\Delta-\omega)^2 + g^2 N}. \tag{7.47}$$

Note the increased splitting between polaritons caused by the bosonic enhancement of the interaction $\sqrt{N}g$. In addition to the polariton states, the Hilbert space is completed with $N-1$ *dark states,* excitations that live in the qubits and do not "see" the cavity. For instance, if we have $N=2$ qubits, the only dark state is the qubit singlet $\frac{1}{\sqrt{2}}(\sigma_1^+ - \sigma_2^+)\,|\tilde{0}\rangle$.

7.3.7 Off-Resonant Qubits and Dispersive Coupling

Many experiments work with qubits and cavities in the non-resonant or *dispersive* regime. In this regime, the qubit and the photons have very different frequencies, with a separation or *detuning* so large that it cannot be bridged by the qubit–photon coupling $\delta := |\Delta - \omega| \gg \langle \bar{n} \rangle\, g$, suppressing the exchange of excitations between the qubit and the resonator.

Why is this regime interesting? One reason is that we minimize the qubit decoherence by placing its frequency deep in the bandgap created by the resonator. As explained in near the beginning of Section 7.3, away from the cavity resonance, the spectral function is smooth and very small $J^{\mathrm{QO}}(\omega) \simeq 4\kappa |g|^2/\delta^2$, signaling a low probability that the cavity extracts energy from the qubit and dumps it into the waveguide.

The second reason has to do with the possibility that qubits exchange excitations via the resonator. The energy difference between a qubit and the cavity makes it unfavorable to transfer excitations from a qubit to the cavity and vice versa. However, it is still possible for excitations to jump from qubit to qubit via the cavity, if the qubits are close in frequency. As we will soon see, these *virtual processes* are terms that appear in perturbation theory and induce both exchange interactions $\sigma_i^+\sigma_j^-$ as well as cavity-dependent frequency shifts of the qubits $\sigma^z \hat{a}^\dagger \hat{a}$.

The dispersive regime is best studied using degenerate perturbation theory via the Schrieffer–Wolff transformation. Following Section A.3, we separate our Hamiltonian into a free term and an interaction $\hat{H} = \hat{H}_0 + g\hat{V}$. Our perturbation is the Tavis–Cummings term $g\hat{V} \sim \sum_i (\sigma_i^+ \hat{a} + \hat{a}^\dagger \sigma_i^-)$. We compute the Schrieffer–Wolff generator on the bare basis of the qubits and the cavity:

$$g\hat{S} = \sum_i \frac{g_i}{\Delta_i - \omega} \left(\sigma_i^+ \hat{a} - \hat{a}^\dagger \sigma_i^- \right). \tag{7.48}$$

The complete effective model is

$$H_{\mathrm{eff}} = \hbar \left(\omega + \sum_i \frac{g_i^2}{\Delta_i - \omega} \sigma_i^z \right) \hat{a}^\dagger \hat{a} + \frac{\hbar}{2} \sum_i \left(\Delta_i + \frac{g_i^2}{\Delta_i - \omega} \right) \sigma_i^z \tag{7.49}$$

$$+ \sum_{i \neq j} \frac{\hbar g_i g_j}{\Delta_i - \omega} \left(\sigma_i^+ \sigma_j^- + \sigma_i^- \sigma_j^+ \right).$$

The first term has been written in a way that evidences a shift of the resonator frequency. The direction or sign of this shift depends on the state of the qubit σ^z and on the detuning. We will see in Section 7.4.2 how this shift can be used to perform a nondemolition measurement of the qubit's state. The dispersive coupling also introduces a constant dc-Stark shift on the qubits and, when there are two or more qubits, an *exchange interaction* $\sigma_j^+\sigma_i^-$ mediated by virtual transitions through the resonator.[15] These exchange terms can be used to generate entangled states, simulate spin Hamiltonians, and implement two-qubit quantum gates, among other applications.

7.4 Circuit-QED Control

Circuit-QED setups are the foundation on which many other experimental ideas are built. Cavities, waveguides, and qubits are the main components of quantum optics, quantum simulation, and quantum computing setups. It is therefore important to pair the theory from earlier sections with the actual experiments to understand how these setups are operated, and also as preparation for later chapters.

7.4.1 Direct Cavity Spectroscopy

Our theoretical treatment of the cavity-QED setup has put great emphasis on the combined dynamics of the qubit and the resonator. However, the first evidence of strong coupling between a qubit and a cavity was given by spectroscopy (Wallraff et al., 2004), a technique that does not require measuring the qubit. Figure 7.9

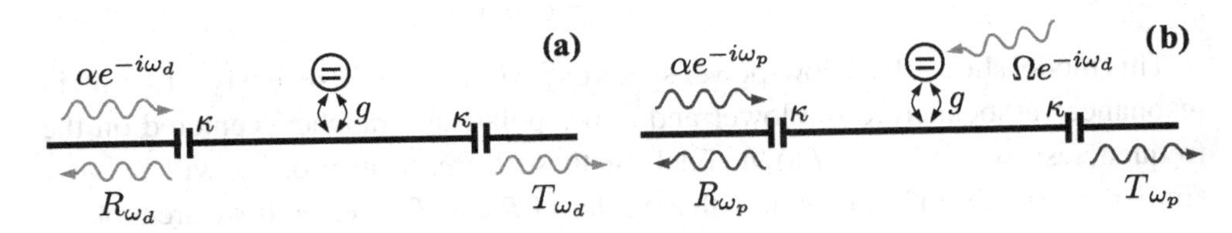

Figure 7.9 Spectroscopy schemes for a qubit in a cavity. (a) Single-tone spectroscopy, consists in driving the cavity with a coherent state at a frequency ω_d and studying the transmitted and reflected light. (b) Two-tone spectroscopy consists in probing the cavity transmission at a frequency ω_p, while driving the qubit with a different frequency ω_d. The driving of the qubit may take place through a separate antenna or – if enough power is available to compensate for the attenuation – through the antenna that drives the cavity itself.

[15] The mediating role of the cavity is evident when you analyze the second-order Schrieffer–Wolff correction $\frac{1}{2}g^2[S,V] \propto \sigma_j^+\hat{a}\hat{a}^\dagger\sigma_i^-$ which is responsible for these terms.

sketches the two most common setups for spectroscopy of a cavity–qubit system. The simplest method, in Figure 7.9a, drives the resonator with a monochromatic microwave at different frequencies ω_d, studying how much power is transmitted and reflected by the cavity.

The analysis of single-tone spectroscopy mimics the spectroscopy of linear resonators from Section 5.5.3. We first establish a relation between the output fields that are reflected/transmitted by the cavity, and the operator of the cavity itself. In the transmission case, all the signal is produced by the emission from the cavity, $\langle \hat{b}^{\text{out}} \rangle = -i\sqrt{\kappa}\,\langle \hat{a}(t)\rangle$. We may compute this expectation value using our master equation for the cavity-QED setup (7.34), with an added term $\Omega(t)\hat{a}^\dagger + \Omega(t)^*\hat{a}$ that accounts for the probe drive $\Omega(t) = \Omega_0 \exp(-i\omega_d t + i\phi)$.

There is a simple trick that works when the strength of the drive is weak enough that it can only excite transitions from the unique ground state – $|g,0\rangle$ in this case – to a finite set of excitations that are disconnected from each other – in this setup, the two polaritonic branches $E_{1,\pm}$ with states $|1_\pm\rangle$. In this limit, we approximate our system as a collection of harmonic oscillators, with the vacuum being the ground state and one oscillator mode for each excited state at frequencies dictated by the allowed transitions – $|\text{vac}\rangle \sim |g,0\rangle$ and $\hat{b}_\pm^\dagger\,|\text{vac}\rangle \sim |1_\pm\rangle$ the two polaritons, and $\hbar\omega_\pm = E_{1\pm} - E_0$. Each of the oscillators is driven by the microwave with a reduced amplitude $\Omega(t)\beta_\pm\hat{b}_\pm^\dagger$ + H.c. and experiences also a reduced decay rate $\kappa_\pm = |\beta_\pm|^2\kappa$, where $\beta_\pm \sim \langle 1_\pm|\hat{a}^\dagger|\text{vac}\rangle$ is the matrix element of the cavity operator. Under these conditions, the output field becomes a linear superposition of the fields absorbed and reemitted through the different allowed transitions:

$$T(\omega_d)\Omega_0 e^{i\phi} = \sum_{s=\pm} \frac{\kappa_s}{2} \frac{1}{i(\omega_s - \omega_d) + \kappa_s/2} \beta_s \Omega_0 e^{i\phi}. \tag{7.50}$$

This model shows that a low-power spectroscopic signal will reveal two Lorentzian resonances associated to the lower and upper polariton branches, centered on the frequencies $\omega_\pm = (E_{1\pm} - E_0)/\hbar$. Each of these Lorentzian profiles will be very similar to the ones in Figure 7.4, with the roles of R and T reversed if we are looking at a transmission spectrum. The height of the resonance is the proportional to the matrix element $\beta_\pm$ and is therefore dependent on how much of a photon component $|g,1\rangle$ is present in each of the eigenstates. Conversely, the width of the Lorentzian depends on the lifetime of these polaritons: The faster that these states decay due to κ or γ, the broader the resonance.

In Figure 7.7, we plotted the transmission spectrum of a setup where the gap of the qubit is tuned using the transverse field $\varepsilon\sigma^x$. When the qubit is far away from the cavity resonance, only one of the two polariton branches has sufficient overlap with the excited cavity mode $|g,1\rangle$. This mode allows the transmission of photons around the cavity frequency $\omega_d = \omega$, with a linewidth given by κ. When the qubit approaches the cavity, we begin to see the vacuum Rabi splitting in action. Straight on resonance, $\Delta^2 + \varepsilon^2 = \omega^2$, the two polariton resonances $\omega_d = \omega \pm g$ split with a gap $2g$. In the strong coupling regime, the decay rate of the cavity and of the qubit are smaller than this splitting, and both resonances can be resolved.

Low-power spectroscopy was also the fundamental tool in identifying the ultrastrong coupling regime, both in the works by Niemczyk et al. (2010) and Forn-Díaz et al. (2010). Note that even in the USC regime, the approximation of independent resonances works, although the computation of the matrix elements β_n is complicated by the strongly hybridized and qubit–photon entanglement in the low-energy excitations. In Figure 7.7b,c, we reproduce a transmission spectrum from Niemczyk et al. (2010), illustrating a qubit that comes to resonance with the third mode of a transmission line resonator. The coupling strength between the qubit and the resonator is very large, $g/\omega \simeq 12\%$, thanks to a clever design where the qubit is embedded in the central conductor of the resonator. As anticipated in Section 7.3.5, the spectrum from Figure 7.7b is only fitted properly when we consider the full Rabi model (7.32), with both rotating and counterrotating terms.

7.4.2 Qubit Dispersive Measurement

When we send a signal against a microwave resonator, it is reflected or transmitted with an amplitude *and a phase shift*. Both quantities depend, among other things, on the detuning between the driving field and the resonator. We saw this in Section 5.5.2 for a linear cavity, and it happens again when we have a qubit. The difference now is that the features of this resonance will be affected by what is inside the cavity.

This idea inspired the *quantum dispersive measurement of a superconducting qubit*, developed by Wallraff et al. (2005), to detect the state of a qubit without destroying it. The technique assumes a qubit and a resonator in the dispersive regime, far away from each other in frequency space. In this limit, the cavity resonance experiencies a shift (7.49) that depends on the state of the qubit. More precisely, as described by the model

$$H_{\text{eff}} = \hbar\left(\omega + \chi\sigma_i^z\right)\hat{a}^\dagger\hat{a} + \frac{\hbar}{2}(\Delta + \chi)\sigma^z, \tag{7.51}$$

the frequency of the resonator is now $\omega + \chi\sigma^z$, with $\chi = g^2(\Delta - \omega)^{-1}$. If $\Delta \ll \omega$, the resonance is shifted upward when the qubit is in the excited state and downward otherwise.

Let us assume that we have a microwave cavity sitting at the end of a waveguide, such that it cannot leak energy elsewhere. This corresponds to a setup like the one in Figure 7.9 where we eliminate the T_{ω_d} port, a configuration also studied in Section 5.5.3. If we send a microwave drive at any frequency through that guide, we expect all the field to be reflected, possibly affected by a small delay that changes the phase of the reflected field:

$$R_{\omega_d} = \frac{i(\omega - \omega_d) - \kappa/2}{i(\omega - \omega_d) + \kappa/2} \sim e^{i\phi(\omega_d)}. \tag{7.52}$$

When we drive on resonance $\omega_d = \omega$ and there is no qubit, the reflected field experiences a π shift and $R_{\omega_d} = -1$. However, if there is a qubit, the reflected field on resonance will produce a slightly different shift:

$$R_{\omega_d} = \frac{i\sigma^z\chi - \kappa/2}{i\sigma^z\chi + \kappa/2} \sim e^{i(\pi + \delta\phi\sigma^z)}, \tag{7.53}$$

which depends on the qubit's state $\sigma^z = \pm 1$:

$$\tan(\pi + \delta\phi\sigma^z) \simeq \frac{\chi\sigma^z}{\kappa}. \tag{7.54}$$

Detecting this phase is a process that takes some time, during which we integrate the quantum field reflected by the cavity, using the techniques from Section 5.5.7. This measurement attempts to distinguish the coherent state created by the microwave drive inside the cavity, and that depends on the state of the qubits $|\alpha_\pm\rangle := |\alpha e^{i(\pi \pm \delta\phi)}\rangle$. A proper discrimination of these states requires us to strike several balances.

- The power of our coherent microwave probe must be low enough not to create large numbers of photons inside the cavity – which would invalidate the dispersive approximation – but also large enough so that the two coherent states are distinguishable. In other words, we want a small overlap $\langle \alpha e^{+i\delta\phi} | \alpha e^{-i\delta\phi}\rangle \propto e^{-|2\alpha\delta\phi|^2} \sim 0$, while keeping $|\alpha|^2 \lesssim 1$.
- Similarly, we want the coupling to be large and the detuning to be small, in order to maximize the dispersive shift χ, achieve a large phase shift, and enable a strong distinguishability between states. However, χ must be small enough that the corrections to perturbation theory may be ignored, and the dispersive Hamiltonian (7.49) remains an accurate description.
- During the integration of the reflected field, we obtain an estimate of the coherent state with an uncertainty that decreases with the integration time $\sim 1/T$. This is both because of the accuracy in the integration of the oscillating field, and also because the cavity requires some time to reach the asymptotic states $|\alpha_\pm\rangle$.
- At the same time, we cannot make the measurement too long, because during measurement the qubit is affected both by its intrinsic relaxation and by additional decoherence channels enabled by the coupling to the cavity, all of which add up to $\gamma_{\text{eff}} \sim \kappa |g/(\Delta - \omega)|^2 + \gamma$.

The dispersive measurement is a type of generalized measurement where we put our the qubit into contact with a cavity, let them interact, and finally deduce the state of the qubit from that of the cavity. Since $\mathrm{Im}\alpha_\pm = |\alpha|\sin(\delta\phi)\sigma^z$, such discrimination can be performed by inspecting the sign of a quadrature $\hat{P} = \frac{1}{\sqrt{2}}(\hat{a}^\dagger - \hat{a})$, to deduce the sign of $\sigma^z = s$. Ideally, this measurement has two outcomes $s = \pm 1$, obtained with probabilities $p_s = \mathrm{tr}(\varepsilon_s(\rho))$, from a positive map ε that combines the qubit–cavity interaction U with the projection Π onto the right sign of the quadrature $\hat{P}$:

$$\varepsilon_s(\rho_{\text{qubit}}) = \mathrm{tr}_{\text{cavity}}\left(\Pi_{\text{sign}(\hat{P})=-s} U\, \rho_{\text{qubit}} \otimes |\text{vac}\rangle\langle\text{vac}|\, U^\dagger\right), \quad \text{with} \tag{7.55}$$
$$U \simeq |g\rangle\langle g| \otimes D(\alpha_-) + |e\rangle\langle e| \otimes D(\alpha_+).$$

In this limited framework, the measurement is specified by two probabilities:

$$\begin{aligned} M_p &= \mathrm{tr}(\Pi_{-1}\,|\alpha_-\rangle\langle\alpha_-|) = \mathrm{tr}(\Pi_{+1}\,|\alpha_+\rangle\langle\alpha_+|), \\ M_n &= \mathrm{tr}(\Pi_{+1}\,|\alpha_-\rangle\langle\alpha_-|) = \mathrm{tr}(\Pi_{-1}\,|\alpha_+\rangle\langle\alpha_+|) = (1 - M_p). \end{aligned} \tag{7.56}$$

The smallest probability M_n measures how much of the $|\alpha_-\rangle$ coherent state lies in the region $\hat{P} > 0$, and vice versa, how much of the $|\alpha_+\rangle$ lies below $\hat{P} < 0$. It quantifies the mistake in the discrimination of σ^z, so that the combined measurement reads

$$\begin{aligned} \varepsilon_{-1}(\rho) &= (1 - M_n)\mathrm{tr}(\rho_{\text{qubit}}\,|g\rangle\langle g|) + M_n \mathrm{tr}(\rho_{\text{qubit}}\,|e\rangle\langle e|), \\ \varepsilon_{+1}(\rho) &= M_n \mathrm{tr}(\rho_{\text{qubit}}\,|g\rangle\langle g|) + (1 - M_n)\mathrm{tr}(\rho_{\text{qubit}}\,|e\rangle\langle e|). \end{aligned} \tag{7.57}$$

From this derivation, we see that in the limit of good discrimination, $M_n \ll 1$, the measurement is close to an ideal projective measurement of the qubit on the basis $\{|g\rangle, |e\rangle\}$. This is a type of *quantum nondemolition* or *QND measurement* that can be repeated multiple times, obtaining almost always the same value. It is also the type of measurement that we want to have in quantum computers, both for the application of quantum algorithms as well as for error correction.

The fidelity of the dispersive QND measurement is limited by the discrimination error M_n and by the decoherence of the qubit – which was not considered in the previous reasoning, and which modifies the maps $\varepsilon_{\pm 1}(\rho)$. We can improve both errors, allowing for longer measurements with lower decoherence, by adding *Purcell filters* to the measurement device (Reed et al., 2010). The Purcell filter is an additional cavity that sits between the qubit's cavity and the waveguide, further restricting the photons that can leak out of the original setup. The filter must be broader than χ so that it can discriminate the two displaced frequencies, but it must be narrow enough so that the qubit's frequency Δ falls outside the filter's linewidth. This way, the effective decay rate of the qubit is lowered from the value $\kappa|g/\delta|^2$ predicted by (7.31), to a much smaller number. At the same time, the setup allows for stronger drives, faster measurements, and reduced errors, which as of the time of writing this book lay around 1% (2%) in 88 ns (48 ns) for state-of-the-art setups (Walter et al., 2017).

7.4.3 Two-Tone Spectroscopy

Having access to the qubit's instantaneous state of the qubit is a powerful tool that facilitates a new type of spectroscopy, where we can probe the qubit's response to external fields, calibrating its Hamiltonian independently of that of the cavity. Sketched in Figure 7.7b, *two-tone spectroscopy* uses two driving fields, one that will excite the qubit ω_d, and a probe drive that is resonant with the bare cavity frequency $\omega_p = \omega$. The transmission of the probe beam is continuously monitored, measuring the phase shift experienced by the photons. Since this is a continuous process, where we are at the same time directly influencing the qubit, the study of the phase shift only provides an *average* of the cavity shift and, by extension, of $\langle\sigma^z\rangle$.

As in usual spectroscopy, this procedure is repeated for a broad range of qubit drives ω_d. When ω_d is far from resonance, the qubit will remain in its approximate ground state and the phase shift will have a fixed, nonzero value. However, when ω_d approaches the renormalized frequency of the qubit, $\Delta + g^2/(\Delta - \omega)$, the qubit will begin to oscillate. In this case, $\langle\sigma^z\rangle$ will average to zero and the photons will not experience any phase shift.

In Figure 7.10, we show an excerpt of the results by Schuster et al. (2005). Note how the method reveals the qubit's hyperbola with great clarity and contrast. The two-tone spectroscopy is complementary to the single-tone spectroscopy from Section 7.4.1. In the latter method, we only get a good signal from the qubit when it is strongly hybridized with the cavity. In contrast, the two-tone spectroscopy works best when the qubit is in the dispersive regime and not hybridized at all.

7.4.4 Single-Photon Generation

Once we have a superconducting qubit inside a microwave resonator, we can apply different microwave and qubit controls to engineer quantum states of the qubit, the cavity, or both. The simplest application of such protocols is to engineer a one-photon state in the cavity. For this, the simplest method is to start with a qubit gap Δ strongly detuned from the resonance frequency ω of the cavity where the qubit resides. The energy separation should be large enough $|\omega - \Delta| \gg |g|$ that we can neglect the qubit–cavity coupling. We assume that the qubit starts in this situation and in the $|g\rangle$ state. The qubit is then driven resonantly to acquire one excitation $|e\rangle$, and we rapidly shift its frequency to bring qubit and cavity into resonance $\Delta = \omega$ for

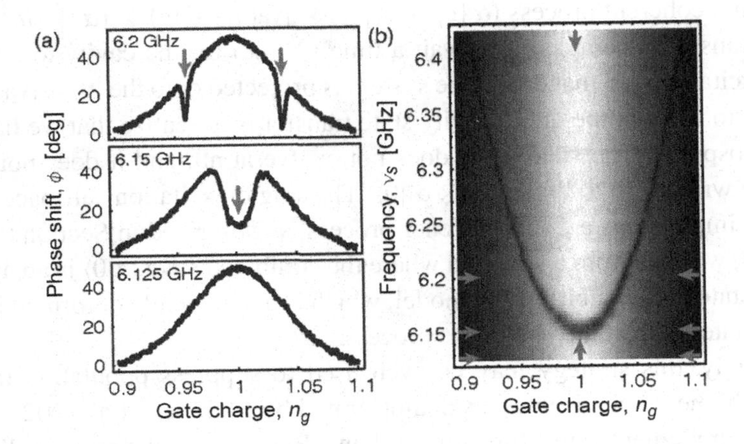

Figure 7.10 Two-tone spectroscopy of a charge qubit. (b) shows the phase shift experienced by the transmitted photons as the qubit energy is moved around using an external voltage and driven with some frequency, ν_d. As shown in (a), when the qubit is appropriately excited, the average phase drops significantly to zero. Figure reproduced from Schuster et al. (2005).

half a Rabi period. This way, the qubit and the cavity exchange excitations, evolving from $|e, 0\rangle$ to $|g, 1\rangle$. If at this point we rapidly bring the qubit out of resonance again, we will have created a single-photon state.

As discussed in Section 5.4.6, this idea was demonstrated by Eichler et al. (2011) in a seminal experiment that not only showed the generation of the photons, but also demonstrated the nature of the quantum state by doing Wigner function tomography of photon's wavefunction as it left the cavity.

7.4.5 Qubit Reset

Many experimental protocols require that we have qubits in well-defined $|g\rangle$ or $|0\rangle$ states. This requisite may be hard to obtain experimentally. First, if we rely on cooling, the quality of the states we prepare is limited by the temperature of the experiment, which may not be enough if the qubit's frequency is modest. Second, cooling may take too long a time, especially when we strive to prepare qubits with long coherence times T_1 of hundreds of microseconds or even longer.

Since most experiments place qubits close to different types of cavities, including the ones for readout, we can use those cavities to enhance the qubit's relaxation and prepare almost perfect zero states. The protocol begins with the qubit in a situation of low gap detuning $\Delta < \omega - |g|$. The gap of the qubit is slowly changed until it lands above the resonator's frequency $\Delta > \omega + |g|$. If this shift is done at an *adiabatic* pace $\mathrm{d}\Delta/dt \ll |g \times \Delta|$, the qubit will transfer its quantum state to the resonator in a coherent process $(\alpha\,|g\rangle + \beta\,|e\rangle) \otimes |\text{vac}\rangle \rightarrow |g\rangle \otimes (\alpha + \beta a^\dagger)\,|\text{vac}\rangle$. Once the transfer is complete, we wait a time $\mathcal{O}(1/\kappa)$ for the cavity to relax and emit any excitations, so that the whole system is projected onto the $|0\rangle \otimes |\text{vac}\rangle$ state.

The way to analyze the qubit–cavity state transfer is to realize that we have two separate subspaces. The state $|g, 0\rangle$ does not evolve at all, and it does not matter whether we write $\alpha\,|g\rangle \otimes |0\rangle$ or $|g\rangle \otimes \alpha\,|0\rangle$. The single-excitation subspace, on the other hand, implements a Landau–Zener process, as described in Section 9.1.1. In this process, we start from a situation where the combined state $|e, 0\rangle$ is an approximate eigenstate of the qubit–cavity model, which is adiabatically deformed into the final eigenstate $|g, 1\rangle$ at the end of the process.

Interestingly, this strategy may be even used to suppress population that has leaked into higher excited states, as demonstrated by McEwen et al. (2021). If the dynamics is implemented in a time longer than a Rabi period, the reset quality over all energy levels can be very high $\mathcal{O}(10^{-4})$, well below what can be experimentally detected.

7.4.6 Cavity Fock States Superpositions

Using our controls on the qubits, the cavities, and the qubit–cavity coupling, we can engineer very sophisticated protocols to create arbitrary states of either system, or both. We have already described how to inject a single photon into the cavity. As explained by Law and Eberly (1996), we can generalize this idea to create superpositions of N Fock states in the cavity:

$$|\Psi_N\rangle = \sum_{n=0}^{N} c_n\,|g, n\rangle\,. \tag{7.58}$$

The idea is technically complex, although it can be understood qualitatively. One just needs to realize that, given a state with N photons

$$|\chi_N\rangle = \sum_{n=0}^{N-1} \left(\alpha^N_{e,n-1}\,|e, n\rangle + \alpha^N_{g,n}\,|g, n\rangle\right), \tag{7.59}$$

we can create a state $|\chi_{N-1}\rangle$ that has one less excitation. For this, we use two operations. The first one, Q_n, consists in bringing the qubit into resonance with the cavity $\Delta = \omega$ and wait a time t_N sufficient for all population from $|g, N\rangle$ to be transferred to $|e, N-1\rangle$. The second operation is a rotation of the qubit C_N, transforming $|e, n\rangle \leftrightarrow |g, n-1\rangle$, which converts $|e, N-1\rangle$ into $|g, N-1\rangle$ and leaves us with a $|\chi_{N-1}\rangle$ state.

Using this technique, we can compute a sequence of gates that undoes the desired state $C_0 Q_0 \cdots C_N Q_N\,|\chi_N\rangle = |0\rangle$. If we apply the reversed sequence $C_0^\dagger$, $Q_0^\dagger \cdots Q_N^\dagger$, we will be able to construct the Fock superposition state experimentally, leaving the qubit unentangled from the cavity. This technique was demostrated by Hofheinz et al. (2009) with up to $N = 5$ photons, in a very interesting work that not only created the state but used Wigner function tomography to reconstruct the complete density matrix of the cavity.

7.4.7 Cavity Schrödinger Cats

A much simpler protocol can be used to engineer the Schrödinger cat states from Section 5.4.4, using the dispersive coupling regime (7.51). For simplicity, we will study the dynamics of the system in a rotating frame where we eliminate the qubit's and cavity's frequencies, working with the Hamiltonian

$$H = \chi \sigma^z a^\dagger a. \tag{7.60}$$

Assume we prepare the cavity in a coherent state $|\alpha\rangle$ and simultaneously excite the qubit into a superposition $\frac{1}{\sqrt{2}}(|g\rangle + |e\rangle)$. After a time t, the combined state will have evolved into

$$\frac{1}{\sqrt{2}} e^{-i(\Delta-\chi)t} |g\rangle \left|e^{-i\chi t}\alpha\right\rangle + \frac{1}{\sqrt{2}} e^{-i(\Delta+\chi)t} |e\rangle \left|e^{+i\chi t}\alpha\right\rangle. \tag{7.61}$$

We will choose $\chi t = \pi/2$, so that both coherent states have opposite signs, and further displace the cavity along $-i\alpha$, obtaining the state

$$\frac{1}{\sqrt{2}} e^{-i(\Delta-\chi)t} |g\rangle |-2i\alpha\rangle + \frac{1}{\sqrt{2}} e^{-i(\Delta+\chi)t} |e\rangle |0\rangle. \tag{7.62}$$

Interestingly, we can now use the dispersive shift of the qubit's frequency to implement a rotation that only is effective in the vacuum state. In particular, we drive the qubit with a long pulse, resonant with the frequency $\Delta + \chi$, which only excites the transition $|g,0\rangle \leftrightarrow |e,0\rangle$, leaving all $|g,n > 0\rangle$ and $|e,n > 0\rangle$ approximately unaltered. The outcome of this rotation is a new state where the qubit is decoupled from the cavity:

$$|g\rangle \otimes \frac{1}{\sqrt{2}} (|-2i\alpha\rangle + |0\rangle). \tag{7.63}$$

Using further displacements, we can bring this into a more symmetric superposition $|-i\alpha\rangle + |i\alpha\rangle$. Note that we can also change the relative phase of the coherent states by playing with the single-qubit rotation's phase. Moreover, due to free evolution with the resonator's Hamiltonian, the states $\pm i\alpha$ rotate in phase space so that we effectively can prepare superpositions along different directions.

This type of coherent state superpositions has been demonstrated by Vlastakis et al. (2013) with coherent states that involve up to $|\alpha|^2 \sim 100$ photons. Such superpositions constitute the basis for very interesting encodings of quantum information and even error-corrected quantum computing protocols (Mirrahimi et al., 2014), which unfortunately we do not have space to discuss here.

Exercises

7.1 Show how (7.13) can be obtained by projecting both sides of the Schrödinger equation $i\partial_t |\psi(t)\rangle = \hat{H} |\psi(t)\rangle$ onto the basis states $|1, \text{vac}\rangle$ and $\hat{a}_k^\dagger |0, \text{vac}\rangle$.

7.2 **Markovian kernel.** Let's have a look at the qubit's dynamics:

$$\partial_t c_1 = -i\Delta c_1 - \frac{i}{2\pi}\int_0^t \int J(\omega) e^{-i\omega(t-\tau)} c_1(\tau) \mathrm{d}\omega\, \mathrm{d}\tau, \tag{7.64}$$

assuming that the spectral function is constant $J(\omega) = 2\pi\alpha$ over the whole real line, $\omega \in \mathbb{R}$. Prove that the differential equation is exactly Markovian and only depends on the value of $c_1(t)$. What are the decay rate γ and the Lamb shift?

7.3 **Qubit–photon entanglement.** We can use (7.24) to estimate the amount of entanglement between the qubit and the emitted photon. The von Neumann entropy of the reduced density matrix ρ_{qb}

$$S = -\text{tr}(\rho_{\text{qb}} \log(\rho_{\text{qb}})) = -\sum_i \lambda_i \log(\lambda_i) \tag{7.65}$$

is a function of the eigenvalues λ_i of the density matrix, and it is also a quantifier of the entanglement between our qubit and the photonic environment. Use the Pauli expansions and diagonalization methods from (2.21) to compute $\lambda_\pm$. Show that this entanglement is zero at $t = 0$ and $t = +\infty$ and find its maximum.

7.4 **Zero-temperature master equation.** Assume a qubit in a generic mixed state, placed in an empty transmission line, $\rho = \rho_{\text{qb}} \otimes |\text{vac}\rangle\langle\text{vac}|$.

(1) Show that ρ can be written as $a\,|\Psi_1\rangle\langle\Psi_1| + b\,|\Psi_2\rangle\langle\Psi_2|$, where $|\Psi_i\rangle = |\psi_i\rangle \otimes |\text{vac}\rangle$ and $|\psi_i\rangle$ are the eigenstates of ρ_{qb}.

(2) Show that we can study the evolution of $|\Psi_1\rangle$ and $|\Psi_2\rangle$ separately using the formalism from Section 7.2.

(3) Show that the total density matrix obeys the relaxation equation (6.19) and that this expression is a solution of (6.17) with $\bar{n}(\Delta) = 0$.

7.5 A **chiral interaction** is one that breaks time-reversal symmetry. In our context of light–matter interactions, a qubit interacting with a chiral medium would couple differently to photons moving along different directions $g_{+k} \neq g_{-k}$. Let us assume that our waveguide is chiral because the $k < 0$ modes are totally absent. Take the following steps to understand how this changes light–matter interactions:

(1) Assuming $g_k \sim g\mathrm{d}k^{1/2}$, integrate the qubit dynamics and compute the relation between g and the spontaneous emission rate, γ.
(2) Write the new input–output relations: How many fields do you have to consider? What is the new prefactor in front of $c_1(t)$?
(3) Repeat the derivation of the scattering coefficients for an incoming photon. Show that $R_\omega = 0$, $|T_\omega| = 1$ and that the photon only experiences a phase shift.
(4) Explain when the chiral model can be used to study the scattering of photons by a qubit that is placed at the end of a semi-infinite transmission line.

7.6 Consider a multilevel circuit with N excited levels, interacting with the transmission line. We assume that each level only couples to the ground state via the propagating photons:

$$H = \sum_{i=1}^{N} \Delta_i \, |i\rangle\langle i| + \sum_k \omega_k \hat{a}_k^\dagger \hat{a}_k + \sum \beta_i \sigma_i^+ \sum_k \left(g_k \hat{a}_k + g_k^* \hat{a}^\dagger \right). \quad (7.66)$$

Here $\sigma_i^+ = |i\rangle\langle 0|$ and the β_i measure the relative strength of the different $0 \leftrightarrow i$ transitions.

(1) Derive the evolution equations for the single-excitation ansatz:

$$|\psi\rangle = \left(\sum_{i=1}^{N} c_i \sigma_i^+ + \sum_k \psi_k \hat{a}_k^\dagger \right) |0, \mathrm{vac}\rangle . \quad (7.67)$$

(2) Integrate the photons out and apply the Markovian approximation to obtain N equations for the c_i coefficients. How does the excited state $|i\rangle$ evolve in time?
(3) Generalize the input–output relations to consider the N transitions.
(4) Compute the scattering amplitudes T_k and R_k for a single incoming photon. What does the spectrum look like?

7.7 **Lambda system and dark states.** Let us study the following energy-level structure, with two degenerate ground states, $|a\rangle$ and $|b\rangle$, and one excited state:

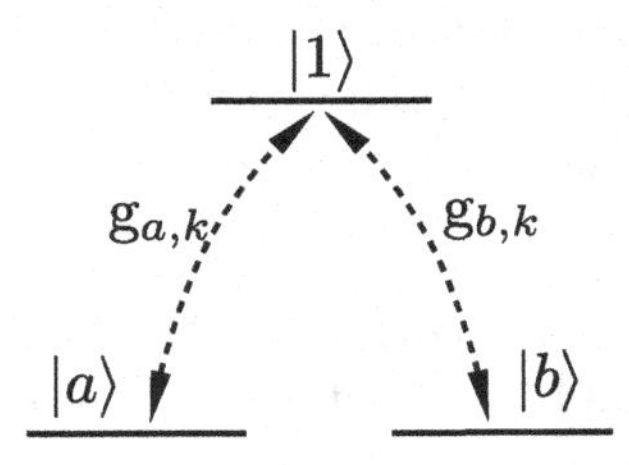

What can you say about the spontaneous emission from state $|1\rangle$ and the interaction with propagating photons? Hint: Show that this problem is equivalent to a two-level system plus one noninteracting *dark state* for any (complex) values of the two coupling strengths, $g_{a,bk}$.

7.8 **Photodetection.** Let us now devise a simple photodetector using the a few-level system that can switch the state through the absorption of a photon. A two-level system is not a valid photodetector because even if the photon is absorbed, the state $|1\rangle$ will subsequently decay. Instead, we introduce the following three level system:

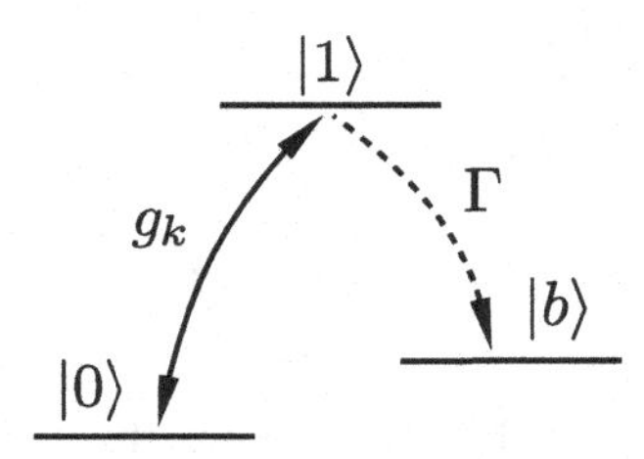

In this scheme, an incoming photon couples to the qubit and excites the transition $0 \leftrightarrow 1$. At the same time, there is an incoherent transition $1 \rightarrow b$ that depletes the excited state population with rate Γ and deposits it irreversibly onto the state $|b\rangle$, thus signaling the presence of a photon. We are going to model this system with the same equations as for the two-level system. The only change is the addition of an ad hoc incoherent channel:

$$i\partial_t c_1 = (\Delta' - i\gamma)c_1 - i\Gamma c_1. \tag{7.68}$$

(1) Assuming an adiabatically switched on wavepacket, compute the probability of landing onto the state $|b\rangle$

$$P_b = 1 - |c_1|^2, \tag{7.69}$$

by integrating the dynamics of the qubit under this driving. Show that it is always below 50%.

(2) What are the transmission and reflection amplitudes of the photon, T_ω and R_ω? How does the photodetection process modify the probabilities $|T|^2$ and $|R|^2$?

7.9 Given (7.42), prove that the probability of having N excitations, given by $P(N,t) = \mathrm{tr}(\rho_N(t))$ decays exponentially. More precisely, prove the equation $\frac{\mathrm{d}}{\mathrm{d}t}P(N) \leq -(\kappa(N-1)+\gamma)P(N)$.

7.10 Prove the expressions for the dispersive measurement (7.57) in the ideal case where decoherence can be ignored.

8

Quantum Computing

8.1 Quantum Circuit Model

The term *quantum computer* originated in the 1980s, although with different meanings and perspectives. At this time, computers were on their way to becoming a commodity and computer science was no longer an emergent field, but there was already a growing concern about the energy requirements and the efficiency of those first computers. Not long before, Landauer had connected information processing and thermodynamics, establishing a lower bound to the energy that is consumed whenever a bit of information is created or discarded: $k_{\mathrm{B}}T \log 2 \simeq 0.017\mathrm{eV}$.

Computers of that era were very inefficient, and scientists were looking for alternative computation models that could attain Landauer's limit. Some focused on the hardware, looking at computer designs that use superconducting circuits or nonlinear optical elements, two paradigms that have been ultimately abandoned. Other, more theoretically minded researchers worked on *reversible computation.* These researchers looked at the two-bit gates that are used in all computers – NAND, XOR, etc. – and realized that each gate drops a bit of information whenever it is executed. Connecting this idea with Landauer's principle, they argued that we could look at reversible operations, such as the Toffoli gate in combination with the NOT gate, as a universal framework for computation that would be more energy efficient.

You may already realize that quantum mechanics was a natural candidate for reversible computation. After all, it is a physical framework built on the premise of unitary and thus reversible time evolution. In 1980, physicist Paul Benioff formalized this notion, establishing a connection between the Turing machine, a universal model of computation, and the evolution of an idealized quantum physical system (Benioff, 1980). In other words, he found a quantum Hamiltonian that implements the Turing machine.

Around the same years, we also find celebrated physicist Richard Feynman looking at the relation between computers and quantum mechanics from a different angle. Coming from a theoretical physics background, and having worked on the application of computers to solving theoretical physics problems, he challenged the capabilities of classical computers (both real and abstract, such as the Turing machine) to simulate the evolution of quantum systems. In a well-known talk, Feynman (1982), he illustrated the exponential complexity of simulating the Hilbert space of even very simple quantum models – particles jumping on a lattice – and introduced the idea of *quantum simulation,* engineering a quantum physical device to reproduce the dynamics of a complex quantum model that we do not know how to solve.

Benioff's work and Feynman's ideas converge in the work by Deutsch (1985), a seminal paper that introduced the concept of a *quantum Turing machine*. This is a device that not only can process classical information – and therefore reproduce a *classical* Turing machine – but also information stored in generic quantum states, including quantum superpositions and entangled states. Deutsch's quantum Turing machine provided the theoretical foundation for what we now understand as a *universal quantum computer.*

Despite its theoretical value, both the classical and the quantum Turing machines are inconvenient designs that do not inform the real architecture of computers. Instead, quantum computers are developed in the lab using more scalable and highly parallelizable frameworks. In this chapter, we focus on the particular framework of *quantum circuits* introduced also by Deutsch (1989). The quantum circuit model assumes a quantum register formed by near-ideal qubits. The register acts both as a memory and as a processor.[1] Quantum computations are implemented by modifying this register through a composition of qubit measurements with unitary operations that are extracted from a universal set.

The quantum circuit model includes also a very elegant graphical representation. The algorithm's operations and measurements are placed on a set of horizontal lines, similar to a musical "pentagram." Each horizontal line addresses one qubit. A box in this pentagram is a unitary operation acting on the qubits that it is connected to. The complete circuit can be read as a set of gates and measurements to be executed, either in parallel or sequentially, as we read the diagram from left to right.

As an example, Figure 8.1a shows a quantum algorithm to recreate and measure a Schrödinger cat or Greenberger–Horne–Zeilinger (GHZ) state $\frac{1}{\sqrt{2}}\,|00000\rangle + \frac{1}{\sqrt{2}}\,|1111\rangle$. In this example, we begin with a set of qubits initialized to a zero state $|00000\rangle$. The first operation is a single-qubit gate, a Hadamard gate H (cf. Table 8.1)

[1] In some sense, one could argue that such a design is closer to an *abacus* than to a Turing machine (a head moving over an infinite, 1D stream of data) or a von Neumann computer (with CPU, control unit, memory, external storage, and I/O).

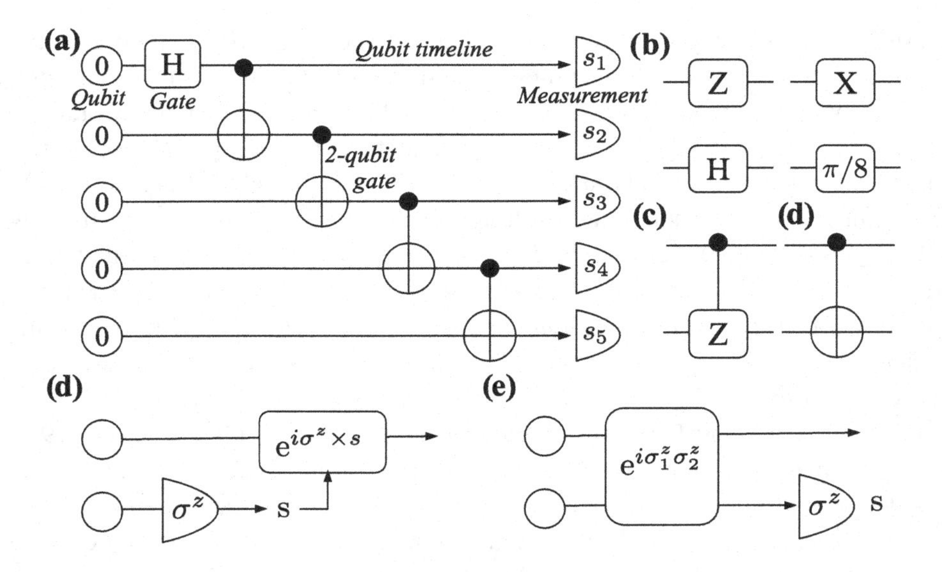

Figure 8.1 (a) Quantum circuits are schematics of quantum algorithms and protocols that combine (i) qubits, represented by circles with horizontal lines; (ii) single-qubit operations that intersect only one such lines; (iii) multiqubit operations; and (iv) measurements. (b) Single-qubit operations, as denoted in Table 8.1. (c) Two-qubit conditional operations for a CZ and CNOT, respectively. Quantum measurements for feedback operations (d) can be moved to the end of an algorithm, using two-qubit gates (e).

that sets the first qubit into a superposition state, $\frac{1}{\sqrt{2}}(|0\rangle + |1\rangle) \otimes |0000\rangle$. Subsequent two-qubit operations, called CNOT's, copy the state of the first qubit to all other qubits, creating a large entangled GHZ state.

The power of the quantum circuit model does not lay in this nice, prescriptive, and visually appealing representation. The circuit model gives us a minimal set of logical ingredients that, combined together, can reproduce any classical or quantum computation we may think of. This includes Shor's celebrated algorithm for factoring, the phase and amplitude estimation algorithm, the Harrow–Hassidim–Lloyd (HHL) algorithm for inverting matrices, and many modern applications to real-life problems in chemistry, finance, logistics, etc.

DiVincenzo (2000) formalized the requirements of the quantum circuit model in a celebrated paper. The *DiVincenzo criteria* are a set of five requirements for scalable quantum computation plus two for quantum communication. Cited verbatim, these elements are the following:

(1) A scalable physical system with well-characterized qubits
(2) The ability to initialize the state of the qubits to a simple fiducial state, such as $|000\ldots\rangle$

Table 8.1. *Most popular single-qubit and two-qubit unitaries, with their names and their generators – i.e., Hermitian operator O such that $U = \exp(iO + i\phi)$ with some irrelevant global phase ϕ.*

Name / generator	Matrix	Name / generator	Matrix
Hadamard $\frac{\pi}{2}\sigma^y$	$H = \frac{1}{\sqrt{2}}\begin{pmatrix}1 & 1\\ 1 & -1\end{pmatrix}$	Pauli-X $\frac{\pi}{2}\sigma^x$	$X = \begin{pmatrix}0 & 1\\ 1 & 0\end{pmatrix}$
Pauli-Y $\pi\sigma^y$	$Y = \begin{pmatrix}0 & -i\\ i & 0\end{pmatrix}$	Pauli-Z $\frac{\pi}{2}\sigma^z$	$Z = \begin{pmatrix}1 & 0\\ 0 & -1\end{pmatrix}$
Phase $\frac{\pi}{4}\sigma^z$	$S = \begin{pmatrix}1 & 0\\ 0 & i\end{pmatrix}$	"$\pi/8$" $\frac{\pi}{8}\sigma^z$	$T = \begin{pmatrix}1 & 0\\ 0 & e^{i\pi/4}\end{pmatrix}$
Control-NOT (CNOT) $\frac{\pi}{2}(\sigma_1^z + 1)\sigma_2^x$	$\begin{pmatrix}1 & 0 & 0 & 0\\ 0 & 1 & 0 & 0\\ 0 & 0 & 0 & 1\\ 0 & 0 & 1 & 0\end{pmatrix}$	Control-Z (CZ) $\frac{\pi}{4}(\sigma_1^z - 1)(\sigma_2^z - 1)$	$\begin{pmatrix}1 & 0 & 0 & 0\\ 0 & 1 & 0 & 0\\ 0 & 0 & 1 & 0\\ 0 & 0 & 0 & -1\end{pmatrix}$
iSWAP $\frac{3\pi}{4}(\sigma_1^+\sigma_2^- + \text{H.c.})$	$\begin{pmatrix}1 & 0 & 0 & 0\\ 0 & 0 & i & 0\\ 0 & i & 0 & 0\\ 0 & 0 & 0 & 1\end{pmatrix}$	$\sqrt{\text{SWAP}}$ $\frac{\pi}{8}\boldsymbol{\sigma}_1 \cdot \boldsymbol{\sigma}_2$	$\begin{pmatrix}1 & 0 & 0 & 0\\ 0 & \frac{1+i}{2} & \frac{1-i}{2} & 0\\ 0 & \frac{1-i}{2} & \frac{1+i}{2} & 0\\ 0 & 0 & 0 & 1\end{pmatrix}$

(3) Long relevant decoherence times, much longer than the gate operation time
(4) A "universal" set of quantum gates
(5) A qubit-specific measurement capability
(6) The ability to interconvert stationary and flying qubits
(7) The ability to faithfully transmit flying qubits between specified locations

These criteria provide an "engineering checklist" that applies to all platforms – trapped ions, superconducting circuits, quantum dots, etc. If all first five criteria are met, and if the decoherence times lay above a certain threshold, we can provably build arbitrarily large quantum computers, capable of any quantum computation we can think of, using error correction and fault-tolerant principles to make those computations with arbitrary precision.

In this chapter, we will go through the elements in DiVincenzo's checklist. We will use the tools and knowledge gathered in Chapters 6 and 7, and complete them with additional techniques for implementing gates, architectural decisions, characterizing coherence and fidelity, etc. The roadmap is as follows. In Section 8.2, we introduce the most common forms of quantum registers, how they are measured and reset, thus covering checklist items (1), (2) and (5). Section 8.3 discusses item (4) and how arbitrary computations can be approximated using a small set of quantum operations, a universal gate set. We show that those gates include single-qubit operations – already covered in Chapter 6 – and at least one universal two-qubit operation, of which we offer various designs. Section 8.4 discusses the characterization of errors in the quantum computer, with an improved version of item (3). Section 8.5 provides some insight on how having low error rates allows us to implement error-correction strategies, opening the door to arbitrarily large and arbitrarily complex quantum computations.

8.2 Quantum Registers

We introduced in Section 6.1.1 the idea of a physical qubit as a real system with two long-lived quantum states that can exist in an arbitrary superposition of those states. A quantum register is a collection of qubits – physical if we refer to an experiment, or logical if we are focused on the algorithmic representation – that can be subject to the unitary operations and measurements of a quantum circuit model.

A physical implementation of a quantum register is conditioned by various experimental choices:

- The choice of physical system (i.e., circuit) that implements a qubit
- How those qubits are locally addressed and measured
- The quantum operations that we can implement on those qubits
- The topology of those qubits in the experimental setup

The choice of qubit is particularly important, because it conditions all other options in the computer design. One could, for instance, think about using the flux qubit, a device with strong anharmonicities, the potential for very strong interactions and complex qubit–qubit interaction geometries (see Section 9.4.1). However, at the moment this book is being written, all superconducting quantum computers are based on the transmon qubit or variations thereof.

The transmon has three nice properties: It has long coherence times, it is a very reproducible qubit, and it has a flexible design that adapts very well for implementing large quantum registers of 50 and more qubits (Arute et al., 2019), with varying topologies and designs. Figure 8.2a illustrates a quantum computer with nine X-mon qubits, a variation of the transmon, so-called because of their shape. Notice that the qubits are never alone: they are accompanied by lines, microwave resonators, and shielding ground planes, which are needed to protect, manipulate, and interrogate the register. Where to place those elements is an important architectural decision, which these days is coming closer to an industrial secret.

8.2.1 Measurements

The election of the transmon qubit naturally imposes one type of measurement scheme: We must adopt the cavity-mediated, QND measurement that was presented

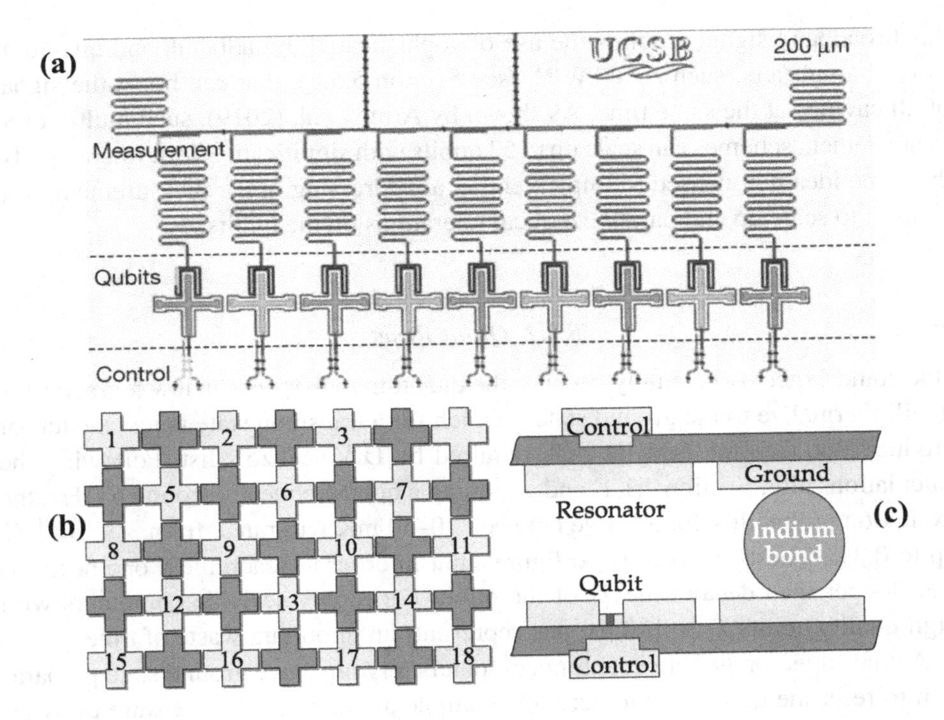

Figure 8.2 (a) Superconducting qubit quantum register built from nine X-mon qubits, a variant of the transmon qubit with a cross shape. The qubits are close to each other for nearest-neighbor interactions; they are connected to individual cavities (wiggly lines on the top) for dispersive qubit readout, and they are tuned by small flux lines coming from the bottom of the picture. Modified from Barends et al. (2014) with permissions. (b) Alternative qubit topology, resembling the work by Arute et al. (2019), with similar qubits, but higher connectivity, and using auxiliary qubits to mediate interactions. (c) Three-dimensional sandwich packing of two chips, joined by superconducting indium bonds. One chip contains resonators and readout channels structures, while the other one packs qubits.

in Section 7.4.2. As shown in Figure 8.2a, each qubit is attached to a readout resonator, which is connected to the outer world. The microwave cavity acts as a narrowband filter that isolates the qubit from the readout cables. However, when we send a resonant microwave pulse through those cables, the cavity reflects the light with a phase that carries information about the state of the qubit. This is a good, nondestructive POVM measurement of the qubit, with a small uncertainty of a few percent, ∼1–3% in current experiments.

If all the resonators that talk to the qubits have slightly different frequencies, we can multiplex the readout phase. Instead of sending a monochromatic pulse that is resonant with *one* cavity, we send a broadband signal with the frequencies of all resonators. Analyzing the reflected signal allows us to gather information about the state of all qubits that were addressed by the probe signal. The readout of this broadband signal requires the use of sophisticated, broadband, and quantum-limited amplifiers, such as a TWPA (see Section 5.5.6), that can boost the signal of all cavities at the same time. As shown by Arute et al. (2019), such multiplexed measurement schemes can scale up to 53 qubits with significant effort. Interestingly, the same idea that is used to implement local addressing in the measurements can be used to scale up and parallelize local operations on the qubits.

8.2.2 Qubit Reset

One could expect that, simply because the quantum register is at a low temperature, it will thermalize to the ground state – which for a transmon state is just the tensor product $|000\ldots\rangle$, the fiducial state required by DiVincenzo's list. Following the calculations after Section 4.1.1, and assuming a qubit frequency around 6 GHz, the excitation probability for a fridge between 10–50 mK will range from $3 \times 10^{-11}\%$ up to 0.3%. These are very good figures, but in order to reach them one needs to wait longer than decay time T_1 of the qubit. For highly scalable computers with high-quality qubits $T_1 \sim 100$ μs, that represents an important waste of time.

Actual superconducting quantum computers rely on active ground state preparation to reset the qubit state to zero. One simple alternative is to measure the state of the qubit. If the qubit is in the desired state $|0\rangle$, one does nothing; otherwise, we apply a π pulse to flip the state from $|1\rangle$ to $|0\rangle$. This scheme does not require any additional tools, but the fidelity of the resulting state is limited by the quality of our measurement, which in existing setups is not very good.

Other methods rely on the readout cavities, using them to shorten the lifetime of the qubit T_1. By bringing the qubit in resonance with the cavity, we enable the transfer of excitations from the $|1\rangle$ state to the resonator, where they may decay into the readout cables, accelerating the reset time. This is the mechanism used by Arute et al. (2019) in their 53-qubit quantum processor, where they achieved a reset time of 10 μs, consistent with the resonator bandwidth. There are other alternatives that use microwave drives to enable the exchange of excitations between the transmon and the readout cavity, even in situations when they are off-resonance (Egger et al., 2018). This type of method achieves good preparation fidelities, around 1% in times below a microsecond (Egger et al., 2018; Magnard et al., 2018).

8.2.3 Architectural Decisions

Even after we have the qubit, readout, and control mechanisms, there are many important choices regarding the physical layout and organization of the quantum register. A very important one concerns the *layout* of the circuit – how qubits, resonators, and control lines are fabricated onto the chip. This layout will inform the *topology* of the circuit, establishing which qubits can interact with others, directly or through mediating circuits.

Take for instance the one-dimensional layout from Figure 8.2a. Qubits are placed on a row and are in close proximity to each other. Each qubit has two nearest neighbors, which are capacitively coupled to it. Resonators are moved away to a separate row and are collectively addressed, and there is a third row of direct controls on the qubits. This arrangement is clean, provides direct access to all qubits, and minimizes the number of qubits that see each other, which is good to lower cross-talk.

Unfortunately, the one-dimensional design does not scale well in applications. Quantum algorithms require interactions between arbitrary pairs of qubits. These interactions can be decomposed into other gates, but the decomposition becomes very costly in one-dimensional topologies. For instance, we can make a quantum operation between qubits 1 and 9 only by inserting gates between 1 and 2, 2 and 3, and so on. If we follow this approach, it becomes clear that we are constrained by the *taxi-driver metric*[2] of how far two arbitrary qubits lie in a given lattice.

There are also other considerations. For instance, quantum error correcting architectures have been designed with certain lattice topologies in mind. Figure 8.2b illustrates an arrangement of physical qubits on a square lattice, where each qubit has four neighbors. This topology is required for error correcting algorithms based on the surface code, as we explain in Section 8.5.4. However, moving from a simple planar architecture (cf. Figure 8.2a) to a dense lattice (cf. Figure 8.2b) raises the question of where to place all other components – resonators, control lines – that do not fit in the space between qubits.

The solution to this problem comes from an early work by O'Brien et al. (2017), which looked at how to shield qubits and protect them from decoherence. This work introduced a sandwich architecture, where two chips are bonded together using tiny indium contacts. One chip contains the qubits, the other one acts as a shield against electromagnetic fields, providing better coherence. Since Rigetti's work, there has been steady progress in using the same sandwich architecture, but transferring resonators and control lines to the top chip.

The idea is sketched in Figure 8.2c, using IBM's patent as inspiration (Brink et al., 2019). The qubit is fabricated on one slab, together with ground planes and control lines – which also need some 3D fabrication, because they would not fit among qubits. Resonators and readout lines are constructed on a different chip. Both chips are attached mechanically to each other, and contacted through a small "ball"

[2] The taxi-driver metric evaluates the distance between two nodes in a lattice by counting the smallest number of edges that make a path connecting both nodes.

of indium. This superconducting ball deforms and sticks to the aluminum, establishing an electrical bridge. In this sketch, the indium bond assists in creating a common ground plane for both chips, but it is thick enough that keeps both layers well separated – creating what is known as an *air gap*.

8.3 Gate Toolbox

In the quantum circuit model, a computation is decomposed into three phases: register preparation, unitary evolution, and measurement. We have seen how the first and last step are done. We will now explain how the most complicated phase, an arbitrary unitary evolution, is decomposed into smaller, manageable operations that can be implemented in the lab.

8.3.1 Universal Set of Gates

A general quantum computation may require the implementation of an arbitrary unitary operation in the Hilbert space of the quantum register. Just from the shape of the unitary matrix itself, we appreciate the exponential number of parameters that inform the operation – 2^{2N} complex amplitudes for N qubits! These arbitrary numbers are produced by the quantum computing expert almost at random, drawing from the set of algorithms, without connection to any experimental protocol, and with no idea of how to implement them in the laboratory.

Fortunately, the circuit model builds on a solid mathematical foundation, which takes care precisely of this question: the decomposition of arbitrary computations into elemental gates. This is made possible by the notion of a *universal set of quantum gates*. We call by this name any finite set of unitary operations $\{U_1, U_2, \ldots, U_M\}$ with which we can approximate global unitary W transforming the N-qubit register space to arbitrary precision $W \simeq \prod_{n=1}^{K_\varepsilon} U_{\sigma_n} + \mathcal{O}(\varepsilon)$.

Fortunately for us, there exists a very powerful theorem that provides two instances of a universal gate set. This theorem states that any unitary transformation W acting N qubits can be approximated to arbitrary accuracy using a *universal set of quantum gates* made from the following:

(1) Arbitrary unitary rotations of individual qubits
(2) One universal two-qubit gate acting on any pair of qubits, such as the CNOT, CZ, iSWAP, and $\sqrt{\text{SWAP}}$ from Table 8.1

The procedure, which can be found in the book by Nielsen and Chuang (2011), may be further simplified, replacing the first step with

(1) Hadamard gates and $\pi/8$ gates acting on individual qubits

Sometimes, these decompositions are easy to find, with patience and a bit of practice. Take for instance the algorithm in Figure 8.1a for constructing a five-qubit gigahertz state $\frac{1}{\sqrt{2}}(|00000\rangle + |11111\rangle)$, or look at the well-known algorithms for a quantum Fourier transform, phase estimation, factoring, etc. More often, however, if we try to apply these ideas to general problems, such as simulating the evolution of a quantum material or a molecule, we may not have a practical decomposition to use. This may easily happen because we cannot explicitly compute the matrix elements of the unitary to simulate, or because the decompositions we find require an exponentially growing number of operations.

With these caveats in mind – which simply highlight the need for more research in the theory of quantum algorithms – the fact is that we have decomposed DiVincenzo's fourth requirement into two very well-defined tasks: implementing single-qubit operations and finding a protocol to create a two-qubit universal gate. We will now explore the theoretical and experimental solutions to these problems.

We do not need to spend too much time discussing the design of local operations. This matter was largely addressed in Chapter 6. In Section 6.1.4, we introduced the continuum of single-qubit transformations in the Bloch sphere (6.8), and how it can be implemented using an external microwave drive (6.6) with tunable amplitude, frequency, and phase. Varying these parameters, and adjusting the duration of the pulse, we showed how to implement the Z, S, T, and H gates from Table 8.1, and provided evidence that *any other local gate* can also be implemented.

Single-qubit gates are the fastest and most precise operations that are performed in superconducting quantum circuits, with fidelities around 0.09% in state-of-the-art experiments (Arute et al., 2019) – which are impressive, but are still above the 2×10^{-5} fidelities achieved in trapped ions by Brown et al. (2011).

Two-qubit gates, on the other hand, are not so simple and also not so precise. There are many choices of universal two-qubit gates, some of which are shown in Table 8.1. We will discuss three designs, which are based on different ways to make the qubits interact, or to control their dynamics.

8.3.2 Two-Qubit Exchange Gates (iSWAP)

In this section, we discuss the two-qubit gates generated by off-diagonal dipolar interactions, such as those found in capacitively coupled transmons (see Section 6.5.2). We start from the Ising model with transverse coupling, shown in (6.67). When the interaction term is usually much weaker than the qubit gap,

$|J| \ll |\Delta_i|$, an RWA-type argument allows us to replace this Hamiltonian with a simple exchange model (see Problem 6.16):

$$H_{\text{eff}} = \frac{\Delta_1}{2}\sigma_1^z + \frac{\Delta_2}{2}\sigma_2^z + J(\sigma_1^+\sigma_2^- + \sigma_1^-\sigma_2^+) \tag{8.1}$$
$$= \begin{pmatrix} \frac{\Delta_1+\Delta_2}{2} & 0 & 0 & 0 \\ 0 & \frac{\Delta_1-\Delta_2}{2} & J & 0 \\ 0 & J & -\frac{\Delta_1-\Delta_2}{2} & 0 \\ 0 & 0 & 0 & -\frac{\Delta_1+\Delta_2}{2} \end{pmatrix}.$$

The two eigenstates, $|0,0\rangle$ and $|1,1\rangle$, are preserved throughout evolution, accumulating the same local phases they would acquire in absence of interaction. When $\Delta_1 = \Delta_2$, the one states $|0,1\rangle$ and $|1,0\rangle$ are resonant and can exchange their only excitation:

$$U(t) = e^{-iH_{\text{eff}}t} = e^{-it\Delta(\sigma_1^z+\sigma_2^z)} \begin{pmatrix} 1 & 0 & 0 & 0 \\ 0 & \cos(Jt) & -i\sin(Jt) & 0 \\ 0 & -i\sin(Jt) & \cos(Jt) & 0 \\ 0 & 0 & 0 & 1 \end{pmatrix}. \tag{8.2}$$

Evolution with this Hamiltonian for a brief period of time will result in the joint action of three commuting operations: local rotations with the phases Δt, an exchange of probability between the $|0,1\rangle$ and $|1,0\rangle$, and an interaction phase in the single-excitation subspace, generated by $\sigma_1^z\sigma_2^z$. This combination of operations is particularly useful to simulate fermionic models in the quantum computer, such as in quantum chemistry applications. A particular instance of this evolution is $Jt = 3\pi/2$, which produces a gate equivalent to the iSWAP gate, up to local transformations:

$$U(\pi/2J) = e^{i\pi\Delta(\sigma_1^z+\sigma_2^z)/2J}\, U_{\text{iSWAP}}. \tag{8.3}$$

The iSWAP is a universal operation – two iSWAP gates can build a CNOT gate (Williams, 2011) – although it is less used in practical applications.

These exchange-type gates appear in all setups that allow for off-diagonal dipolar interactions. They can be implemented in quantum computers with directly coupled transmons – as in Figure 8.2 – but they also appear in devices with circuit-mediated interactions. The work by Majer et al. (2007) used the dispersive coupling between qubits in a superconducting cavity from Section 7.3.7 to implement the whole family of rotations enabled by the exchange term, including the iSWAP. More recently, we find these gates in the Google Sycamore quantum processor, which relies on detuned transmons to mediate the interaction between qubits in the quantum register.

Their setup can recreate various dipolar interactions, including the exchange term from (8.1), which they used to implement iSWAP two-qubit gates, as a foundation for their 53-qubit quantum supremacy experiment (Arute et al., 2019).

8.3.3 Two-Qubit Tunable Frequency CZ Gate

Even though the iSWAP is a very natural choice for superconducting circuits, we need to explore other gates that are more commonly used. One of them is the CZ, control-Z, CPHASE or control-phase operation, shown in Table 8.1. This gate and the CNOT are equivalent up to two local rotations (see Problem 8.2). The CZ belongs to a the family of *two-qubit phase gates,* which are diagonal gates defined by four phases $\theta(s_1, s_2)$:

$$U\,|s_1, s_2\rangle = e^{i\theta(s_1,s_2)}\,|s_1, s_2\rangle\,. \tag{8.4}$$

A two-qubit phase gate is a maximally entangling and universal gate when

$$\Delta\theta = \frac{1}{4}\{\theta(1,1) + \theta(0,0) - [\theta(0,1) + \theta(1,0)]\} = \frac{\pm\pi}{4} \bmod 2\pi\,. \tag{8.5}$$

All these two-qubit phase gates are equivalent to each other, up to single-qubit transformation and global phases. A popular gate in this set is the CZ or control-Z operation from Table 8.1 (cf. Exercise 8.4):

$$U_{\mathrm{CZ}} = \sum_{s_1,s_2} e^{\pm i\pi s_1 s_2}\,|s_1, s_2\rangle\langle s_1, s_2| = \exp\left[\frac{i\pi}{4}(\sigma_1^z + 1)(\sigma_2^z + 1)\right].$$

The CZ may be generated by evolution with an interaction of the form $\sigma_1^z\sigma_2^z$. Unfortunately, the coupling terms that we have found in Section 6.5 are not diagonal in the qubit basis and cannot be directly used for this purpose.

A common way to implement phase gates is to do an *adiabatic passage* or adiabatic deformation of the Hamiltonian that enhances the strength of interactions for some period of time. Our system is built with a control parameter ω, which tunes the Hamiltonian $H(\omega)$. We assume that the control begins and ends at the same point, $\omega(0) = \omega(T) = \omega_0$, at which $H(\omega_0)$ approximates to two noninteracting qubits with a well-defined basis $|s_1, s_2\rangle$. We design the passage so that around the middle point $\omega(T/2)$, the interaction between qubits is dominant.

If the states of the two qubits are never degenerate when we change ω and if we modify the control $\omega(t)$ slowly (adiabatically) enough, every two-qubit eigenstate $|s_1, s_2\rangle$ of $H(\omega_0)$ will be mapped to the instantaneous eigenstates $|n_\omega\rangle$ of the Hamiltonian and return back to their original configuration. Those states only experience some dynamical and geometric phases $|s_1, s_2\rangle\, e^{i\theta(s_1,s_2)}$, which can be derived from the instantaneous eigenenergies they experienced during the passage:

$$\theta(s_1, s_2) = -\frac{1}{\hbar}\int_0^T \mathrm{d}t\, E(n_{s_1,s_2}, \omega(t)). \tag{8.6}$$

This idea was first used by DiCarlo et al. (2009) to implement a CZ gate for transmon qubits in a cavity. It was later perfected by Barends et al. (2014) for transmons that interacted directly with each other, in the design from Figure 8.2. The CZ adiabatic gate requires two qubits with slightly different frequencies, $\Delta_2 > \Delta_1$, and our control is the frequency of the second qubit $\Delta_2 = \omega(t)$. The transmons have an always-on exchange interaction (cf. Section 6.5), either direct or mediated by the cavity. This interaction is weaker than the anharmonicities $\alpha_{1,2}$ of both transmons along the passage.

As explained in García-Ripoll et al. (2020), we can analyze the gate in the reduced space $\{|0,0\rangle, |1,0\rangle, |0,1\rangle, |1,1\rangle, |0,2\rangle\}$ of the qubit basis plus a state $|0,2\rangle$ with two excitations in the second transmon. The effective Hamiltonian in this subspace reads

$$H_{\text{eff}}(t) = \begin{pmatrix} 0 & 0 & 0 & 0 & 0 \\ 0 & \hbar\Delta_1 & J & 0 & 0 \\ 0 & J & \hbar\Delta_2 & 0 & 0 \\ 0 & 0 & 0 & \hbar\Delta_1 + \hbar\Delta_2 & \sqrt{2}J \\ 0 & 0 & 0 & \sqrt{2}J & 2\hbar\Delta_2 - \hbar\alpha_2 \end{pmatrix}. \tag{8.7}$$

The gate protocol starts at a point of large detuning $\Delta_2 - \Delta_1 \gg |J|$, shown in Figure 8.3. The qubits almost do not see each other, we can neglect the couplings

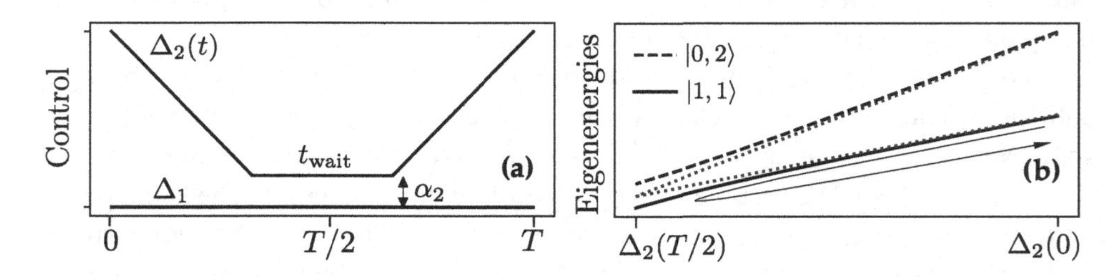

Figure 8.3 Phase gate with two transmons with different energy gaps. (a) We tune the frequency of the second transmon $\omega_2(t)$, from a largely detuned value down to the resonance $\omega_2 = \omega_1 + \alpha_2$ between the $|0,2\rangle$ and $|1,1\rangle$ states. (b) Eigenenergies of the tuned Hamiltonian, for levels that begin and end at the states $|1,1\rangle$ (solid) and $|0,2\rangle$ (dashed). The $|1,1\rangle$ state is pushed downward from the eigenenergies of the noninteracting model ($J = 0$, dotted) by an amount $\delta E_{\text{int}}(t)$.

J, and the qubit basis is almost an eigenstate of H_{eff}. The gate proceeds by lowering down Δ_2 close to the point $\Delta_2 = \Delta_1 + \alpha_2$ and back to its original value (cf. Figure 8.3a). If the tuning of the frequency is adiabatic enough and $|J| \ll \Delta_2 - \Delta_1$, states $|0,0\rangle$, $|1,0\rangle$, $|0,1\rangle$ remain approximately invariant throughout the gate.

The qubit register $|1,1\rangle$, on the other hand, experiences an interaction with the $|0,2\rangle$ state that cannot be neglected. Throughout the evolution, $|1,1\rangle$ is therefore mapped to an adiabatically deformed eigenstate, with a trajectory shown in Figure 8.3b as a solid line. Halfway through the passage, it reaches the superposition state $\frac{1}{\sqrt{2}}(|1,1\rangle - |0,2\rangle)$ around the degeneracy point $\Delta_2 = \Delta_1 + \alpha_2$. This deformation also changes the instantaneous eigenenergy, which is shifted downward $\hbar\Delta_1 + \hbar\Delta_2 - \delta E_{\text{int}}(t)$, with a maximum value $\delta E_{\text{int}} \sim 2\sqrt{2}J$ right at the avoided level crossing.

The instantaneous eigenenergies of the various adiabatic passages manifest in different dynamical phases for each of the two-qubit states. The three lowest energy states acquire phases that are noninteracting in nature:

$$\theta(s_1, s_2) = -\int_0^T (\Delta_1 s_1 + \Delta_2(t) s_2)\mathrm{d}t. \tag{8.8}$$

These are phases similar to the ones generated by the single-qubit Hamiltonian $\frac{1}{2}\hbar\Delta_1\sigma_1^z + \frac{1}{2}\hbar\Delta_2(t)\sigma_2^z$. The $|1,1\rangle$ state also acquires this local phase rotation, but it also experiences the energy shift caused by the interaction:

$$\theta(1,1) = \theta(0,1) + \theta(1,0) - \frac{i}{\hbar}\int_0^t \delta E_{\text{int}}(t)\mathrm{d}t. \tag{8.9}$$

By adjusting the detuning $\Delta_2(t)$ (total duration, waiting time around degeneracy, ramp profile, etc.), one may engineer a very good CZ gate that satisfies (8.5). In addition to this, it is also possible to tune the rotation angle to produce other control-phase gates, which may be useful in algorithms such as the quantum Fourier transform (see Problem 8.4).

As shown experimentally (Barends et al., 2014) and theoretically (García-Ripoll et al., 2020), this gate can be very fast. It does not need to be implemented adiabatically. The only requirements for the gate to succeed are (i) minimizing leakage and (ii) ensuring a large interaction phase. For the first condition, we need to ramp down the qubit slow enough as to prevent transitions outside the two-excitation subspace. Fortunately, this is limited by the anharmonicity, and happens already at the speed of local gates ($T \gg 1/|\alpha|$). The second condition only involves tuning the waiting time, so that the integral $\int \delta E_{\text{int}}$ becomes large enough. Once more, this also happens in a relatively short time, of the order of the energy splitting $2\sqrt{2}J$, with gates as short as tens of nanoseconds.

8.3.4 Two-Qubit Tunable Coupling CZ Gate

The tunable-frequency CZ gate has been successfully demonstrated by various laboratories, but it has three important drawbacks, which arise from our use of the qubit detuning to modulate effective interactions. The first problem is the frequency crowding phenomenon discussed in Section 6.5.3, which makes it very hard to have a large number of qubits, parked at sufficiently distanced frequencies with negligible interactions. The second problem is that, even for weak couplings, those off-resonant interactions manifest as residual couplings that fundamentally limit the achievable fidelity. The final problem is that during the frequency ramp, qubits may accidentally cross degeneracies with other neighbors, not the ones we want them to interact with, a phenomenon known as *cross-talk*.

An elegant solution to this architectural nightmare is the use of tunable couplers, devices that mediate the interaction and allow us to switch it on and off, without the need of moving the qubit's frequencies (see Section 6.5.4). In the work by Chen et al. (2014), a SQUID mediates the coupling between nearby transmons, in a scheme that allows not only for a dipolar coupling, but also for canceling spurious residual interactions.

Using this setup, it is possible to implement a CZ gate that builds on the $|11\rangle$–$|02\rangle$ interaction, as before. Before and after the gate, the coupling is effectively canceled, so that the qubit frequencies do not really matter, and can be optimized for minimum decoherence (Arute et al., 2019). However, during the operation of the gate, the qubits will be parked at resonant frequencies, $\Delta_2 = \Delta_1 - \alpha_2$, and the coupling will be switched on and off, from $J(0) = 0$ up to a maximum coupling $J_{\max} \ll \alpha_2$ and back.

The model at hand is identical to the one before (8.7). We neglect the interaction on the states $|0,0\rangle$, $|0,1\rangle$, and $|1,0\rangle$, which remain instantaneous eigenstates of $H_{\text{eff}}(t)$ and only acquire local phases (8.8). In the two-excitation subspace, however, the instantaneous eigenstates are different:

$$|\pm\rangle = \frac{1}{\sqrt{2}}|1,1\rangle - \frac{1}{\sqrt{2}}|0,2\rangle . \tag{8.10}$$

The state $|1,1\rangle$ suffers a different, but still quasi-adiabatic dynamics, that depends on the energy splitting between these eigenstates $2 \times \delta E_{\text{int}}(t)$:

$$U(t)\,|1,1\rangle = \frac{1}{\sqrt{2}}e^{-i(\omega_1+\omega_2)t} \times \left(e^{-i\theta_{\text{int}}(t)}\,|1,1\rangle - e^{+i\theta_{\text{int}}}\,|0,2\rangle\right), \tag{8.11}$$

$$\theta_{\text{int}}(t) = \int_0^t \delta E_{\text{int}}(\tau)\mathrm{d}\tau .$$

If we tune the passage so that $\theta_{\text{int}}(T) = \pm\pi \bmod 2\pi$, we obtain a CZ gate. As demonstrated by Chen et al. (2014), the gate can be implemented very rapidly, in a time $1/J \sim 30$ ns, with very small leakage outside computational space and reasonable fidelity.

8.4 Tomography and Error Characterization

We have discussed four out of the five essential ingredients in our shopping list for a quantum computer, but we have not elaborated much on how errors in those items impact the scalability of the computer, or even how to detect those errors. In this section, we are going through that list again, identifying sources of imperfection. Some of them have already been brought up in earlier chapters, but we will now also derive general mathematical models for those errors and tools to characterize them.

8.4.1 Classes of Errors

DiVincenzo's third requirement is actually insufficient. The checklist refers to "long decoherence times," assuming that all errors in the design come from decoherence. Instead, we must consider four types of errors that pop up in a circuit model quantum computer.

- **State preparation.** These are the errors we make when initializing the quantum register. As mentioned earlier, thermalization by itself is not fast enough, and we need to couple our qubits to other items (cavities, controls, measurement devices) and apply complex protocols to ensure a fast reset. State preparation errors account for deviations from the ideal zero state $|0\rangle^{\otimes N}$, which, as we saw earlier, can be around 1–2%.
- **Measurement errors.** The precision of measurements in quantum computers is limited. Limited integration time, measurement induced decoherence, and amplifier and detector noise are some circumstances that require us to model measurements as POVMs with intrinsic uncertainties. The *measurement fidelity* is the probability that, given a certain outcome of our device, the measured qubit was in that particular state. Circuit-QEDs' fast dispersive qubit measurements seem to have relatively good fidelities, with around ~99–99.8% fidelity or success probability (see Barends et al., 2014; Jeffrey et al., 2014; Walter et al., 2017). This is good, but not yet so good that we can use unlimited repeated measurements on the same qubit.
- **Qubit decoherence**. We have studied the loss of coherence and quantum superpositions due to environment-induced relaxation, heating, and dephasing.

The models from Sections 6.1.6 and 6.1.5 are good descriptions when these process are uncorrelated – i.e., when qubits talk to separate environments in Markovian processes that degrade the purity of the states over time. Those models are simply characterized by the relaxation and dephasing times, T_1 and T_2, for each qubit. However, sometimes this characterization is insufficient, because decoherence is correlated – e.g., fluctuations of a surrounding magnetic field or control cables affect various qubits similarly – or because there is residual *cross-talk*, unwanted interactions between the qubits underlying this loss of information.

– **Gate imperfections.** Single-qubit and two-qubit unitaries are implemented using microwave drives in combination with magnetic fields that tune the frequencies and couplings of qubits. The operation of these gates is affected by intrinsic errors as well as by our limited capacity on stabilizing and crafting the controls. A π-pulse flip of the qubit will have errors if we fail to give the microwave pulse the right amplitude, phase, or duration, and also if the frequency of the qubit fluctuates slightly due to environment-induced dephasing. The quality of our gates is measured in terms of a *gate fidelity,* which is the probability that we implement the right operation or, alternatively, the mathematical overlap between the experimental and the desired outcome.

There is nothing fundamentally wrong in quantum states being subject to errors and imperfections. Our day-to-day electronics and devices are subject to similar issues. Some causes are fundamentally unavoidable, such as the soft errors induced by cosmic rays that hit our computers every few hours. However, engineers have learned to detect and correct those errors, making semiconductor chips a reality with immense scaling power. It is now well understood, and one of the fundamental pillars of the quantum computation theory, that we can follow a similar roadmap: (i) understanding and modeling the errors that take place in our computer, (ii) quantifying the intensity of those errors, and (iii) developing error-correcting and fault-tolerant operation strategies.

8.4.2 Error Models: Completely Positive Maps

We use the formalism of *trace-preserving completely positive maps* to represent the errors and imperfect operations in a quantum computer and any quantum information process. We introduced this representation when studying qubit dephasing (Section 6.1.5) and relaxation (Section 6.1.6). These maps are linear transformations $\varepsilon(\rho)$ of density matrices into density matrices (cf. Section B.1), and thus represent the most general physical operations that are allowed by quantum mechanics.

Table 8.2. *Completely positive maps for useful error models.*

Bit flip	$\varepsilon(\rho) = p\rho + (1-p)\sigma^x \rho \sigma^x$
Phase flip	$\varepsilon(\rho) = (1-p)\rho + p\sigma^z \rho \sigma^z$
Bit flip	$\varepsilon(\rho) = p\rho + (1-p)\sigma^x \rho \sigma^x$
Depolarizing channel	$\varepsilon(\rho) = (1-3p/4)\rho + p/4\sigma^x \rho \sigma^x$
Totally depolarizing channel (asymmetric)	$\varepsilon(\rho) = \left(1 - \sum_{\alpha=x,y,z} p_\alpha\right)\rho + \sum_{\alpha=x,y,z} p_\alpha \sigma^\alpha \rho \sigma^\alpha$
Totally depolarizing channel	$\varepsilon(\rho) = (1-p)\,\rho + p\frac{\mathbb{1}}{\mathrm{d}}$
Amplitude damping	$A_1 = \lvert 0\rangle\langle 0\rvert + \sqrt{1-\gamma}\,\lvert 1\rangle\langle 1\rvert,\ A_2 = \sqrt{\gamma}\,\lvert 0\rangle\langle 1\rvert$
Phase damping	$A_1 = \lvert 0\rangle\langle 0\rvert + \sqrt{1-\gamma}\,\lvert 1\rangle\langle 1\rvert,\ A_2 = \sqrt{\gamma}\,\lvert 1\rangle\langle 1\rvert$

A trace-preserving completely positive map is a linear superoperator that transforms density matrices into normalized density matrices – i.e., probability is conserved, and we do not discard events – which admits a decomposition into Kraus operators:

$$\varepsilon(\rho) = \sum_i A_i \rho A_i^\dagger. \tag{8.12}$$

The Kraus operators satisfy $\sum_i A_i A_i^\dagger = \mathbb{1}$, because they produce normalized density matrices, and we may have up to 2^N of them.

In the ideal operation of a quantum computer, a positive map can be just a unitary transformation $\varepsilon(\rho) = U\rho U^\dagger$, describing a quantum gate with a single Kraus operator $A = U$. In the most general case, a physical map will describe an imperfect gate, decoherence of a quantum register, or even the loss of information during measurements.

Quite often, we make physically reasonable assumptions about the underlying errors. A common one is to separate the ideal gates from the errors, as if they were independent processes[3] $\varepsilon_{\text{error}}(U\rho U^\dagger)$. Quite often, we also assume that those errors decompose into *spatially uncorrelated* errors acting on the idle qubits, and some average error that is *independent of the operation,* acting only on the qubits that are manipulated. Moreover, in the description of the spatially uncorrelated errors, we also resort to a family of models, summarized in Table 8.2, that are physically appealing and have simple parameterizations.

These simpler models and views are needed to develop intuition, understand the experiments, and make progress in the development of error-correction strategies. Without simplification, completely positive maps are humongous objects: For a

[3] For unitary operations, this is always possible, at least formally.

system of N qubits, the most general map is described by a matrix with about $d^4 = 2^{4N}$ parameters. To see that, let us examine the Bloch sphere representation of a density matrix, introducing a complete set of observables formed by all tensor products of Pauli matrices[4] $S_\alpha \in \{1, \sigma^x, \sigma^y, \sigma^z\}^{\otimes N}$. The 2^{2N} operators form a complete orthogonal basis with respect to the matrix scalar product $\mathrm{tr}(S_\alpha^\dagger S_\beta) = d\delta_{\alpha,\beta}$. This allows us to expand any density matrix:

$$\rho = \mathbf{s}^T \cdot \mathbf{S} = \sum_{\alpha=1}^{d^2} s_\alpha S_\alpha, \quad \text{with } s_\alpha = \frac{1}{d}\mathrm{tr}\left(\frac{1}{d} S_\alpha^\dagger \rho\right) = \langle S_\alpha \rangle_\rho . \tag{8.13}$$

A general completely positive map $\varepsilon(\rho)$ is a linear transformation of the generalized Bloch vector, represented by a matrix $M \in \mathbb{R}^{4^N \times 4^N}$ with about 16^N independent coefficients:

$$\varepsilon(\rho) = (M\mathbf{s})^T \cdot \mathbf{S} = \frac{1}{d} \sum_{\alpha,\beta=1}^{d^2} M_{\alpha,\beta} \mathrm{tr}(S_\alpha^\dagger \rho) S_\beta. \tag{8.14}$$

There are other alternative representations for the completely positive maps in an experiment. The χ matrix (Gilchrist et al., 2005; O'Brien et al., 2004) expresses the superoperator in a complete basis S_α:

$$\varepsilon(\rho) = \sum_{\alpha,\beta=1}^{d^2} \chi_{\alpha\beta} S_\alpha \rho S_\beta^\dagger. \tag{8.15}$$

For physical maps, the matrix χ is positive semidefinite and can be diagonalized $\chi = VV^\dagger$, producing one of many equivalent definitions for the Kraus operators $A = \sum V_\alpha S_\alpha$ (Neeley et al., 2008). This representation is linearly related to the Pauli transformation matrix M:

$$M_{\alpha\beta} = \sum_{\gamma,\delta=1}^{d^2} \frac{1}{d} \mathrm{tr}(S_\alpha S_\gamma S_\beta S_\delta) \chi_{\gamma\delta}, \tag{8.16}$$

and equivalent from any formal or operational point of view.

A third representation is based on the *Choi–Jamiolkowski isomorphism* between quantum processes and quantum states. The *Choi matrix* for a quantum process $\varepsilon(\rho)$ is obtained by applying the positive map onto one subsystem of a maximally entangled state $|\Phi\rangle = \sum_i \frac{1}{\sqrt{d}} |ii\rangle$:

$$\rho_\varepsilon := (\mathbb{1} \otimes \varepsilon)(|\Phi\rangle\langle\Phi|) = \sum_{\alpha,\beta} M_{\alpha\beta} S_\beta^T \otimes S_\alpha = \sum_{\alpha,\beta} \chi_{\alpha\beta} |\alpha\rangle\langle\beta| . \tag{8.17}$$

[4] For two qubits, for instance, this would be formed by all operators in the set of 16 combinations: $\{\mathbb{1}, \sigma_1^x, \sigma_1^y, \sigma_1^z, \sigma_2^x, \sigma_2^y, \sigma_2^z, \sigma_1^x\sigma_2^x, \sigma_1^x\sigma_2^y, \sigma_1^x\sigma_2^z, \sigma_1^y\sigma_2^x, \sigma_1^y\sigma_2^y, \sigma_1^y\sigma_2^z, \sigma_1^z\sigma_2^x, \sigma_1^z\sigma_2^y, \sigma_1^z\sigma_2^z\}$. Note how these operators are both Hermitian and unitary.

It is linearly reated to two other positive map representations, M (8.14) and χ (8.15), through the convenient entangled basis $|\alpha\rangle = \sum_i (\mathbb{1} \otimes S_\alpha) |ii\rangle / \sqrt{d}$.

Quantum process tomography is the reconstruction of the superoperator associated to a physical map, $\varepsilon(\rho)$ or M. The tomography protocol uses a complete set of observables S_α together with a complete set of states constructed from them:[5] $\rho_k = B_{k\alpha} S_\alpha$. The protocol involves preparing the 2^{2N} or more input states ρ_k and, for each possible input, measuring the 2^{2N} observables S_α multiple times.

The outcome of a tomography experiment is a set of real numbers that are unbiased estimators of the expectation values $\lambda_{\alpha,k} \simeq \langle S_\alpha \rangle_{\varepsilon(\rho_k)}$ that must be related to the positive map ε. If there was no statistical uncertainty, we establish a linear relation between the estimators and the matrix M:

$$\lambda_{\alpha,k} = \mathrm{tr}(S_\alpha \varepsilon(\rho_k)) = M_{\alpha,\beta} B_{\beta,k}. \tag{8.18}$$

In practical situations, this equation cannot be solved, because the statistical uncertainties in $\lambda_{\alpha,k}$ create instabilities that produce unphysical maps.

A more robust strategy is to realize that for M to be physically acceptable, the Choi matrix $\rho_\varepsilon := \sum_{\alpha,\beta} M_{\alpha,\beta} S_\beta^T \otimes S_\alpha$ must be positive semidefinite (Merkel et al., 2013). The technique of *maximum likelihood estimate* (MLE) is a convex optimization method that finds the superoperator M that best approximates all experimental estimates – e.g., the expectation values $\lambda_{\alpha,k}$ – while satisfying the physical constraints for this matrix (see supplementary material for Chow et al., 2012).

Standard quantum process tomography is a costly and delicate process. It has been used to study the process tomography of free evolution and local gates for one or two qubits (Chow et al., 2009, 2012; Yamamoto et al., 2010), using both the M or χ matrix representation. In either case, the tomography does not scale well for larger setups. It involves multiple repetitions of 2^{4N} experiments, and a very delicate handling of uncertainties and systematic errors. Moreover, as shown by Merkel et al. (2013), *state preparation and measurement* (SPAM) errors can make the reconstruction process unstable, or at least produce unreasonable estimates of the tomographic errors. This motivates the search for alternative characterization of errors, which are less general and do not care so much about the error model, but focus on the intensity or strength of those errors.

8.4.3 Error Quantification: Fidelity

One way to quantify the relevance of noise and decoherence in an experiment is to use a metric that tells us how far our states and quantum processes are from

[5] For instance, a qubit has $S \in \{\mathbb{1}, \sigma^x, \sigma^y, \sigma^z\}$, and we could use the overcomplete set of pure states $\rho_k \in \{\frac{1}{2}(\mathbb{1} \pm \sigma^z), \frac{1}{2}(\mathbb{1} \pm \sigma^x), \frac{1}{2}(\mathbb{1} \pm \sigma^y)\}$ as reconstruction basis (Chow et al., 2012).

their ideal targets. The natural measure for quantum states is the *fidelity*, a number between 0 and 1 determining the probability that we prepared the desired quantum state or reference. Given any two states, described by density matrices ρ and σ, their mutual fidelity is

$$F(\rho,\sigma) = \left[\mathrm{tr}\sqrt{\rho^{1/2}\sigma\rho^{1/2}}\right]^2 \in [0,1]. \tag{8.19}$$

Note how this expression simplifies to the overlap squared for pure states $F(|\psi\rangle, |\phi\rangle) = |\langle\psi|\phi\rangle|^2$. It is clear from this particular case that a perfect fidelity $F = 1$, read as "100% fidelity," is only achieved for identical states, up to global phases. The *infidelity* $\epsilon = 1 - F(\rho,\sigma)$ measures the error in approximating σ by ρ and is related to their separation.[6]

The notion of fidelity can be extended to work with quantum processes in different ways. For instance, say that we wish to compare an ideal unitary operation U to the actual process that takes place in the lab $\varepsilon(\rho)$. Formally, we could apply $\varepsilon(\rho)$ to different pure states $|\psi\rangle$ and estimate the average probability that we obtain the right answer – i.e., the state fidelity between $U\,|\psi\rangle$ and $\varepsilon(|\psi\rangle\langle\psi|)$. This leads to the *average gate fidelity*:

$$F_{\mathrm{avg}}(\varepsilon, U) = \int \langle\psi|U^\dagger\varepsilon(|\psi\rangle\langle\psi|)U|\psi\rangle\, \mathrm{d}\psi. \tag{8.20}$$

This integral is computed by averaging over all possible pure states, sampled with a uniform (Haar) distribution.

The average gate fidelity works with general positive maps, but as argued by Gilchrist et al. (2005), it lacks stability under composition with other trivial operations.[7] The solution to this problem is given by the *process fidelity* or *entanglement fidelity*, a quantifier based on the *Choi–Jamiolkowski* isomorphism between processes and states (8.17).

We introduce the process fidelity of two completely positive maps, using their Choi matrices as proxies for the comparison:[8]

$$F_{\mathrm{proc}}(\varepsilon_0, \varepsilon_1) := \mathrm{tr}\big(\rho_{\varepsilon_0}^\dagger \rho_{\varepsilon_1}\big) = \frac{1}{d^2}\mathrm{tr}\big(M_0^\dagger M_1\big) = \mathrm{tr}\big(\chi_0^\dagger \chi_1\big). \tag{8.21}$$

[6] For normalized pure states, $\|\psi - \phi\|^2 = 2(1 - F^{1/2}) \simeq \epsilon$ for small errors. Note that traditional definitions in the literature chose the geometrical convenience (see Nielsen and Chuang (2011)) $F^{1/2} = |\langle\psi|\phi\rangle|$ as fidelity for pure states. You find this convention in early works in quantum technologies (Steffen et al., 2006), but it has since been abandoned.

[7] Imagine that we put together our system and an uncoupled experiment where we do nothing. We would expect that the auxiliary system should not modify the average of the fidelity, $F_{\mathrm{avg}}(\mathbb{1}\otimes\varepsilon_0, \mathbb{1}\otimes\varepsilon) = F_{\mathrm{avg}}(\varepsilon_0, \varepsilon)$, but this is not true in general!

[8] Originally, the process fidelity was introduced as $F_{\mathrm{proc}}(\varepsilon, U) = \langle\Phi|U^\dagger\varepsilon(|\Phi\rangle\langle\Phi|)U|\Phi\rangle$, to compare positive maps ε with ideal unitary operations (Gilchrist et al., 2005). This formula measures the fidelity of the Choi matrix for the positive map, with respect to the pure state associated to U.

Interestingly, the average gate fidelity and the process fidelity are related:

$$F_{\text{avg}} = \frac{dF_{\text{proc}} + 1}{d+1} = \frac{\text{tr}(M_0^\dagger M_1) + d}{d^2 + d}. \tag{8.22}$$

If an experiment is designed for full process tomography, we can recover both F_{proc} and F_{avg}. This idea has been applied to the study of a qubit under no operations (Neeley et al., 2008) to characterize single-qubit operations (Chow et al., 2009) and to study universal two-qubit gates (Bialczak et al., 2010). Unfortunately, it is difficult to scale up this quantifier to larger numbers of qubits due to exponential growth in experimental setups required for complete process tomography $\sim d^4 \sim 2^{4N}$. This limitation has motivated the quest for simpler experimental protocols that exploit the power of randomized operations to estimate the fidelities of state preparation, measurements, and single-qubit and two-qubit gates.

8.4.4 Randomized Benchmarking

The term *randomized benchmarking* (RBM) denotes a family of experimental methods that estimate the average errors of quantum operations in a quantum register, as well as the combined SPAM errors. There are various incantations of this technique, all of them based on these two principles:

- We can approximate the average over states in (8.20) by perturbing a state $|\psi\rangle$ randomly with Pauli operations that act locally on each qubit – e.g. the set $\{S_\alpha\}$ we introduced before – and averaging over many realizations of these perturbations.
- If there are no systematic errors, our experimental realization of the elementary single- and two-qubit gates is an approximate combination of the ideal gate and an average error map $\varepsilon_{\text{avg}}(U\rho U^\dagger)$. The average error does not depend on the gate, but may be different for single- and two-qubit gates.

A randomized benchmarking protocol uses random Pauli sampling to provide an estimate of the average error or average infidelity associated to a set of gates, $F_{\text{avg}}(\varepsilon_{\text{avg}})$. There are many variations of randomized benchmarking, but most focus on the *Clifford group* $\mathcal{C}$. This is the group that transforms the tensor products of nonidentity Pauli matrices onto other products of Pauli matrices, up to signs and global phases. The group is generated by Hadamard and $\pi/2$ phase rotations on each qubit, plus CNOT gates on any pairs of qubits (Gottesman, 1998). It is of particular interest because it contains the gates that are used in stabilizer-state quantum error correction, but this group is not, by itself, sufficient for universal quantum computation.

A typical RBM protocol follows this algorithm:

(1) Select a set of natural numbers $\mathcal{L} = \{l_1 < l_2 < \cdots < l_m\}$.
(2) For each integer $l \in \mathcal{L}$, generate a random sequence of l gates in the set $\{g_1, g_2, \ldots, g_l\} \in \mathcal{C}$.
(3) Repeat these steps N_p times, to gather sufficient statistics:
    (a) Create a random sequence of Pauli operations $\{S_1, S_2, \ldots, S_l\}$.
    (b) Compute the state $\rho_x = g_l S_l g_{l-1} S_{l-1} \ldots g_1 S_1 |0\rangle$.
    (c) Determine a Pauli S_{l+1} and an additional gate g_{l+1} such that $\rho_f = g_{l+1} S_{l+1} \rho_x$ is an eigenstate of the σ_i^z operators.
    (d) Create experimentally the state ρ_f and measure whether σ_i^z operators correspond to the prediction. If so, increment the counter C_l.
(4) Estimate the probability of success for each length l, using the outcomes of these experiments, as $P_l = C_l / N_p$.

Typically, the probability of success P_l can be fitted to an exponentially decaying function:

$$P_l = A_0 p^l + B_0. \tag{8.23}$$

The average error map therefore behaves as the completely depolarizing model from Table 8.2, with the average depolarization parameter p. The fitting parameters A_0 and B_0 then include contributions from preparation and measurement errors. If $d = 2^N$ is the dimension of the Hilbert space for a register with N qubits, the average fidelity and the average error for a single quantum operation can be estimated as

$$F_{\text{avg}} = \frac{p}{d} + (1 - p), \qquad \epsilon = 1 - F_{\text{avg}} = \frac{d-1}{d}(1 - p). \tag{8.24}$$

These are averaged quantities, not only over input states, but also over all the possible Clifford gates that our quantum computer could implement. We expect that this double-averaging justifies even more the assumptions listed at the beginning of this section.

The same protocol can be extended to calibrate one specific gate, G, with some ad hoc changes: First, we insert the benchmarked gate after each Clifford operation:

$$\rho_f = g_{l+1} S_{l+1} \prod_{i=l}^{1} G g_l S_l |0\rangle . \tag{8.25}$$

This produces a similar exponential fit (8.23), with an updated depolarization parameter p_G. The *average error per gate* for G is estimated as

$$\epsilon_G = \frac{d-1}{d}\left(1 - \frac{p_G}{p}\right). \tag{8.26}$$

The operation G may be as simple as idling or waiting for a time t after each Clifford gate. When doing so, we are calibrating the coherence of our qubits, either individually – when the experiment is done on each qubit separately – or as a whole – when we consider the Clifford group on more than one qubit. O'Malley et al. (2015) has taken this idea a step further, developing an error metrology protocol that combines randomized benchmarking with Ramsey-type pulses. The result is a very sensitive analysis toolbox to explore the decoherence and dephasing properties of the qubits, including a careful study of the phase fluctuations that bound deviations from the exponential decay model.

Randomized benchmarking was introduced by Knill et al. (2008) and later generalized to multiple qubits by Magesan et al. (2011) and Gaebler et al. (2012). It exploits important connections between the performance of quantum computers in large, arbitrary calculations and the quality of the ingredients we built into the computer. In particular, it demonstrates in an intuitively appealing way the exponential accumulation of tiny errors in the quantum processors. For instance, a failure rate as small as 0.1% error per gate transforms into a success probability of $P_{20} = 80\%$ after 20 gates, $P_{40} = 66\%$ after 40, and $P_{160} = 20\%$ after 160, and so on. Considering that useful quantum algorithms must be much longer to compete against classical methods, it becomes obvious that we need to find ways to cope with this rapid deterioration of the quantum register, incorporating some type of error correction.

One may argue that randomized benchmarking provides a standard, low-cost,[9] and reproducible strategy to map out the quality of a quantum computer's components. In particular, RBM is a very generic algorithm that applies to all quantum computers based on the circuit model. It therefore enables a systematic comparison of setups and platforms, such as superconducting circuits versus trapped ions – Magesan et al. (2012) versus Gaebler et al. (2012) – which is not very different from how we compare the quality of high-end versus low-end electronics – e.g., using global figures such as the failure rates of their hard disks, memories, and processors. However, keep in mind that by doing so we might be dropping relevant information about the nature and correlations among errors, which is crucial for the development of error correction and fault-tolerant computation strategies.

8.5 Fault-Tolerant Quantum Computers

Imagine a hard disk that can store classical bits, with a low probability of corruption ε after a time T. The probability that one bit is faulty after all this time grows exponentially $p(1,L) = \varepsilon(1-\varepsilon)^L\binom{L}{1} \sim \mathcal{O}(\varepsilon L^L)$ with the number of bits L. To avoid

[9] At least when compared with full tomography.

this scenario, we can use a simple error-correction strategy, which is to copy the information redundantly, using *N physical bits* to encode one *logical bit*, our unit of information. Because the setup is classical, we may inspect the states of these logical bits periodically, to see whether one or more bits have changed their state and fix them.

In this classical scenario and encoding, a simple error-correction strategy is to implement a majority vote, which copies to all physical bits the state that is occupied by at least $N/2$ of them. This majority vote strategy works, because the probability that n out of N bits have been corrupted decreases exponentially with n. If we sum the probability that $n > N/2$ bits have been corrupted – in which case the majority vote would be wrong! – we obtain the probability that our error-correction fails. This probability scales with the redundancy as $\mathcal{O}(N^N \varepsilon^{N/2})$. The error – correction strategy therefore works provided $\varepsilon \ll 1/N$, which we may achieve if we interrogate the physical qubits with fast enough frequency.

Unfortunately, this error-correction scheme does not immediately work for quantum information. The *no cloning theorem* (Dieks, 1982; Wootters and Zurek, 1982) implies that we cannot copy an arbitrary quantum state of one qubit onto other qubits, going from $|\phi\rangle\,|0\rangle^{N-1}$ to $|\phi\rangle^N$. But, more fundamentally, there is no way that we can measure a qubit without destroying the quantum superpositions that might be stored in them. We need other encodings and other strategies that are well adapted to the quantum world.

8.5.1 Local Errors and Global Qubits

All quantum or classical correction strategies need some assumptions about the errors that we need to protect ourselves from – the *error model*. This concerns the type of perturbations that our system will experience, and the probability that those perturbations occur in a given time. In our preceding example, we worked with *local* bit flips, taking place at rate $\sim \epsilon/T$. We adopted a *global* strategy, the majority vote, which was able to find the original information in a global degree of freedom of our bits.

This distinction – local errors versus globally stored information – maps rather well to the quantum treatment of information. We typically assume that quantum *errors happen locally* on each qubit. The physical intuition is that each qubit interacts with its immediate environment, which is sufficiently large, complex, and Markovian, that it cannot correlate this qubit without far away qubits in the register. In this scenario, which may be confirmed experimentally with RBM or similar techniques, the error models are compositions of spatially uncorrelated incoherent transformations – e.g., any error model from Table 8.2, from depolarization to pure dephasing.

If this assessment is correct, we can develop a redundant encoding of quantum information into *logical qubits*, using *collective degrees of freedom* that are only mildly affected by the local errors. In this encoding, the Hilbert space of N qubits is mapped to a collection of states $|\mathbf{O},\alpha\rangle$. The labels $\mathbf{O}$ run over 2^m collective states that encode the Hilbert space of m logical qubits.

One expectation is that, by using collective or redundant degrees of freedom, we will reduce the probability that information is completely lost. Then, depending on the approach to error correction we take, we may assume that the additional degrees of freedom α will either absorb the entropy arising from decoherence, or we will expect to be able to detect, from changes $\alpha \to \alpha'$, which errors took place – and in that case, undo the error in the relevant degrees of freedom O with appropriate unitary transformations $U_{\alpha\to\alpha'}$. In both cases, our information and quantum superpositions will survive, because we never look into the logical quantum register!

8.5.2 Passive versus Active Error Correction

There are two big types of error correction: There is *intrinsic fault tolerance*, and there is *active error correction and fault-tolerant operation.* We say that a quantum system is intrinsically resilient to errors – *intrinsically protected* or relies on *passive error correction* – when the energy level structure protects the logical qubits from external perturbations.

Ideally, the logical qubit degrees of freedom $\hat{O}_n$ should commute with the operators that couple the quantum register to the environment $[\hat{O}_n, \hat{g}_m] = 0$, so that the logical states are not affected by decoherence. More generally, we will find that the environment is unable change the state of the logical qubits $\langle O|\hat{g}_{m_1}\hat{g}_{m_2}\cdots\hat{g}_{m_k}|O'\rangle = 0$, unless we have a large number $k \geq k_{\text{min}}$ of coupling operators. By considering the effects of the environment perturbatively (see Section A.3.2), we will confirm an exponential suppression of logical perturbations that could be sufficient for practical computations – specially if k_{min} grows with the size of the system!

One example of this approach is the Majorana excitations in a hybrid superconductor–nanowire interface. These are excitations of one delocalized particle onto two spatially separated regions of the wire that cannot be distinguished locally. Quantum computation is still possible, by braiding Majorana excitations, in a model of *topological quantum computation* that is *intrinsically fault tolerant* – not only the logical qubits, but all operations take place in the the logical space and are immune to errors.

Intrinsic protection is rare outside the world of topologically protected models and topological quantum computation. A more general approach is to engineer *active error-correction methods*, using the building blocks of the circuit model – single-qubit gates, two-qubit gates, and measurements – to correct errors in the information-free degrees of freedom $|\alpha\rangle$, before they accumulate and leak into the logical qubit space. Shor (1995) first suggested this possibility for a small set of errors, using a very wasteful repetition code. Later, Steane (1996) and Calderbank and Shor (1996) introduced a scalable approach to error correction that spreads the information of a logical qubit over multiple physical qubits with the help of entanglement.

The works by Steane, Calderbank, and Shor also introduced explicit error models – i.e., they decomposed the expected errors into fundamental transformations of the physical qubits – and studied rigorously at which rates those errors could be corrected. However, their approach to error correction poses difficulties in the generalization to arbitrary error models, growing to arbitrary code sizes, and implementing completely fault-tolerant quantum operations – i.e., implementing quantum computations in the presence of errors, not just correcting the decoherence in an imperfect qubit. Nowadays, both theoreticians and experimentalists have shifted their focus to other schemes that are qualitatively simple to understand and build at scale using the architectures we have discussed.

8.5.3 Stabilizer Codes

The *stabilizer formalism* is a particular approach to error correction that attempts to stabilize a quiescent state $|\alpha\rangle = |0\rangle$, using a complete set of measurements on the information-free degrees of freedom. The method identifies a large enough set of *stabilizer operators* P_k that commute with each other $[P_k, P_{k'}] = 0$. Typically, the stabilizers have eigenvalues ± 1, and the quiescent state is the only eigenstate that satisfies $P_k |0\rangle = + |0\rangle, \forall k$.

The *error-detection phase* consists in measuring all observables P_k to obtain the *syndrome* $\{p_k\}$, a collection of measurement outcomes. When any of these values is different from +1, we know that an error-occurred and mutated the quiescent state $|0\rangle \rightarrow |\alpha'\rangle$.

The *error correction phase* investigates the values of the syndrome and, through a classical algorithm, proposes with a unitary transformation to undo the errors $U |\alpha'\rangle \rightarrow |0\rangle$. As in the majority vote method, there is a maximum amount of errors that a stabilizer method can correct. Above a certain number, we can no longer determine which state $|\alpha'\rangle$ we have and how to take it back to the quiescent state. But even below that threshold, the *classical* computation of U may be extremely hard when there are many changes in the syndrome. To prevent either problem, we must repeat the error-detection and correction stages frequently enough to reduce the average number of perturbations in the syndrome.

The good thing about stabilizer formalisms is that they can be implemented using the tools from the circuit model. Both the measurements P_k and the corrections U are small quantum algorithms that can be compiled into elementary single- and two-qubit gates and single-qubit measurements, possibly with some auxiliary qubits to assist the syndrome measurements.

Most importantly, the types of stabilizer codes that we care about also allow the implementation of *fault-tolerant computations.* These are unitary gates that act on the logical qubit space without perturbing the stabilizer space or the syndrome. Such gates are called "fault tolerant" because they work even if we use imperfect components to reproduce them – i.e., single qubit gates with some residual errors, imperfect measurements, etc. When we implement the fault-tolerant gate, errors manifest as perturbations of the syndrome that can be corrected, producing the ideal operation.

Let us emphasize that we can only implement fault-tolerant gates when the syndrome does not show too many changes in between correction steps. To put limits on this, we set the *fault-tolerance threshold,* a bound in the infidelity of any operation from our quantum toolbox – including state preparation, measurement, and gates,[10] all of which can be characterized via RBM. In theory, when a platform remains within the specified thresholds, we can create fault-tolerant computers working with quantum registers and quantum algorithms of arbitrary size. In practice, this is largely dependent on assumptions such as noise models, code sizes, and the specific correction protocols.

8.5.4 Surface Code

There are various stabilizer codes in the literature, but two topological models have significantly driven change in the quantum computing world: these are the *surface code* by Bravyi and Kitaev (1998) and the *color code* by Bombin and Martin-Delgado (2006). Both codes have similar properties:

- They are lattice-based codes. They assume that qubits are laid out on a planar topology, with nearest-neighbor interactions.
- They are stabilizer codes, where the stabilizer operators are multiqubit observables defined by the plaquettes and edges of the lattice.
- Both codes support fault-tolerant computations, with universal sets of single-qubit and two-qubit gates.

[10] Including periods in which a qubit is idle, which are treated as "identity" operations.

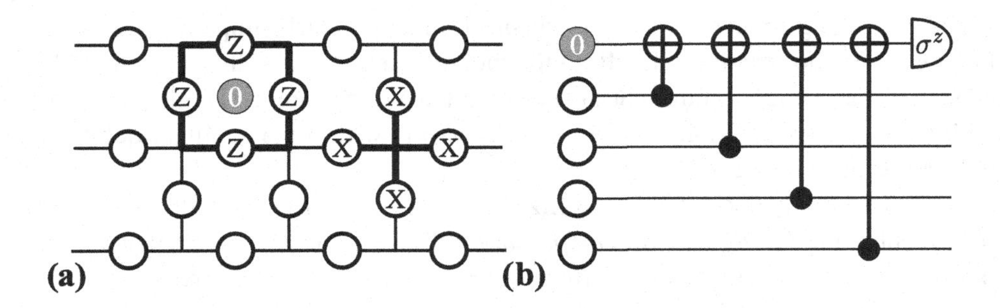

Figure 8.4 (a) Surface code layout corresponding to the qubit arrangement from Figure 8.2b. We show a plaquette and a vertex stabilizer operator, and one of the ancillary qubits. (b) Quantum circuit associated to the measurement of a $\sigma_1^z\sigma_2^z\sigma_3^z\sigma_4^z$ plaquette stabilizer, using the central auxiliary qubit for the measurement.

- They can encode any number of logical qubits, provided the lattice sizes are large enough.
- The codes can be implemented in a formally simple way, using the single- and two-qubit operations from the circuit model.

We will now work out the ingredients of the surface code formalism, using a simple lattice as an example. For further reference, we recommend the review articles by Fowler et al. (2009, 2012).

Stabilizer Operators

Figure 8.4a shows a toy realization of a surface code, where the qubits match the numbered transmons in the quantum register from Figure 8.2b. The surface code operates on a square lattice, but the qubits sit on the edges, not on the vertices of the lattice. A whole plaquette contains four edges and four qubits; we also accept open plaquettes formed by just three edges. The example in Figure 8.4a has four complete plaquetes and $2 + 2$ incomplete ones, distributed along the two *rough edges* on the sides.

There are two sets of stabilizers on this lattice. One set is formed by products of σ^z operators on the qubits of each plaquette; the other set is constructed by products of four σ^x operators from the qubits surrounding a vertex. Figure 8.4a illustrates two stabilizers: $Z_2 = \sigma_2^z\sigma_5^z\sigma_6^z\sigma_9^z$ and $X_6 = \sigma_7^x\sigma_{10}^x\sigma_{11}^x\sigma_{14}^x$, assuming we order qubits from left to right, top to bottom, as in Figure 8.2b.

Logical Qubits and Local Gates

Logical qubits are best described using the collective observables that act as the qubit's Pauli matrices – e.g., they can be used to expand the density matrix using a

Bloch vector. These operators implement the usual pseudospin algebra $[\Sigma_j^x, \Sigma_k^y] = 2i\Sigma^z\delta_{jk}$, $[\Sigma_j^y, \Sigma_k^z] = 2i\Sigma^x\delta_{jk}$, and $[\Sigma_j^z, \Sigma_k^x] = 2i\Sigma^x\delta_{jk}$. They also commute with all stabilizers $[\Sigma^{x,y,z}, Z_n] = [\Sigma^{x,y,z}, X_n] = 0$, because the stabilizers operate in the independent degrees of freedom $|\alpha\rangle$.

Logical qubits in the surface code are created by the boundaries and the holes in the lattice. Each defect decreases the number of commuting stabilizers, leaving undetermined degrees of freedom to be used for storing quantum information. The logical qubit operators for these degrees of freedom are constructed from products of Pauli matrices of physical qubits across the lattice, along the edges that connect defects (holes) and borders. One may create an arbitrary number of qubits in large lattices, provided they accommodate enough independent rough edges and holes (Fowler et al., 2009, 2012).

The example from Figure 8.4a hosts 18 qubits and 17 stabilizers – eight stabilizers of type Z and nine of type X. Since the stabilizers can take two values $\{+1, -1\}$ and are linearly independent, they jointly determine that the dimension of the subspace $|\alpha\rangle$ is 2^{17}. The remaining degrees of freedom can be grouped in a logical Hilbert space of dimension $2^{18-17} = 2$, one qubit, described by three Pauli-like operators $\{\Sigma^{x,y,z}\}$. The diagonal Pauli operator for the logical qubit is any product of σ^z along a string that connects both rough edges – e.g., $\Sigma^z = \sigma_8^z\sigma_9^z\sigma_{10}^z\sigma_{11}^z$. The real off-diagonal operator string operator is given by $\Sigma^x = \sigma_2^x\sigma_9^x\sigma_{16}^x$, or any topologically equivalent string that crosses plaquettes from top to bottom. With these two transformations, we also have $\Sigma^y = -i\Sigma^z\Sigma^x$, totaling three logical gates $\Sigma^{x,y,z}$, that we can implement by manipulating the physical qubits.

Error Detection and Correction

The error model for the surface code is based on detecting two elementary types of errors: spin-flips and phases. Both errors correspond to Pauli unitary transformations, as in $\sigma^x\rho\sigma^x$ or $\sigma^z\rho\sigma^z$. By extension, this error model also detects Pauli Y errors, which can be decomposed as a phase in combination with a spin-flip, $\sigma^y\rho\sigma^y = \sigma^x\sigma^z\rho\sigma^z\sigma^x$. Consequently, the surface code may correct many of the noise models from Table 8.2, from pure dephasing, to completely depolarizing channels.

As mentioned earlier, these errors are detected as changes in two or more values of the syndrome measurements. Take for instance the ninth qubit on our lattice and imagine it experiences a bit flip. This perturbation affects two stabilizers that overlap on this qubit, Z_2 and $Z_6 = \sigma_9^z\sigma_{12}^z\sigma_{13}^z\sigma_{16}^z$. Their expectation values are transformed by the unitary perturbation U as $\langle O,\alpha|U^\dagger Z_{2,6}U|O,\alpha\rangle$, which means that in the Heisenberg picture the new stabilizers have changed sign $\sigma_9^{x\,\dagger} Z_{2,6}\sigma_9^x = -Z_{2,6}$. If the error is a π phase, then the affected stabilizers will be $X_4 = \sigma_5^x\sigma_8^x\sigma_9^x\sigma_{12}^x$ and

$X_5 = \sigma_6^x \sigma_9^x \sigma_{10}^x \sigma_{13}^x$, which will also change sign, $\sigma_9^{z\,\dagger} X_{4,5} \sigma_9^z = -X_{4,5}$. And finally, Pauli Y errors on this qubit manifest as joint phase and spin-flip errors, which affect all four stabilizers that overlap on the affected qubit $Z_{2,6}$ and $X_{4,5}$.

We build our our *error-detection phase* on the capacity of the surface code to detect all these single-qubit transformations – σ^x, σ^y, and σ^z – in isolated or multiple qubits. Qualitatively, we rely on the expansion of a general completely positive map as a linear combination of products of Pauli matrices – the S_α operators from Section 8.4.2. Following this expansion, we interpret generic errors as statistical superpositions of single-qubit transformation, which can be detected.

The error detection works when changes on the syndrome can be mapped, unequivocally, to the actual Pauli transformations. For this, we need the expansion to involve products of few operators, and that these products are localized enough as not to change the values of one or more logical qubits.[11] We achieve this safe regime by assuming architectures with spatially uncorrelated errors and ensuring that error monitoring takes place faster than the rate of errors. In this situation, we can identify not just the errors, but a successful *error-correction strategy* to bring back the syndrome to the desired quiescent state $X_k, Z_j = 0\ \forall k, j$.

The measurement of the stabilizers, like all other operations in the surface code, can be implemented the quantum circuit toolbox. The stabilizer operators such as Z_1, Z_2, or X_4 are three- or four-qubit observables that measure the parity[12] of those qubits in either the σ^z or σ^x basis. This *parity measurement* requires a single auxiliary qubit and four two-qubit operations. The algorithm shown in Figure 8.4b maps the Z-parity of four qubits – i.e., a plaquette – onto an auxiliary qubit that is placed at the center of it. The extra qubit begins on the $|0\rangle$ state and suffers as many spin-flips as neighboring qubits are in the $|1\rangle$ state. The outcome is $|0 \oplus P\rangle$, where $P = \frac{1}{2}(1 + Z)$ is 1 if the parity is odd on the plaquette, or 0 otherwise. A similar algorithm can be constructed for X-parity measurements, with auxiliary qubits placed on the vertices.

Since the parity measurements demand auxiliary qubits, it is quite common that these qubits be included in the architecture of the quantum register. For instance, the surface-code lattice with 18 qubits from Figure 8.4a could be constructed using the white qubits from Figure 8.2b, with the additional 18 qubits as targets for the parity measurement.

[11] Imagine that the transformations happen to connect the boundaries, as in $\Sigma^x = \sigma_2^x \sigma_9^x \sigma_{16}^x$. This error is actually flipping the logical qubit, and is no longer localized to the degrees of freedom $|\alpha\rangle$ that we can safely manipulate.

[12] Take a set of bits $s_1, s_2, \ldots, s_n \in \{0, 1\}$. The parity is even or odd depending on whether the number of ones is even or odd. This definition extends to a qubit basis by determining how many qubits return a positive result $+1$ when measuring σ^z or σ^x.

The error-correction phase, in which we bring the syndrome back to the original value, is implemented as a collection of single-qubit quantum gates acting on strings of physical qubits. These strings of Pauli matrices run along the edges that connect vertices or plaquettes with incorrect stabilizer values, correcting all of the errors simultaneously. The identification of the actual errors and the unitary operations that undo them is a computationally intensive stage that is still subject to investigation and improvement.

As an illustration of the difficulty in finding the right error-correction strategy, it is illustrative to read the article by Kelly et al. (2015). This experiment demonstrated the operation of a repetition code – a simpler 1D stabilizer code that works with the parities of neighboring qubits $Z_i = \sigma_i^z \sigma_{i+1}^z$ – discussing strategies for the detection of errors. More recently, Chen et al. (2021) demonstrated this code working with up to 21 qubits and 50 layers of correction using Google's Sycamore quantum processor, in a work that also showed an implementation of the surface code with four qubits, for a more limited number of operations.

Universal Gates

From the previous discussion, we could infer how to implement three logical single-qubit gates, Σ^x, Σ^z, and Σ^y, and any logical-qubit measurement, in a fault-tolerant way. The first step is to implement the gate or measure the collective observable, as prescribed by the definition of the logical operator. The second step involves measuring the syndrome and correct any errors – which in the case of the measurement might involve updating the measured value to reflect the changes in the physical qubits.

In practice, the $\Sigma^{x,y,z}$ gates are not needed. As explained by Fowler et al. (2012), we only need to care about building a universal set of gates with the H, S, T, and CNOTs operating on the logical qubits. The $\Sigma^{x,y,z}$ will be "moved" out of the algorithm and converted in corrections of the measurement outcomes.[13] However, this does not really simplify the task of explaining how the new gates are implemented.

Qualitatively, a standard implementation (Fowler et al., 2012) relies on the possibility to manipulate the qubit lattice in topologically nontrivial ways. First of all, to create the surface code lattice, we need to measure the stabilizers of a set of qubits – the ones that form the lattice – correcting those values to the desired outcome. This already suggests that we can engineer holes or defects, or cut a lattice into two separate pieces, by ignoring certain qubits, not considering them part of any stabilizer. Conversely, we can merge lattices by stabilizer measurements

[13] Consider a stupid algorithm that combines X_1, Z_2, and a $CNOT$ acting on both qubits C_{12}. We can transform $(X_1 \otimes Z_2)C_{12} = (\mathbb{1} \otimes Z_2)C_{12}(X_1 \otimes X_2) = C_{12}(X_1 \otimes (-Z_2 X_2))$. These ideas allow us to move the Pauli matrices to the end, where they only affect the final measurement outcome.

along their boundaries, and include new qubits or fill holes by including those qubits in the definition of the stabilizer.[14]

If we define the logical qubits as strings that connect holes in the lattice, we can implement CNOT gates in a completely fault-tolerant operation by braiding holes around each other. The Hadamard gates H are also implemented in a fault-tolerant way that involves isolating the region of the lattice that contains the qubit, doing some key measurements, and merging this patch back to the lattice. In both cases, errors introduced during either the CNOT or the H are self-corrected by the repeated measurement and corrections of the stabilizer group.

Unfortunately, the S and T gates are more complicated. They must be implemented using quantum algorithms that rely on H, CNOT, and auxiliary logical qubits prepared in a given state. The preparation of these auxiliary states is a delicate operation not covered by the stabilizer formalism, and which may introduce errors. To make these errors as small as possible, we rely on quantum information recipes that fall outside the code itself. In particular, we use state distillation, which combines multiple imperfect copies of a state to produce one with higher fidelity. All this makes the S and T operations the most costly of the universal and fault-tolerant gate set – and a valuable resource that is thoroughly analyzed when designing a quantum algorithm.

8.5.5 Fault-Tolerant Thresholds and Outlook

The discussion of stabilizer codes, and the surface code in particular, is much richer than this summary. We are intentionally leaving out important topics – e.g., what happens when we "lose" a qubit, or when errors move, or how many errors a code supports, or how large are the overheads, etc. – to simplify the conversation and provide you with some qualitative ideas that drive the architectural decisions on superconducting quantum computing platforms.

One first takeaway message is the fact that there exists a robust formalism for encoding logical qubits in the quantum states of extended lattices, which are simple enough to be implemented with superconducting circuits. The second takeaway is that this encoding provides us with error-correction and fault-tolerant operation schemes, to store and process quantum information for arbitrarily long periods of time, provided that error rates are low enough. The key concept here is the term *low enough,* which translates into existence of one or more *thresholds for fault-tolerant operation,* as mentioned before.

[14] In the process, a stabilizer that contained three qubits may pass to contain four, as the lattice has one more edge.

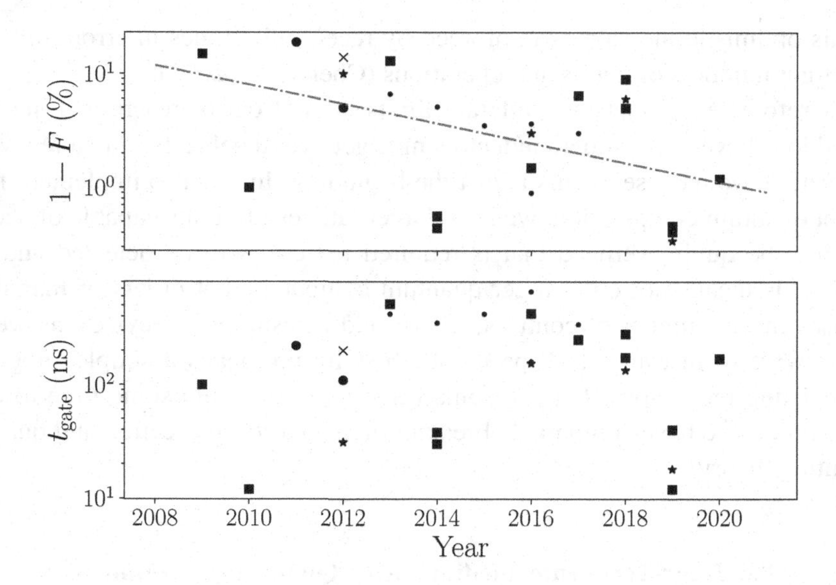

Figure 8.5 (a) Two-qubit gate errors and (b) two-qubit gate speeds for transmon qubits. The type of marker denotes the type of gate: CZ (square), CNOT (dot), cross-resonance (circle) (Chow et al., 2011), iSWAP (star).

There is a growing literature on the conditions for fault-tolerant computation on different quantum technologies. In particular, the surface code has been subject to extensive analysis. Some optimistic results by Fowler et al. (2009) establish an *error threshold* of 1% on all operations – single- and two-qubit fidelity and measurements – for building arbitrarily large codes. Other works set the boundary at even higher levels, when errors are not generic but are biased toward certain models.

One could argue that we are very close to those thresholds. Single-qubit gates have very good fidelities, with errors below 0.1% in most works. We also know how to improve state preparation using active controls. Measurement errors are still above desired thresholds, but can be improved using Purcell filters and parametric amplification (Walter et al., 2017). Regarding two-qubit gates, Figure 8.5 shows a collection of data points from over 30 works with transmon qubits. There are already several realizations of fast CZ and iSWAP two-qubit gates with less than 1% error (Barends et al., 2014; Arute et al., 2019). The overall tendency is also favorable, with a 20% decrease in two-qubit error gate every year[15] over all gates,

[15] Note that qubit coherence already follows a Moore's type law – with an 80% increase in coherence time every year (cf. Section 6.6) – this naturally leads to a decrease in gate errors. This said, gate operations are more complicated than just "waiting," as they are also sensitive to the quality of the controls, distortions in the cables that bring the control signals, and many other technical sources of errors that need to be overcome.

and this optimistic message is reinforced by recent milestones in error correction with larger numbers of qubits and operations (Chen et al., 2021).

It therefore seems that fault-tolerant limits are not out of reach, and, just like trapped ions have done, superconductors may get over the threshold in a few years. However, as we have seen, this is just the beginning. In order to implement fault-tolerant quantum computation, we need larger lattices of qubits capable of accommodating the qubit overhead that is required for errors to be detected and corrected. This means not only larger quantum computers, but orders of magnitude increases in the number of controls, filters, and measurement devices, as well as the hardware to instrument all protocols. It is by no means a simple "upgrade" from existing technology, but a roadmap that will take quite some time to complete – unless we have comparable breakthroughs in error correction and quantum computing theory!

8.6 Near-Term Intermediate Scale Quantum Computers

8.6.1 What Is NISQ?

This book has been written in what Preskill (2018) called the noisy interme-diate-scale quantum (NISQ) era. There exist already working quantum computers ranging from a few to 50-something qubits. Some of them are being developed in public laboratories (Rol et al., 2019; Krinner et al., 2020), while others are being developed in big companies – Google (Arute et al., 2019), IBM (Cross et al., 2019) – and startups – e.g., Rigetti (Otterbach et al., 2017) – but none of these machines is capable of scalable, error-corrected fault-tolerant computations.

This is an era in which the control of large quantum mechanical systems is a reality, but it is far from perfect. State-of-the-art decoherence times, gate and SPAM fidelities, and system sizes are not yet within the specifications for building a robust and scalable quantum computer. It is also likely that we will be unable to achieve scalable and fault-tolerant operation in the very near future, due to technical demands in many aspects of superconducting and microwave technology. These involve material science, quantum control, error correction, and electronic engineering as some of the areas where there is immediate work to do, but it is also safe to assume that we need important breakthroughs in other areas to move forward.

In this interim time, we are stuck with devices that lack error correction, but are steadily growing in size and quality by contrasted metrics. These devices are by no means "useless." They are covering important milestones in the path toward scalable quantum computation. They are also a platform on which we steadily learn important lessons on what works and what does not – even when these challenges become more and more difficult over time!

For instance, we already have first evidence of quantum devices and experiments that can beat what can be simulated with classical computers (Arute et al., 2019) while ensuring full control of the experimentalist. In this rewarding time, we are seeing new developments in the field of quantum algorithms, adopting new paradigms where the quantum computer works hand in hand with classical algorithms and postprocessing techniques. This approach has illuminated new solutions to canonical (toy) problems in quantum chemistry (Kandala et al., 2019; Arute et al., 2020), even in the presence of significant errors and decoherence! We expect that the same platforms will continue to support scientists in studying such interesting concepts in theoretical physics and quantum many-body physics as quantum transport, many-body localization, and time crystals (Mi et al., 2022; Neill et al., 2021).

8.6.2 Hybrid Quantum Computers

From our first collective steps into the NISQ era, there are already some interesting lessons learned. First, the circuit model for quantum computation has consolidated as a useful abstraction for interfacing with the quantum computer. The qubit reset, measurements, single-qubit operations, and universal two-qubit gate of each platform become the new assembly language with which all computations can be constructed. This language is understood by the quantum computer's *firmware*. This is a set of classical computers, microwave generators, and FPGAs for data collection, which converts those instructions into actual microwave pulses, interprets the signals from the quantum device as binary measurement outcomes, and performs other low-level tasks, such as qubit calibration.

On top of this layer, we find higher-level interfaces, such as IBM Qiskit, Google's Cirq, Rigetti's Pyquil, etc. These are classical frameworks, written in friendly programming languages (Python, C#, etc.) or even in graphical interfaces (IBM's Quantum Experience[16]), with which we can write quantum algorithms and operate the quantum computer. These frameworks do not need to be attached to the quantum computing infrastructure. They may run separately, sending remote orders to one or more quantum computers, in our more or less common assembly language, for the actual firmware to process and execute.

Interestingly, this is not a limitation. We can run very low-level operations this way – for instance, we can engineer a complete randomized benchmarking protocol using entirely the high-level interface. We can also benefit from the possibilities of combining quantum and classical computations, in a *hybrid quantum-classical computing* model. This means relying good classical algorithms for certain tasks (e.g., optimizers, compilers) that are easier for large silicon-based computers, and

[16] https://quantum-computing.ibm.com/.

delegating to the Hilbert space of the quantum computer other taks that might be too large or too slow on an ordinary computer – for instance, creating and manipulating the wavefunction of a complex material, or the complex multivariate probability from a sophisticated risk model.

8.6.3 Quantum Volume

This level of abstraction means that computations may be, to some extent, agnostic to the actual hardware on which it runs. There is a natural competition between platforms – trapped ions, circuits, quantum dots, and photonic systems – for offering the most powerful backend. In this scenario, an important question arises: How do we compare those devices? A partial answer could be the one from Section 8.4, which involves creating a map of the coherence times and gate fidelities of the device.

Unfortunately, this approach is insufficient, because it does not take into account that the same algorithm may be compiled very differently on different platforms. For instance, to make gates between arbitrarily separated qubits may be more costly on a one-dimensional register than on a 2D lattice. And different algorithms may have different assembly representations depending on whether the computer uses an iSWAP or a CNOT or can support a continuum of two-qubit gates!

A partial solution to this problem is the *quantum volume,* a metric introduced by Bishop et al. (2017) and perfected by Cross et al. (2019). In the spirit of RMB, the quantum volume studies random circuits, computing the largest integer N such that we can execute quantum algorithms involving N qubits and N layers of generic two-qubit gates[17] and still have a decent fidelity. Once we have that N_{max}, the quantum volume is the size $2^{N_{\text{max}}}$ of the Hilbert space that can be encoded in it. As of this writing, the largest quantum volume 32 for superconducting circuits indicates that we can execute a random circuit using five qubits and five layers of one- or two-qubit operations and be safe about the outcome.

8.7 Outlook

In this chapter, we have seen how quantum computers are built and operated. The hardware is conceptually simple. It builds on the tools developed in Chapters 6

[17] By generic, we mean any rotation in $SU(4)$ covering all of the Hilbert space for two qubits. We also mean that those gates can take place among any pair of qubits. In practice, the arbitrary separation means that quantum gates must be decomposed – *transpiled* – into a much larger number of assembly gates (Cross et al., 2019). This is really what makes the quantum volume a challenging metric.

and 7, with additional techniques for two-qubit quantum gates and for characterizing the computer's performance.

It is clear we need further developments in error correction and fault-tolerant computation models. Existing paradigms impose significant overheads – number of measurements, postprocessing of data, and rapid feedback controls – and demand orders of magnitude more qubits than what we are likely to supply in the following decade. We also need a better understanding of quantum circuits, algorithms, and compilation strategies.

However, there are also many architectural, design, and theoretical challenges open. We can highlight the following ones:

- We need better qubits, with longer coherence times. There are hints that this can improve with new materials and the application of state-of-the-art surface treatment and fabrication (Place et al., 2021). But maybe we also need different types of qubits, with intrinsic topological protection!
- Complex algorithms and error correction require us to pack more qubits into the chip. This is an important challenge, because the use of microwave technology sets a lower bound on the size of qubits and control devices. Maybe the route to success lays in going beyond existing sandwich architectures to fully three-dimensional integrated devices (Brecht et al., 2017, 2016).
- If we increase the number of qubits means, we need to increase commensurately the number of readout and control lines. This requires important developments in the control and microwave electronics – which already has shifted to customized FPGA circuitry (Arute et al., 2019) – and clever strategies to optimize the limited number of microwave lines that can be inserted into the fridge. If at some point even this becomes an unavoidable limitation, we might need to transition to distributed quantum computers connected by quantum links, as explained in Section 7.2.5.
- We need better controls and higher-quality gates. This challenge may be addressed at the hardware level, looking for new interconnects between qubits, producing different, more scalable or stronger couplings. Or it may be addressed at a quantum control level, improving on the design of quantum gates, making them more robust against external perturbations, or even against the obvious distortions suffered by microwaves when they are injected into a chip.

These are very important and very difficult challenges. I hope that the information in this book, combined with existing literature and your ingenuity, will stimulate you to address them, joining many of the excellent teams working in the field worldwide, or working to create your own!

Exercises

8.1 Prove that a Hadamard gate acting on each qubit creates a quantum superposition of all possible binary numbers in a quantum register:

$$H^{\otimes N}|00\ldots 0\rangle = \frac{1}{2^{N/2}}\sum_{b=0}^{2^N-1}|b\rangle\,. \tag{8.27}$$

Discuss how to implement this gate for N identical flux qubits.

8.2 **Unitary equivalence of universal gates.** Prove that the CZ gate and the CNOT gates in Table 8.1 are equivalent up to local transformations. In other words, there are single-qubit unitary operations U_i and W_i such that $(U_1 \otimes U_2)U_{\text{CZ}}(W_1 \otimes W_2) = U_{\text{CNOT}}$.

8.3 **Cavity-mediated exchange interactions.** Assume two qubits interacting with the same cavity:

$$H = \sum_i \left[\frac{\Delta_i}{2}\sigma_i^z + g_i(\sigma_i^+ a + a^\dagger\sigma_i^-)\right] + \omega a^\dagger a. \tag{8.28}$$

(1) Using second-order perturbation theory in the off-resonant limit, $|\omega - \Delta_i| \gg |g_i|$, obtain a dispersive coupling model. Show that this Hamiltonian in the subspace of no photons is equivalent to

$$H_{\text{eff}} = \sum_i \left(\frac{\Delta_i}{2} - \frac{2g_i^2}{\delta_i}\right)\sigma_i^z - \frac{g_1 g_2}{2}\left(\frac{1}{\delta_1} + \frac{1}{\delta_2}\right)\sigma_1^x\sigma_2^x. \tag{8.29}$$

(2) Solve the dynamics induced by the effective Hamiltonian $\exp(-iH_{\text{eff}}t)$ after a time t. In which limit does this operation become an iSWAP gate?

(3) Suppose that the cavity is in a thermal state with temperature $T = 0.050$ K, for $\omega \sim 6$ GHz. How does the excitation probability of the cavity affect the two-qubit exchange operation? Estimate the infidelity of the gate.

8.4 **Phase gate**. Take the model for the two interacting transmons (8.7).

(1) Diagonalize the Hamiltonian for all values of J and ω_2 and plot the eigenenergies as a function of $\omega_2/\omega_1 \geq 1$.

(2) Show that there are two avoided energy level crossings at $\omega_1 = \omega_2$ and $\omega_1 + \alpha = \omega_2$.

(3) Show that in the limit $\omega_2 \gg \omega_1$, the qubit basis is mapped to four eigenstates of this problem. Which energy levels do those states follow as we adiabatically change ω_2?

(4) Assume that we can perform a perfect adiabatic passage, from a large ω_2 down to resonance $\omega_2 = \omega_1 + \alpha$, where each two-qubit state $|s_1, s_2\rangle$ is mapped to an instantaneous eigenstate of $H(\omega_2)$. Compute formally the accumulated phases by all four two-qubit states along the round trip.

(5) Prove that it is possible to recreate any control-phase gate, with any angle $\exp(i\Delta\theta\sigma_1^z \otimes \sigma_2^z)$, assuming arbitrarily long adiabatic passages with no decoherence. Hint: Show that for a large enough passage time T, there is a nonzero interaction phase $\Delta\theta$. Then prove that the value of $\Delta\theta$ can be enlarged arbitrarily by deforming the adiabatic passage $\omega_2(t \times T'/T)$, with $T' > T$.

(6) Find out the algorithm for the quantum Fourier transform. What phases $\Delta\theta$ are required in that algorithm?

8.5 We want to simulate time-evolution with the Ising model, implementing the unitary operation

$$U(t) = \exp\left(-it\sigma_1^z\sigma_2^z\right), \tag{8.30}$$

but we only have a toolbox with control-NOTs and arbitrary single-qubit operations. Devise a protocol to approximate $U(t)$ using such gates. Hint: Investigate what happens when you sandwich a local operation in between two CNOTs.

8.6 Useful positive maps. We now study a set of very common positive maps.

(1) **Completely depolarizing channel**. Provide a physical interpretation for the completely depolarizing channel in Table 8.2. Show that the symmetric version is recovered from the one in that table by setting $p_x = p_y = p_z = p/3$.

(2) **Random flip.** Write down a completely positive map to describe this physical operation: With probability p, preserve the state and elsewhere flip the state of the qubit, exchanging $|0\rangle$ and $|1\rangle$.

(3) **Dephasing**. Show that the dephasing of a qubit (6.14) may be written as a channel that with probability p preserves the state, and elsewhere keeps only the diagonal elements:

$$\varepsilon(\rho) = p\rho + \frac{1-p}{2}\left[\mathrm{tr}(\rho)\mathbb{1} + (1-p)\mathrm{tr}(\sigma^z\rho)\sigma^z\right].$$

(4) In all previous maps, determine the Kraus operators and show that the map preserves the trace of a density matrix: $\mathrm{tr}\varepsilon(\rho) = \mathrm{tr}\rho$.

8.7 Prove that the set of operators formed by all tensor products of Pauli matrices $S_\alpha \in \{\mathbb{1}, \sigma^x, \sigma^y, \sigma^z\}^{\otimes N}$ (cf. Section 8.4.2) forms a basis of linearly independent and orthogonal elements in which all matrices can be expanded. Hint: Reinterpret matrices as column vectors, show that $(A, B) = \mathrm{tr}(A^\dagger B)$ is a good scalar product for the matrix space, and prove the expansion (8.13).

8.8 Derive the Choi matrices (8.17) using the M and χ representations. Hint: Show that $|\alpha\rangle = \frac{1}{\sqrt{d}}\sum_{i=1}^{d} |i\rangle \otimes S_\alpha |i\rangle$ forms an orthonormal basis.

8.9 Deduce the form of the matrices M and χ for a perfectly coherent positive map, $\varepsilon(\rho) = U\rho U^\dagger$, in terms of the unitary operation U. Particularize it for a single-qubit rotation $U = \exp(-i\theta\vec{n} \cdot \boldsymbol{\sigma})$ and a CNOT.

8.10 Compute the process matrices M and χ for the single-qubit error map from (6.19). Compute the **average fidelity** and the **process fidelity** for this map. How do those figures relate to T_1 and T_2^*?

8.11 Design a **randomized benchmarking protocol** for a single qubit. What gates do you need to use? Simulate this protocol for a qubit that experiences relaxation ($T_1 = 1$ μs) and one that experiences dephasing ($T_\phi = 1$ μs), and has no SPAM errors. In the simulation, assume a fixed duration $\tau \sim 10$ ns for all single-qubit gates, and that the error of the gate can be fully attributed to decoherence, that is, $\varepsilon_U(\rho) = \varepsilon_{\text{decoh}}(U\rho U^\dagger)$. Fit the success probability P_l to the exponential curve (8.23) and extract the parameter p. Compare the error estimate p with the estimated average and process fidelity associated to $\varepsilon_U(\rho)$.

8.12 The average fidelity (8.20) results from the fidelity of a physical process over all pure states. It has been shown by Nielsen (2002) that this definition is equivalent to replacing the map $\varepsilon(\rho)$ with a totally depolarizing channel (see Table 8.2) with depolarization p. Show that the relation between p and F_{avg} is given by (8.24).

9
Adiabatic Quantum Computing

In Chapter 8, we introduced the quantum circuit model for quantum computing, in which we decompose algorithms into elementary operations – resets, gates, measurements – that can be mapped to local controls and qubit–qubit interactions. In this chapter, we introduce a very different computation model, based on the simulation of complex many-body Hamiltonians and the preparation of ground states. This is an extremely appealing idea, which connects the worlds of quantum simulation and quantum computation and which has been rediscovered many times. A particularly relevant work by Farhi et al. (2000) gave it the name of *adiabatic quantum computation,* because of the particular choice for ground state preparation. This work had a tremendous impact in the field and even inspired the creation of large quantum devices with thousands of qubits, D-Wave's *quantum annealer*. We will discuss the ideas in this work, together with the computational power of the Hamiltonian model, and also get the differences between ideal adiabatic quantum computers and the NISQ quantum annealers that are available in the lab. As in the previous chapter, we try to stay as focused on the superconducting circuit hardware as possible, but there are unavoidable theoretical concepts that must be well explained in order to motivate and understand why and how experiments work.

9.1 Adiabatic Evolution

In most of this book, we have studied fast controls and the real-time dynamics they induced in qubits and cavities. The motivation was, of course, to make quantum operations as fast as possible, minimizing the role of decoherence. In this chapter, we switch gears and focus mainly on slow control and *adiabatic passages* that are used to prepare low-temperature states.

Imagine, as sketched in Figure 9.1a, a complex physical system with a Hamiltonian that can be manipulated. The control, denoted ε, represents some generic parameters of the Hamiltonian that we can change in time, so that

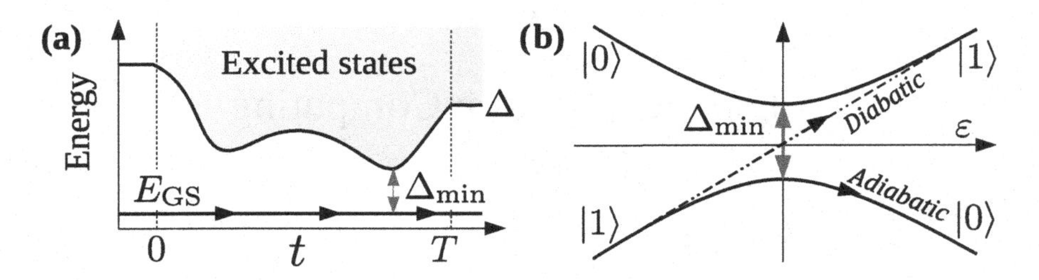

Figure 9.1 (a) Sketch of the energies and gaps throughout an adiabatic passage in which a Hamiltonian $\hat{H}(\epsilon)$ is deformed from $\epsilon(t = 0)$ to time $\epsilon(t = T)$. The Hamiltonian maintains a gap between the excited states and the ground state energy E_{GS}, which is at least $\Delta E_{\min}$. (b) Qubit hyperbola for a flux qubit that begin in its ground state and is subject to a linearly growing bias field, from a very large negative value $\varepsilon \ll |\Delta_{\min}|$ to a very large positive one. In this Landau–Zener process, there is a minimum qubit gap $\Delta_{\min}$, depending on which the process will be adiabatic – the qubit remains in the ground state manyfold and $|1\rangle$ is deformed to $|0\rangle$ – or diabatic – there are transitions to the excited state, leaving some probability in $|1\rangle$ at the end.

$\hat{H}_t = H(\varepsilon(t))$. We start the adiabatic passage from an eigenstate of the Hamiltonian $\hat{H}_0$, which we assume is easy to prepare. We change the control in time $\varepsilon(t)$, expecting that our experiment remains as close as possible to the ground state of the time-dependent Hamiltonian $\hat{H}_t$.

Implicit in the notion of adiabatic passage is the idea that ε can change slow enough for the system to adapt and remain in a quasi-equilibrium state without exciting to higher-energy states. To pose some classical analogs, think of how you can adiabatically stretch a rubber band. If you stretch it too fast, it will acquire too much energy, exciting phonons that show as vibrations of the band. Or think about how you can push a glass full of water across a table. If you push too fast, the surface of the water will deform and even spill out of the glass, but if you push it gently, the surface of the liquid will remain unperturbed.

We now address the idea of adiabatic passage from two points of view. We start with a very specific and simple model, a qubit, where we can solve the problem analytically and even provide a quantitative answer to the shape of the control. This problem will give us an intuition about adiabatic and diabatic processes, which can be applied to many different problems. This intuition then acquires a rigorous mathematical framework through the *adiabatic theorems*, a set of sufficient criteria to ensure an adiabatic passage for every given Hamiltonian. This theorem lays the foundations for the adiabatic quantum computation (AQC) model that we discuss next.

9.1.1 Landau–Zener and Qubit Adiabatic Control

The simplest model in which we can study an adiabatic passage is a qubit. Imagine a flux qubit (cf. Section 6.4) with a tunneling amplitude $\hbar\Delta_{\min}$, controlled by an external bias field $\hbar\varepsilon$. The Hamiltonian for this qubit adopts the familiar form (6.2), which we change basis to become

$$\hat{H}(\varepsilon) = \frac{\hbar\varepsilon}{2}\sigma^z + \frac{\hbar\Delta_{\min}}{2}\sigma^x. \tag{9.1}$$

We can compute the ground states of the qubit for any value of ε, using the Pauli matrix decomposition:

$$\hat{H}(\varepsilon) = \frac{1}{2}\hbar\Delta(\varepsilon)\,\mathbf{n}\cdot\boldsymbol{\sigma}, \quad \text{with} \begin{cases} \Delta = \sqrt{\varepsilon^2 + \Delta_{\min}^2} \geq |\Delta_{\min}|, \\ \mathbf{n} = \frac{1}{\Delta}(\Delta_{\min}, 0, \varepsilon). \end{cases} \tag{9.2}$$

This model is equivalent to a magnetic field $\mathbf{B} = \Delta(\varepsilon)\mathbf{n}$ interacting with a quantum spin $\mathbf{S} = \frac{1}{2}\boldsymbol{\sigma}$.

The ground state eigenenergy $-\frac{1}{2}\hbar\Delta(\varepsilon)$ is separated from the excited state by a gap $\hbar\Delta(\varepsilon)$. The ground state is a pure state with a Bloch vector anti-aligned with the magnetic field $\langle\boldsymbol{\sigma}\rangle = -\mathbf{n}$. It has a wavefunction

$$|\tilde{0}_\varepsilon\rangle = \cos\left(\frac{\theta(\varepsilon)}{2}\right)|0\rangle + \sin\left(\frac{\theta(\varepsilon)}{2}\right)|1\rangle, \tag{9.3}$$

that depends on the polar angle $\mathbf{n} = (\sin\theta, 0, \cos\theta)$. In the limit of large bias $|\varepsilon| \gg |\Delta$, the ground state of the Hamiltonian approximates the two current states, $(0,0,+1)$ or $|1\rangle = |R\rangle$, for very negative field and $(0,0,-1)$ or $|0\rangle = |L\rangle$ for very large positive bias. These are two states that we can prepare, either by waiting for the qubit to relax to the ground state at those extreme points, or using some type of measure-and-reset protocol — e.g., we could dispersively measure the qubit's state and flip it using a microwave if the qubit was in the excited current state.

We now wonder what happens if we start from such a well-prepared ground state and change the control parameter ε, as in Figure 9.1b. Does the qubit remain in the instantaneous ground state or does it get excited? In the first case, which we call the *adiabatic evolution*, the qubit's wavefunction is always the one in (9.3) up to global phases. In the second case, which is called the *diabatic evolution*, the qubit will not adapt, and it will either get locked in the original current state or, more generally, become a quantum superposition of both qubit states.

To answer this question, we assume that $\varepsilon(t)$ is a monotonously growing function, and write the wavefunction on a rotating frame that brings the instantaneous eigenstates of $\hat{H}(\varepsilon)$ to the qubit basis:

$$|\psi(t)\rangle = e^{\frac{1}{2}i\theta(\varepsilon(t))\sigma^y} |\xi(t)\rangle . \tag{9.4}$$

If we start at a large negative value of ε and we are perfectly *adiabatic,* $|\xi(t)\rangle \simeq |0\rangle$, and the probability of *diabatic transitions* $P_{\text{dia}} = |\langle 1|\xi(t)\rangle|^2$ becomes approximately zero.[1] We write the dynamics of $|\xi(t)\rangle$ in a rotating frame, using ε instead of time as coordinate (see Problem 9.1):

$$i\partial_\varepsilon |\xi(\varepsilon)\rangle = \left[\frac{1}{2\dot{\varepsilon}}\Delta(\varepsilon)\sigma^z - \frac{\Delta_{\min}}{2\Delta(\varepsilon)^2}\sigma^y\right]|\xi(\varepsilon)\rangle. \tag{9.5}$$

This Hamiltonian has contributions that describe the *adiabatic* (σ^z) and *counteradiabatic* (σ^y) evolution. In the perfectly adiabatic limit, in which the rate of change $\dot{\varepsilon} \to 0$ is very slow, the adiabatic term dominates the Hamiltonian, and the wavefunction remains an eigenstate up to a dynamical phase $|\xi(t)\rangle = e^{i\phi(t)}|0\rangle$, with $\phi(t) = \int \Delta(\varepsilon)\mathrm{d}\varepsilon/\dot{\varepsilon}$. In this limit, the counteradiabatic term is a weak perturbation that changes more slowly than $4/\Delta_{\min}$, and is therefore incapable of bridging the gap $\Delta/\dot{\varepsilon}$. We can estimate the small corrections introduced by this term as being of order $\dot{\varepsilon}\Delta_{\min}/\Delta^3 < \dot{\varepsilon}/\Delta_{\min}^2$.

When the speed of change $\dot{\varepsilon}$ is constant, the two-level system admits an analytical solution called the Landau–Zener formula (Landau, 1932; Zener, 1932). This formula describes a limit in which we start from the instantaneous ground state $|\psi(-\infty)\rangle = |1\rangle$, and sweep the control all the way to $\varepsilon \to +\infty$. In this problem, the probability of diabatic excitation is

$$P_{\text{LZ}} = \exp\left(-\frac{\pi}{2}\frac{\Delta_{\min}^2}{\dot{\varepsilon}}\right) = \exp\left(-\frac{\pi}{2\bar{v}\tau_{\text{adiab}}}\right). \tag{9.6}$$

This expression contains two timescales: the rate of change of the Hamiltonian's gap $\bar{v} = \dot{\varepsilon}/\Delta_{\min}$, and the smallest energy scale for the qubit along the passage, $\Delta_{\min} = 1/\tau_{\text{adiab}}$. When the rate of change $\bar{v}$ is faster or comparable to the minimum frequency of the qubit, we are in the *diabatic regime,* in which P_{LZ} approaches 1. In the opposite regime, when the rate of change $\bar{v}$ is slow enough, the qubit is in the *adiabatic regime* and the excitation is negligible $P_{\text{LZ}} = 0$.

[1] Note that, due to our unitary transformation, $|0\rangle$ and $|1\rangle$ are no longer current states, but an instantaneous ground state and excited state.

9.1.2 The Adiabatic Theorem

There are many rigorous generalizations of the Landau–Zener problem, all of which receive the name *adiabatic theorem*. They analyze the adiabatic deformation of eigenstates in a tunable Hamiltonian, assuming that a relevant subspace[2] is well separated from the rest of the spectrum by an energy gap. The theorems provide sufficient conditions for an initial state that begins in this subspace, to remain inside it, as it is continuously deformed. These theorems also bound the error of the adiabatic passage – the probability of leaking to other states – by relating the rate of change of the control and the instantaneous gaps of the Hamiltonian, similar to Section 9.1.1.

The study of adiabatic passages is rather old, but it starts to acquire rigorous foundations around year 1950, when Kato (1950) derived the first conditions for the adiabatic deformation of a ground state. Later works have made these results more useful and rigorous using spectral theory. A full demonstration of any of the adiabatic theorems falls outside the scope of this book,[3] but most versions (Kato, 1950; Galindo and Pascual, 1991; Jansen et al., 2007; Lidar et al., 2009) assume similar conditions on the tunable Hamiltonian $H(\varepsilon)$ and the initial state:

(1) The adiabatic passage happens by deforming a Hamiltonian $\hat{H}(\varepsilon)$ through a control $\varepsilon(t)$ that takes a total time T.
(2) The Hamiltonian $H(\varepsilon)$ and the parameter tuning $\varepsilon(t)$ must be differentiable a number of times.
(3) The spectrum of the deformed Hamiltonian is gapped and separates into two sectors, $\sigma(\hat{H}(\varepsilon)) = \sigma_{\text{ground}}(\varepsilon) \cup \sigma_{\text{excited}}(\varepsilon)$, such that
   – it has a ground state manifold with eigenvalues in a set σ_{ground},
   – it has a set of excited states, with eigenvalues σ_{excited},
   – the minimum distance between these sets is $\hbar\Delta_{\text{min}}$.
(4) In many proofs, the Hamiltonian spectrum is upper-bounded, that is, $E_{\text{max}} := \max_\varepsilon\{\max \sigma(\hat{H}(\varepsilon))\} < +\infty$.
(5) The evolution begins on an eigenstate from the ground state manifold, with eigenvalue σ_{ground}.

The conditions are a bit hard to read, but we can map them to our qubit example. The deformed Hamiltonian $\hat{H}(\varepsilon)$ was given by (9.1), and it had one ground state

[2] The theorems work for one state, or for a collective of states that are very close in energy, degenerate, or almost degenerate. We do not care much what happens within this collective of states, but we want to preserve all probability within this subspace.
[3] This said, we encourage the reader to have a look at the cited works, which are long, but relatively easy to follow.

$\sigma_{\text{ground}} = \{-\frac{1}{2}\hbar\Delta(\varepsilon)\}$ and one excited state $\sigma_{\text{excited}} = \{+\frac{1}{2}\hbar\Delta(\varepsilon)\}$, separated by a gap $\hbar\Delta_{\text{min}}$. The Hamiltonian was bounded, $E_{\text{max}} = \max_\varepsilon \frac{1}{2}\hbar\Delta(\varepsilon)$. We used a linearly growing control, which was therefore differentiable $\dot{\varepsilon} = \bar{v}\Delta_{\text{min}}$.

Using these conditions, the adiabatic theorems bound the diabatic excitation or error probability[4] as

$$P_{\text{excitation}} \leq \left(\frac{1}{T} \times \frac{1}{\Delta_{\text{min}}} \times \frac{E_{\text{max}}^{\alpha}}{(\hbar\Delta_{\text{min}})^{\alpha}}\right)^{2\beta}. \tag{9.7}$$

The exponents α, β are integers and depend on the particular proof and assumptions. Traditional demonstrations (Kato, 1950; Galindo and Pascual, 1991; Teufel, 2003) use $\alpha = 0$ and $\beta = 1$ and require only one-time differentiability. In these versions, the maximum energy drops out of the equation and we recover an LZ-like formula:

$$P_{\text{excitation}} \leq \mathcal{O}\left(\frac{1}{T\Delta_{\text{min}}}\right)^2. \tag{9.8}$$

If we increase the differentiability to $k \geq 2$, Jansen et al. (2007) derives a slightly better bound where the error still decays algebraically, but the power has improved: $P_{\text{excitation}} \simeq \mathcal{O}(1/T^2)^2$. Finally, Lidar et al. (2009) showed that sufficient control over the Hamiltonian, introducing $n+1$-times differentiability, allowed improving the exponent arbitrarily, to $\beta \sim 1/n$. This implies that the adiabatic error $P_{\text{excitation}}$ can be made smaller and smaller without significantly changing the total adiabatic time.

The adiabatic theorem is a powerful tool which gives full rigor to the intuitions developed in the study of the Landau–Zener processes, and extends it to arbitrary models, including large setups with thousands of qubits and resonators. This generality implies two limitations: First, the argument is perturbative, not a closed formula like (9.6); second, the theorems rely on global quantities, such as the total adiabatic passage time T and minimum gaps, without regard for the controls. Those limitations are inconsequential, because in many circumstances we do not have access to a detailed knowledge of the spectrum. Instead, we work with bounds for $\hbar\Delta_{\text{min}}$ or E_{max}, and seek a global scaling of the passage time T to ensure adiabaticity.

9.1.3 Circuit-QED Applications of Adiabatic Theorem

The Landau–Zener processes and the adiabatic theorem give us a fantastic insight into the manipulation of quantum systems, illustrating how the energy separation between eigenstates determines the regimes of adiabatic and diabatic evolution. We can use this to better design the circuit-QED and quantum computing experiments from earlier chapters.

[4] Some works study the norm-2 of the error, $|\delta\rangle = |\psi_{\text{ideal}}\rangle - |\psi(T)\rangle$. This is related to the error *probability* $\|\delta\| = \langle\delta|\delta\rangle^{1/2} = P_{\text{excitation}}^{1/2}$ and affects the powers shown in (9.7).

As an example, imagine we wish to bring a qubit in resonance with a cavity, to make them interact with the JC Hamiltonian for a brief period of time. We want this change to be "abrupt" or diabatic. We therefore need to compare the minimum energy splitting in the qubit-cavity Hamiltonian, given by the Rabi splitting $2|g|$, with the rate of change in the qubit's gap. For the control to be diabatic, we find $\dot{\Delta}/g > g$. Moreover, if we use a transmon, a parametric change of $\Delta = \omega_{01}$ introduces some perturbation into the qubit's Hamiltonian $\propto \dot{\omega}_{01}$. The analysis of this perturbation gives us another LZ-like condition, $\dot{\omega}_{01} \ll \omega_{01}^2, \omega_{12}^2$.

Alternatively, the LZ technique also gives us insight in how to adiabatically connect states of a model for different values of a control parameter, to ensure smooth adiabatic transitions. This idea can be used to create single-photon wavepackets, starting from excitations in a qubit. This experiments relies on the avoided level crossing between an excited qubit $|e,0\rangle$ and a cavity with one photon $|g,1\rangle$. The protocol begins with a qubit that is highly detuned from the cavity $\Delta < \omega$, which is adiabatically brought into resonance with the cavity and beyond, until $\Delta > \omega$. If Δ changes at a rate slower than the qubit–cavity coupling, $\dot{\Delta} \ll g^2$, this process preserves the state $|g,0\rangle$ and maps adiabatically $|e,0\rangle \rightarrow |g,1\rangle$. If we calibrate the dynamical phases properly, we can even use this protocol to transfer any quantum superposition from the qubit onto the cavity $(|g\rangle + \beta\,|e\rangle) \otimes |0\rangle \rightarrow |g\rangle\,(\alpha\,|0\rangle + \beta\,|1\rangle)$.

Finally, Zener (1932) showed that the Landau–Zenner formula is an upper bound for the excitation probability, and that other bounds are derived from the instantaneous speed $\dot{\varepsilon}(t)$. This idea has inspired the fast quasi-adiabatic passage (FAQUAD) method, which advocates fixing the rate of change in the Hamiltonian to be proportional to the minimum energy gap in the controlled quantum system $\dot{\varepsilon} \propto \Delta(\varepsilon)$. This technique enables faster controls by speeding up the dynamics in regions where the gap to excited states is still large – e.g., large $|\varepsilon|$ in Figure 9.1b. One application discussed by García-Ripoll et al. (2020) is to use these methods for accelerating the transmon-based two-qubit CZ gates from Section 8.3.3.

9.2 Adiabatic Quantum Computing Model

The adiabatic quantum computation model puts together two ideas:

- All quantum computations may be recasted as finding the ground state of a *gapped*[5] problem Hamiltonian $\hat{H}_p$.
- This ground state can be prepared by adiabatic evolution with a quantum simulator of the Hamiltonian $\hat{H}_p$ or a deformation thereof.

[5] Here "gapped" means that there is an energy gap between the ground state manyfold that encodes the computation and all other states.

The first connection was formalized by different groups, starting from Kitaev et al. (2002) until the more recent work by Biamonte and Love (2008), which creates apparently simple models for arbitrary quantum circuits. We will explore this mapping in the context of Section 9.3, where we worry about which Hamiltonians represent useful and hard computations, and which are physically implementable. But first, let us explore how realistic it is to satisfy the second requirement, which is building the ground state of $\hat{H}_p$.

9.2.1 The Adiabatic Quantum Computing Algorithm

Preparing the ground state of a generic complex quantum system is a difficult task. This is true, even if the simulated Hamiltonian is gapped and we can put the simulator in contact with a bath at a temperature so low, $k_B T \ll \hbar \Delta_{\text{min}}$, that the Boltzmann distribution predicts a macroscopic occupation of the ground state. First of all, the contact with the bath can change the eigenstates of a quantum system, causing it to thermalize into states that are different from the desired ones. Second, even if this incoherent coupling does not change the system, we cannot guarantee that *all degrees of freedom* will interact similarly with the bath. There may be uncountable properties and symmetries that have a vanishing coupling to the thermal environment and cool at impossibly slow rates.

The *adiabatic quantum computation algorithm* was introduced by Farhi et al. (2000) as a coherent protocol to create the ground state of complex Hamiltonians, using a carefully engineered adiabatic passage. The *AQC algorithm* works with a gapped problem Hamiltonian $\hat{H}_p$, whose ground state (or ground states) $|\psi_p\rangle$ we wish to prepare. It assumes the following:

(1) There exists a Hamiltonian H_0 with an easy to prepare ground state $|\psi_0\rangle$.
(2) The interpolating Hamiltonian

$$\hat{H}(\varepsilon) = \hat{H}_0(1 - \varepsilon) + \hat{H}_p\varepsilon \tag{9.9}$$

can be implemented in some hardware – what we call a *quantum simulator* – for any values of the parameter $\varepsilon \in [0, 1]$.
(3) The interpolating Hamiltonian has an energy gap that never drops to zero, $\Delta(\varepsilon) \neq 0$ (cf. Figure 9.1).

The adiabatic algorithm for preparing $|\psi_p\rangle$ consists of the following steps:

(1) Initialize the quantum simulator with interactions $\hat{H}(0) = \hat{H}_0$ and prepare its ground state $|\Psi(0)\rangle = |\psi_0\rangle$.
(2) Tune $\varepsilon(t)$ from $\varepsilon(0) = 0$ to $\varepsilon(T) = 1$ at a speed that satisfies the adiabaticity condition:

$$\frac{1}{\Delta_{\min}} \frac{\mathrm{d}}{\mathrm{d}t} \Delta(\varepsilon_t) \ll \frac{\Delta_{\min}}{\hbar}. \tag{9.10}$$

(3) Measure and characterize the final state $|\Psi(T)\rangle \simeq |\psi_p\rangle$.

The adiabatic algorithm works by smoothly distorting a state that is easy to prepare, typically an uncorrelated product state, through an adiabatic passage that is guaranteed to succeed by the adiabatic theorem. Unlike other quantum computing paradigms, such as the quantum circuits from Chapter 8, the construction of $|\psi_p\rangle$ is a continuous process, not a combination of simple, well-controlled operations.

One important question in the AQC algorithm is the choice of initial Hamiltonian H_0. Ideally, we want a model whose ground state $|\psi_0\rangle$ has some overlap with the final solution, because if we do not achieve this, the adiabatic passage might have a vanishingly small gap. A common choice is to prepare a product state that contains a quantum superposition of all states in the computational basis:

$$|\Psi(0)\rangle = |+\rangle^{\otimes N} = \left(\frac{1}{\sqrt{2}}|0\rangle + \frac{1}{\sqrt{2}}|1\rangle\right)^{\otimes N} = \frac{1}{2^{N/2}} \sum_{\mathbf{s}} |\mathbf{s}\rangle\langle\mathbf{s}|. \tag{9.11}$$

This state can be prepared as the ground state of a noninteracting model, a collection of transverse fields:

$$H_0 = -\sum_{m=1}^{N} \sigma_m^x. \tag{9.12}$$

The adiabatic protocol works by interpolating between both limits, starting with large external fields and small interactions, and gradually increasing the strength of the couplings in H_p.

9.2.2 Resource Accounting

We need a careful and *fair* understanding of the resources needed in a quantum computation to determine whether the problem can be solved with existing resources, and whether the algorithm and the implementation are competitive with classical solutions.

In the quantum circuit model, the resources are *size* and *time*, quantified by the number of qubits and the number of elementary operations (gates and measurements). If these resources grow favorably with the problem size, as it does for Shor's factoring algorithm, we consider that the quantum circuit is a good solution, and we talk about a *quantum advantage*.

In AQC, the accounting of resources also involves *size* and *time*, given respectively by the size of the quantum simulator – e.g., number of qubits required to implement $H(\varepsilon)$ – and the duration of the adiabatic passage. In AQC, the size of the simulator grows gently with the problem size. Time, on the other hand, is usually quantified as the *time to solution*, the duration of an adiabatic passage that guarantees we reach the ground state with a minimum probability p.

As discussed earlier and in the context of the Landau–Zener processes, the time to solution depends on the minimum energy gap of the problem Hamiltonian. This gap typically shrinks with growing problem size, with different scalings depending on the problem's difficulty – exponential $T \sim e^{-\alpha N}$ for difficult NP problems, algebraically for "manageable" problems $T \sim N^{-\alpha}$, as discussed in Section 9.3.4.

The relation between time and gap shows how this resource can be distorted by our choice of Hamiltonian. Imagine we can guarantee an adiabatic passage in time T for a problem Hamiltonian $\hat{H}$ with gap Δ_{min}. If we can implement a simulator for the rescaled Hamiltonian $\hat{H}' = 2\hat{H}, 3\hat{H}$, or even larger, we will arbitrarily increase the gap and arbitrarily contract the time to solution.

In practice, the strength of interactions available in the lab is finite and the total available energy $E_{\max} = \max_\varepsilon \|H(\varepsilon)\| = \max_\varepsilon\{\max\, \sigma(H(\varepsilon))\}$, is bounded. Therefore, when comparing different adiabatic quantum algorithms it is often useful to work with a rescaled dimensionless time $\tau := T E_{\max}/\hbar$, which is free from this uncertainty and scales more adequately with the problem size N. The different accounting in time and size will pop up when we compare AQC's with universal gate-based quantum computers, when we discuss the difficulty of adiabatic computations, and in Section 9.4.3, where we discuss this paradigm critically.

9.3 The Choice of Hamiltonian

The adiabatic algorithm is a global optimizer for problem Hamiltonians. It can therefore solve arbitrary optimization problems $\mathbf{s}_{\text{opt}} = \operatorname{argmin} F(\mathbf{s})$ if we encode the *cost function* into the problem Hamiltonian:

$$\hat{H}_{\text{p}} = \sum_{\mathbf{s}} F(\mathbf{s})\, |\mathbf{s}\rangle\langle\mathbf{s}|\,. \tag{9.13}$$

This suggests that we can explore the relation between problem Hamiltonians and families of computational problems $F(\mathbf{s})$ with different complexity. We will confirm this idea, deriving Hamiltonians for difficult families of classical and quantum problems, called NP-hard and QMA-complete. But first, we need to know these classes.

9.3.1 A Primer on Complexity Classes

In order to make this discussion a bit rigorous, we have to introduce some notions of computational complexity classes. Do not worry! There's not much that needs to be understood to get a grasp of what follows. We begin in the classical world, using the paradigm of Turing machines or universal classical computers, and define (qualitatively) the following three relevant classes:

P This class contains all problems that can be *solved with polynomial resources (time and space) by a classical computer.*

NP This is a larger class of problems *whose solution can be verified on a classical computer with polynomial resources.* Given a solution to an NP problem, testing whether it is correct is a task in P.

NP-complete This subset gathers universal problems in NP. That is, any problem in NP can be translated to or solved with an NP-complete problem. Note that this mapping is efficient (it takes polynomial resoures), but solving the NP-complete problem is not!

NP-hard This set contains all problems that can be used to solve NP-complete problems, but which may or may not be so easily verifiable.

You already know problems that are in P. For instance, arithmetic problems or checking whether a number is prime are tasks that can be done with polynomial resources in simple computers. You may also know some NP-complete problems, such as the map coloring problem or the graph isomorphism problem – finding whether two graphs are identical and can be mapped one to another by just relabeling the vertices. Finally, there are NP-hard problems, such as the traveling salesperson problem – finding the optimal route through all the cities in a graph – to which P and NP-complete problems can be recast, but which cannot be verified so easily.[6]

A similar hierarchy of classes has been developed for problems that can be addressed in a quantum computer:

BQP The Bounded Quantum Polynomial class contains all problems that can be *solved with polynomial resources (time and space) on a universal quantum computer,* with a fixed success probability $p > 1/2$.

QMA Quantum Merlin–Arthur is the equivalent of NP. It includes all problems *whose solution can be verified by a universal quantum computer with fixed success probability,* $p > 1/2$.

QMA-complete As with NP-complete, this class gathers all the QMA problems that can be used to solve *any* other problem in QMA, with at most a polynomial overhead in resources due to the translation.

Note how the BQP class is the quantum analog of P and contains all the quantum circuits that we can write with moderate resources and run in polynomial time. Some

6 Indeed, how do you know if the route is optimal? Note, however, that answering the question, "is there a route of length L?" is NP-complete.

well-known examples are the quantum Fourier transform and Shor's factoring algorithm. From the definitions, using the fact that a quantum computer can reproduce any classical computation, we can infer the following relations:

$$\mathrm{P} \subset \mathrm{NP} \subset \mathrm{QMA}, \quad \text{and} \quad \mathrm{P} \subset \mathrm{BQP} \subset \mathrm{QMA}. \tag{9.14}$$

We believe that BQP contains problems that are not in P. There is evidence of this in the difficulty of simulating the evolution of quantum many-body systems, or even relatively "small" random quantum circuits, where entanglement grows quickly. It is also widely believed in the academic world[7] that NP is not contained in BQP. This means that quantum computers cannot exponentially speed up the solution of those problems, even though one may not rule out other less dramatic accelerations.

In the following two subsections, we will show that both NP and QMA problems can be restated as the task of preparing the ground state of a Hamiltonian. We will identify specific representatives of those Hamiltonians that are *universal,* in the sense that they are NP- or QMA-complete and can be used to solve any problem in those classes. Interestingly, both will turn out to be models of interacting qubits that are good candidates to be implemented in the laboratory.

9.3.2 QUBO and NP-Complete Hamiltonian Problems

The *spin glass* is a deceivingly simple NP-hard problem that we find in the world of condensed-matter physics:

$$H_{\mathrm{p}}^{\mathrm{NP}} = \sum_{i,j} J_{i,j}\sigma_i^z\sigma_j^z + \sum_j h_j\sigma_j^z. \tag{9.15}$$

Barahona (1982) showed that computing the ground state of this model is at least an NP-complete problem, provided the spins are arranged on a planar graph,[8] interactions are antiferromagnetic along the edges that connect qubits $J_{ij} = J > 0$, and we can control the sign of the magnetic fields, which have comparable strength to the interactions $|h_j| \sim J$.

The spin glass in (9.15) is a diagonal Hamiltonian (9.13). Finding the ground state of this problem is thus equivalent to finding the minimum of a classical function. If we move from sign variables $s_i \in \{-1, +1\}$, to boolean variables $x_i = \frac{1}{2}(s_i + 1) \in \{0, 1\}$, the spin glass become an instance of a *quadratic binary optimization* (QUBO) problem:

$$F(\mathbf{x}) = \sum_{ij} x_i Q_{ij} x_j, \quad x_i \in \{0, 1\}. \tag{9.16}$$

This is a family of *NP-hard* problems, to which many other NP problems can be recast.

[7] But apparently not so much in the quantum investment world.
[8] A planar graph is one that can be drawn on a plane without edges crossing each other.

We call *embedding* the transformation of a QUBO problem back to a spin-glass (9.15) simulator, for instance, to solve the problem using AQC. As you may have guessed, the Q matrix of general QUBO problems can be more complex than a planar graph. In this case, the problem is typically mapped in an approximate way, identifying some variables x_i directly to spins, and introducing many other spins and interactions to reproduce the desired energy landscape.

The spin-glass model (9.15) belongs to the kind of Hamiltonians that can be simulated using superconducting qubits and resonators. Indeed, as explained in Section 6.5, the qubit–qubit dipolar interactions and the qubit's external controls may, up to suitable rescalings, imitate any kind of problem of the type (9.15). In Section 9.4, we will discuss the D-Wave quantum annealer, a quantum simulator for Ising-type spin glasses that follows these ideas, adding ingredients such as long-distance tunable couplers and large-scale integration.

9.3.3 QMA-Complete Problems

We have mentioned before that AQC and the quantum circuit model are equivalent. This idea was put forward by Kitaev et al. (2002), engineering a Hamiltonian that encodes the outcome of a quantum computation with N qubits and M gates, using an adiabatic quantum computer with $N \times M$ qubits. Several works after Kitaev's continued the assimilation of the QMA class into Hamiltonians with simpler interactions and lower complexity. Finally, Biamonte and Love (2008) demonstrated that already an apparently simple two-body Hamiltonian with XX and ZZ interactions is QMA-complete:

$$\hat{H}_p^{\mathrm{QMA}} = \sum_i \left(h_i \sigma_i^z + \Delta_i \sigma_i^x\right) + \sum_{i,j} \left(J_{ij}\sigma_i^z\sigma_j^z + t_{ij}\sigma_i^x\sigma_j^x\right). \tag{9.17}$$

The local controls and couplings in $\hat{H}_p^{\mathrm{QMA}}$ can be derived constructively from a quantum circuit. Ideally, we could simulate the XXZZ Hamiltonian in our setup, prepare the ground state using the AQC algorithm, measure its energy – this amounts to measuring Pauli operators, as shown in Problem 9.5 – and verify that energy lies below a prescribed bound, to determine the success of the computation. Moreover, using the same setup, we could also address other NP-complete problems.[9]

In practice, this introduces two technical difficulties. The first one is to ensure a topology of qubits that maps to the problems we wish to solve, or that at least can provide suitable embeddings. The second problem is how to simulate the XXZZ interactions, when the qubits that we have studied only seemed to have one type of dipolar moment.

One solution to this problem is to use flux qubits, and engineer them to enable both inductive *and* capacitive interactions. If we use current superposition as basis

[9] Note how (9.17) contains the NP-hard spin-glass model (9.15) in the limit $\Delta = t = 0$.

states, the magnetic and capacitive interactions create $\sigma^x\sigma^x$ and $\sigma^y\sigma^y$ couplings, respectively. Recent experiments by Ozfidan et al. (2020) have demonstrated simultaneous interactions in rf-SQUID qubits, which are corroborated by numerical simulations of the same setup (Consani and Warburton, 2020). However, it is still unclear to what extent we can control the J and t couplings individually, or whether, as our work suggests (Hita-Pérez et al., 2022), they appear in combination with other couplings, such as $\sigma^z\sigma^z$.

9.3.4 Scaling of Resources

The analysis of time as a resource is a very complicated and controversial matter. Ideally, the adiabatic theorems provide a sufficient condition for the passage to succeed. Equations (9.7), (9.8), or (9.10) give a heuristic scaling of T based on global quantities, such as the minimum gap Δ_{min}. In practice, we rarely know Δ_{min}, but we may calibrate the difficulty and the scaling of time based on a coarse understanding of the interpolating Hamiltonian.

A central idea is that the models $\hat{H}_0$ and $\hat{H}_p$ are so different that the interpolating Hamiltonian $\hat{H}(\varepsilon)$ must experience a *phase transition* as we increase the parameter ε from 0 to 1. For infinitely large systems, the energy gap strictly vanishes at the critical point where we transition between ground states of a different nature. However, for finite systems such as the ones we study in the quantum simulator, we will experience some type of finite-size scaling $\Delta_{\text{min}} = f(N)$.

A lucky situation would be that, when interpolating between (9.12) and (9.15), the lattice of qubits experiences *only* a *second-order phase transition*, in which the gap shrinks algebraically with the problem size $\Delta_{\text{min}} \sim N^{-\alpha}$. Those problems are considered "easy" because the time to solution grows "slowly" with the problem size, as $T \propto N^{\alpha}$.

More generally, for NP-hard problems the spin glass $\hat{H}_{\text{p}}$ will exhibit a collection of almost degenerate ground states that are macroscopically different from each other. In those models, the interpolating Hamiltonian passes through a first-order phase transition, where the gap vanishes exponentially fast $\Delta_{\text{min}} \propto \alpha^{-L}$ and the time-to-solution diverges exponentially $T \propto \alpha^{L}$. This growth is consistent with the hardness of the problem and the scaling of classical algorithms, and reveals that we cannot obtain an exponential advantage by using the quantum algorithm.

This type of physics is so common that it can already be observed in simple one-dimensional problems $H_p = \sum J_i \sigma_i^z \sigma_{i+1}^z$. When the couplings are regular, such as an all-ferromagnetic interaction $J_i > 0$, the interpolating Hamiltonian will cross a second-order phase transition from para- to ferromagnetic ground states. At the critical point on which the transition happens, the gap vanishes algebraically around the transition point, $\Delta_{\text{min}} \simeq N^{-\nu}$, and the passage is experimentally "affordable."

However, if we make the Ising interaction random,[10] then suddenly the phase transition from paramagnet to spin glass will scale much worse, $\Delta_{\min} \propto \exp(-g\sqrt{N})$.

9.4 D-Wave's Quantum Annealer

In 1999, a small team of enterpreneurs and academic researchers founded a company to work on superconducting quantum technologies. This company is known for one of the longest and more focused efforts toward building an adiabatic quantum processor using superconducting circuits. In 2007, the firm made a modest and rather controversial debut introducing a superconducting chip with 16 qubits. After this demonstration and further improvements in the machine's design, the company produced the so-called D-Wave One, a larger processor with 128 qubits, which was later followed by the larger D-Wave Two (2012) with 512 qubits, the D-Wave 2X (2015) with about 1 152 functional qubits, the D-Wave 2000Q (2017) with 2 048 qubits, and a D-Wave Advantage, with 5 640 qubits, already disclosed as of this writing.

In this section, we are going to discuss D-Wave's older architecture. Following the literature, we must call D-Wave's devices *quantum annealers* and not an adiabatic quantum computer, because they are not capable of universal quantum computation – they are not QMA-complete. Instead, D-Wave's efforts focus on NP-hard QUBO optimization problems, implementing powerful quantum simulators for Ising models with transverse magnetic fields:

$$H(t) = \sum_i \Delta_i(t)\sigma_i^x + \sum_{ij} J_{ij}(t)\sigma_i^z\sigma_j^z + \sum_I h_i(t)\sigma_i^z, \tag{9.18}$$

Here, you recognize the spin-glass model (9.15) in combination with the initial bias field $H_0 \propto \Delta(t)$ from which we start the adiabatic passage (9.12). D-Wave's machines can be programmed to change in time all parameters from the effective Hamiltonian (9.18), allowing for a good dynamical range and determination of the sign in all variables. This enables the users to implement an adiabatic passage in which $\Delta(t) \propto (1 - \varepsilon(t))$ is decreased, while the problem Hamiltonian is enhanced $J(t), h(t) \propto \varepsilon(t)$.

The quantum simulation, using superconducting devices, of an NP-hard Hamiltonian with time-dependent quantum fluctuations (9.18) is a great scientific achievement and a great technological challenge worth studying. We will discuss the devices' architecture, how the annealer works in practice, its performance in comparison to classical algorithms, and how "quantum" the D-Wave machine really seems to be.

[10] See section 4.1.2 in Bapst et al. (2013) for an overview.

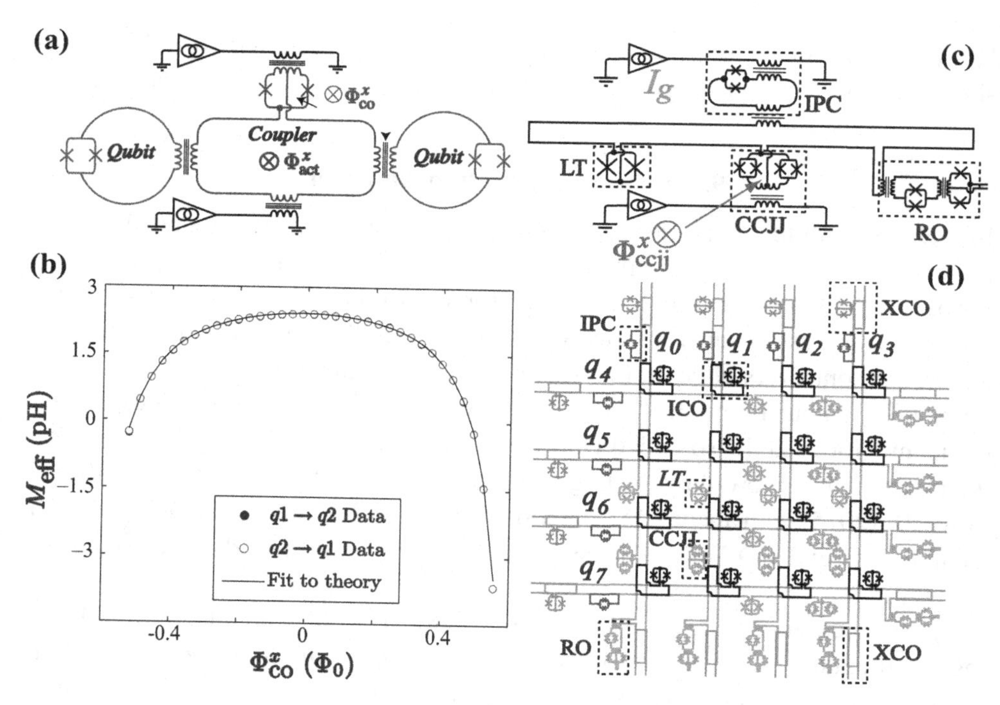

Figure 9.2 (a) Tunable coupling circuit for two flux qubits with an interaction mediated by an rf-SQUID. (b) Experimentally measured mutual inductance between the two flux qubits for this particular circuit. (c) Detail of a flux qubit design, including elements to tune the qubit frequency (LT), tune the qubit gap (CCJJ), induce an external field (IPC) and read-out the qubit state (RO). (d) Architectural design of a cell with eight qubits in the D-Wave chimera graph. All qubits have local controls, tunable interactions and readout, and talk to each other through the ICO couplers. Different cells talk to each other via the XCO couplers. Figure adapted from Harris et al. (2009) and Harris et al. (2010b) with permissions.

9.4.1 D-Wave's Architecture

D-Wave's architecture is summarized in various panels from Figure 9.2. The simulator is made of moderately good flux qubits, with controllable tunneling strength and magnetic bias. Qubits interact with each other through a long-range inductive coupler with tunable inductance to implement matrix J_{ij}. The whole setup sits on top of a sophisticated circuitry – partly superconducting, partly semiconducting digital technology – that provides the dynamical controls of the local and interact on terms. Let us go through the qubits and couplers in detail.

Highly Tunable Flux Qubits

The first ingredient in Hamiltonian (9.18) is a two-level system with adjustable "magnetic fields" along the σ^z and σ^x direction. Until 2017, D-Wave relied on slightly modified rf-SQUIDs to implement these two-level systems (Harris et al., 2010a). Introduced in Section 6.4, rf-SQUID qubits consist of a single Josephson junction and a large inductor. The qubit is crossed by a half-integer multiple of the magnetic flux quantum $\Phi_{\text{ext}} = \frac{1}{2}\Phi_0$, to induce phase frustration. The inductive potential thus acquires a symmetric double-well structure (cf. Figure 6.6b), with minima on two states with opposite current, $|L\rangle$ and $|R\rangle$. We identify these as our qubit states, which are coupled by tunneling Δ_i, as in Hamiltonian (9.18).

The choice of a flux qubit for an Ising-type quantum simulator is very natural because of the following:

- We control the magnetic field $\sim\sigma^z$ through the injected flux.
- We can control the tunneling amplitude $\sim\sigma^x$ if we design the qubit with a tunable Josephson junction, as explained later. This allows us to reduce the qubit's gap Δ_i at the end of the passage, something that would be hard for other qubits.
- Flux qubits have a very large anharmonicity and work very much like a two-level system, without leakage to other states.
- Flux qubits interact inductively. As explained in Section 6.5.4, this type of interaction can be mediated and tuned with SQUIDs or similar devices.
- Because the anharmonicity can be very large, we can afford large qubit–qubit dipolar interactions $J\sigma_i^x\sigma_j^x$. This is required at the end of the phase transition, where the problem Hamiltonian $\hat{H}_p$ dominates.

To implement all these features and simulate the quantum annealing Hamiltonian, D-Wave has introduced changes in the qubit design (Harris et al., 2010a,b). The changes, illustrated in Figure 9.2b, work as follows:

(1) The tunable Josephson junction of the rf-SQUID is implemented by a complicated circuit, called a *compound Josephson junction* (CCJJ). Introduced in Harris et al. (2010a), this circuit works like a dc-SQUID, creating a nonlinear inductive potential $E_J(\Phi^x_{\text{ccjj}})\cos(\phi/\varphi_0)$ for the qubit's flux variable ϕ. The Josephson energy E_J depends on the external flux that passes through the CCJJ. Changing this flux modifies Josephson barrier E_J between potential wells. In other words, the field Φ^x_{ccjj} may increase or decrease the tunneling amplitude $\Delta(\Phi^x_{\text{ccjj}})$ in Hamiltonian (9.18).

(2) Changes in the qubit's Josephson energy affect the critical current in the loop, as well as the inductive coupling to the external fields and to other qubits. The qubit's current is inversely proportional to the inductance of the loop $I_{L,R} \sim \pm\phi/L$. We can adjust and compensate for changes in this value by inserting a dc-SQUID, labeled LT in Figure 9.2c. The dc-SQUID operates in the linear regime as a tunable inductor $L(\Phi)$ (see Section 4.7.2), whose value can be decreased to increase the current, and vice versa.

(3) An adjustable inductor – labeled IPC in Figure 9.2c – mediates the interaction of the qubit with a current source I_g. By adjusting this inductor, we can control how much flux enters the qubit,[11] thereby controlling the effective magnetic field $\frac{1}{2}\hbar h_i(t)\sigma^z$ in (9.18).

(4) Another tunable inductor controls the coupling between the qubit and a readout (RO) device. The RO is a current-biased CCJJ that operates very similarly to a current-biased dc-SQUID (cf. Section 4.7.2). During a measurement, the tunable inductor is activated, so as to inject a small amount of flux into the CCJJ $\delta\phi = \delta\phi_0 + \eta\sigma^z$. If the current bias and η are both large enough, the CCJJ will switch to a voltage state when the qubit is in one of the states $\sigma^z = +1$, remaining inert in the other case. As explained in Berkley et al. (2010), this implements the almost projective measurement described in Section 6.4.4. Naturally, to minimize decoherence, at all other times the mediator is deactivated, eliminating any residual coupling to the RO device ($\eta = 0$), and minimizing decoherence.

We see in this design an abundant use of tunable inductors as control elements for qubit gaps, magnetic fields, qubit potentials, measurement, and decoherence. A simple counting of elements in Figure 9.2c reveals a total of six hardware controls per qubit – one flux control for the inductor, one for the qubit gap, two for the measurement, and two for the bias. Some of these controls are stabilized with the use of the CCJJ, a device that allows better tolerance against fabrication errors – for instance, small variations in the Josephson energies in the coupler circuit – and a perfect cancellation of inductance.

Tunable Qubit–Qubit Coupling

The loop of an rf-SQUID qubit is not required to have any particular shape, other than enclosing the right amount of magnetic flux. This allows experimentalists to stretch the flux qubits so that they overlap with each other, forming a highly connected graph. Figure 9.2d sketches how this is done for a unit cell of a D-Wave quantum processor with eight qubits and 16 crossings.

Wherever two qubits overlap, both qubits experience an enhanced magnetic flux, influenced by the other qubit's current state. Similar to Section 6.5.1, the mutual inductance between qubits creates a $J_{ij}\sigma_i^z\sigma_j^z$ dipolar interaction.[12] The constant J_{ij} is nonzero only for connected qubits – e.g., J_{04}, J_{14}, etc., a total of 16 connections

[11] See Section 6.4.2 and around (6.56), with the convention that now $|L\rangle$ and $|R\rangle$ are the qubit basis states.

[12] Section 6.5 defined the dipolar moment operator in the qubit basis at the degeneracy point. However, in this chapter we have changed basis, and the dipolar moment operator is defined in term of current states $\sigma^z = |L\rangle\langle L| - |R\rangle\langle R|$.

in Figure 9.2. It is, however, static in nature, because the mutual inductance only depends on the topology and geometry of the qubits' intersection.

In order to achieve tunability of the coupling J matrix, D-Wave's researchers added tunable inductors that mediate additional couplings. As sketched in Figure 9.2a and d, D-Wave's mediator is an rf-SQUID that contains a CCJJ circuit instead of a junction. The SQUID implements the interaction mechanism from Section 6.5.4, but now with two controls: the SQUID's internal flux and the amplitude of the Josephson energy.

Figure 9.2b illustrates the dynamical range of the qubit–qubit mutual inductance: One may control not only the amplitude, but also the sign of every qubit–qubit interaction, reinforcing, canceling, or even reversing the geometric inductance previously discussed. As for the orders of magnitude, the mutual inductance shown before ranges around ± 1 pH, a very decent value. Considering that the qubit persistent currents are $|I_p| \simeq 1.0\,\mu\text{A}$ for each of the two possible orientations, this gives a safe estimate for the interaction energy $E = M|I_p|^2 \simeq 10^{-24}$ J, which corresponds to a frequency $\omega \simeq E/\hbar \simeq 2\pi \times$ GHz. This value is consistent with the 4 GHz antiferromagnetic splitting measured spectroscopically for qubits within a single cell of the D-Wave architecture (Berkley et al., 2013; Lanting et al., 2014).

9.4.2 Device Operation

The operation of the flux-qubit quantum simulator has four phases: a *design* phase, in which we find out the parameters of the Hamiltonian that we want to program (9.18); a *programming* stage, in which we schedule the flux biases that will control the local and interaction terms in the Hamiltonian, determining the annealing schedule; an *execution* phase, where the annealing is performed; and a final *measurement* stage, in which we detect the state of each qubit in the simulator.

Design and Embedding

Before using the qubit array to implement an adiabatic passage, we need to design a set of interactions and local fields, and a protocol to change them along the passage, to simulate the interpolated Hamiltonian $\hat{H}[\varepsilon(t)]$. Assuming that $\hat{H}_0$ is just a transverse magnetic field (9.12) with $J = h = 0$, we can focus on determining the couplings J and biases h that implement the problem Hamiltonian at the end of the passage – i.e., the spin glass (9.18) or its equivalent QUBO formulation (9.16).

Unfortunately, as explained before, the topology of the desired problem Hamiltonian, given by the QUBO matrix Q, will not coincide in general with the qubit arrangement in the quantum simulator. In particular, D-Wave's design separates qubits into cells of eight elements, with 16 mutual interactions. Each of these cells then interacts with every neighboring cell via four of its qubits. For instance, in Figure 9.2d, qubits 4, 5, 6, and 7 connect to neighboring cells on the same column, while qubits 0, 1, 2, and 3 propagate interactions along a row. This sparse, nonplanar topology is called the *Chimera graph.*

Our first goal is to find an *embedding* of the problem Hamiltonian H_p as an equivalent model H_p' implemented in the Chimera graph. Usually, the embedding is done by directly identifying the logical qubits of $\hat{H}_p$ as a subset of the *physical qubits* in the destination graph. Other qubits within the graph are used to mediate interactions between the logical bits. The embedding must be such that $\hat{H}_p'$ and $\hat{H}_p$ exhibit similar low-energy eigenenergies, and from every configuration of the embedded model $|\mathbf{S}'\rangle$ we can deduce the logical bits $|\mathbf{s}\rangle$ of the desired model.

In some lucky cases, the graph that we want to simulate can be found as a subgraph – a *minor* – of the Chimera graph. In this case, we need no auxiliary qubits and the interactions can be mapped directly to couplings in the Chimera lattice. For instance, the four-qubit model $H = J\sigma_a^z(\sigma_b^z + \sigma_c^z + \sigma_d^z)$ can be embedded into the unit cell from Figure 9.2d, identifying $(a,b,c,d) \to (q_4, q_0, q_1, q_2)$ in the unit cell, and setting $J_{40} = J_{41} = J_{42} = J_{43} = J$ in the annealer's controls.

Other times we need to copy a qubit, duplicating its value, because there is not an exact one-to-one embedding. For example, think of the triangular lattice cell $H = J(\sigma_a^z\sigma_b^z + \sigma_b^z\sigma_c^z + \sigma_c^z\sigma_a^z)$. We can implement this model with the logical qubit assignment $(a,b,c) \to (q_4, q_0, q_1)$, and copy the value of q_0 to q_5 with a strong ferromagnetic interaction. This means we will have a coupling matrix $J_{40} = J_{41} = J_{51} = J$, and $J_{50} = -J_{\text{extra}}$, with a large constant $|J_{\text{extra}}| \gg |J|$.

More generally, we need to engineer arbitrary interactions. The technique of perturbative gadgets by Kempe et al. (2006) allows the engineering of rather general effective interactions starting from an experimental platform. As before, we identify logical qubits with physical qubits, and add auxiliary qubits and interactions to reproduce the target model. However, this framework is more systematic and powerful, enabling the simulation of arbitrary two-, three- and other many-body interactions between the logical qubits, which are engineered from the simple-looking Hamiltonian in (9.18).

Embedding and gadgets extend the family of models accessible to a given architecture, but they have associated costs. First, the use of auxiliary qubits increases the *size* of the computation. Venturelli et al. (2015) studied the worst-case scenario of a fully connected network and found that the Chimera graph requires $\mathcal{O}(N^2)$ physical qubits to represent N logical ones. Second, the search for a good embedding is itself an expensive classical overhead – minor embedding is an NP-hard problem. Therefore, in many cases we must resort to a suboptimal, but good-enough heuristic (Cai et al., 2014) for the simulation. Third, some gadgets will produce interactions that can be perturbatively small, leading to a decrease in the available energies $E_{\max}$

and an increase of time. Finally, some embeddings require many auxiliary qubits and long chains of interactions. The transverse magnetic field combined with these long chains produces a contraction of the gap[13] $\Delta_{\min}$ and an increase of passage time.

Annealing Schedule and Control

Once we have defined the values of the parameters Δ, J_{ij}, and h_i at the beginning $t = 0$ and at the end $t = T$ of the passage, we need to find a control that interpolates between both limits. In Section 9.2, we suggested a linear ramp of the fields, which decreases the transverse field $\Delta \propto (1 - t/T)$, at the same rate it increases the strength of the problem Hamiltonian $J, h \propto t/T$. It turns out that this is not a natural choice for a flux-qubit design, because the values of Δ, J and h are all related to the critical current of the qubit.

Take, for instance, the tunneling amplitude. It decays exponentially as we increase the Josephson energy of the qubit, $\Delta_i \propto \exp(-\alpha E_{J,i}(\Phi^x_{\text{ccjj},i}))$. We can find a schedule for the control flux $\Phi^x_{\text{ccjj}}(t)$ that imposes the desired decrease in the quantum fluctuations Δ_i. However, that schedule will also change the critical current of the qubit[14] $I_{J,i} \propto E_{J,i}$, enhancing the coupling to the external bias fields $h_i \propto I_{J,i}$ and the qubit–qubit interactions $J_{ik} \propto M_{ik} I_{J,i} I_{J,k}$.

These implicit dependencies may cause some terms, such as the transverse fields, to decrease faster than other terms, as biases, and interactions grow. To some extent, one may compensate these dependences through the control of the mutual inductances M_{ik} or the qubit's injected flux. However, it will quite often happen that both the maximum available energy $\|H_p(t)\| \sim E_{\max}$ and the instantaneous gap Δ will deteriorate, affecting the actual passage speed and conditions of the adiabatic theorem. As example, see figure 9 in Albash and Lidar (2018), which shows the annealing schedules as measured by D-Wave for the 2000Q and 2X devices, along the dimensionless passage time $s = t/T$.

Programming and Measurement Overheads

In D-Wave's devices, the annealing schedule is programmed into a specialized superconducting circuitry, using an embedded digital-to-analog converter and on-board superconducting switches (Johnson et al., 2010). The control circuitry also takes care of the final measurement of all qubits, extracting the information at the end of the passage. The total execution time of an annealing protocol thus includes three times: the programming time, the annealing time, and the measurement time.

[13] Qualitatively, this contraction is caused by the difficulty of transverse fluctuations $\Delta_i \sigma_i^z$ in effecting a change over a string of qubits that want to be on the same state (see Problem 9.6).

[14] See figure 5 in Harris et al. (2010b) and how a change of Φ^x_{ccjj} by an amount $0.02\Phi_0$ causes an increase of I_J from 0 to 1.6 μA, and an exponential decrease in Δ from 10^9 to 10^5 Hz.

The programming and measurement times are constant overheads that do not depend so much on the problem at hand. These can be very large, with respective values around 7 ms and 125 μs reported for a D-Wave 2000Q by Albash and Lidar (2018). The annealing time is problem dependent, but it is usually the shortest value in the whole experiment. Due to strong filtering of controls to limit the influence of high-frequency noise, the shortest annealing time starts at 20 μs in the D-Wave Two device and 5 μs in D-Wave One, Two X, and 2000Q. This minimal passage time can obscure or perturb the interpretation of resource scaling, as explained by Mandrá et al. (2016).

9.4.3 Performance Analysis

D-Wave's quantum optimizer is probably the largest man-made quantum device with independently addressable qubits, tunable local fields, and fully adjustable interactions.[15] It is therefore important to understand how coherent or how "quantum" these devices are, and what intrinsic and extrinsic causes limit their performance. Unfortunately, the architectural choices that facilitate the operation of these quantum annealers – highly filtered and slow controls, slow measurements at the end of the control, no individual qubit addressing or excitation – prevent us from applying sophisticated quantum information methods to analyze coherence, evolution, and entanglement. For instance, the tomography protocols from Section 8.4.2, fidelity measurements, and randomized benchmarks from earlier chapters are out of question. Instead, we obtain information from the device by studying its operation as quantum simulator and through multiple adiabatic passages.

Evidence of Quantum Effects

One way to show evidence of quantum effects is to study the ground state and excitations of the interpolating Hamiltonian (9.18) in different regimes of interactions: from the paramagnetic phase ($J, h = 0$), to the spin-glass phase ($\Delta = 0$), passing through the strongly correlated phase transitions. Unfortunately, D-Wave's chips are protected by strong low-pass filters that eliminate any high-frequency microwave. This means we cannot use microwave spectroscopy (Sections 7.4.1 and 7.4.2) to probe the simulator.

Berkley et al. (2013) designed a method to circumvent this limitation, which they called *qubit tunneling spectroscopy.* This technique relies on isolating a *probe* qubit

[15] In terms of qubits, it may be argued that ultracold atoms in optical lattices have larger systems. However, as of this writing, they have not exhibited a similar control of both local fields *and* separate interactions between neighboring quantum systems.

that interacts with other qubits that implement the desired Hamiltonian (9.18). The setup is described by the effective model

$$\hat{H} = \frac{\Delta_0}{2}\sigma_0^x + \frac{\varepsilon_0}{2}\sigma_0^z + J\sigma_0^z\sigma_k^z + \frac{1}{2}\varepsilon^{\text{comp}}\sigma_0^z + \hat{H}_p. \tag{9.19}$$

We assume that the probe qubit σ_0^z interacts with only one other qubit σ_k^z from the many-body simulator $\hat{H}_p$. We control the probe's bias ε_0 and tunneling Δ, as well as a compensating field acting on the contacted qubit. When this field is $\varepsilon^{\text{comp}} = 2J$, the eigenstates of H split into two manifolds, with different values of $\sigma_0^z = \pm 1$, which are coupled by a weak tunneling Δ_0:

$$\hat{H} = \hat{H}_p + |1\rangle_0\langle 1| \otimes \left(\frac{1}{2}\varepsilon_0 + 2J\sigma_k^z\right) + \frac{1}{2}\Delta_0 \mathbb{1} \otimes \sigma_k^x. \tag{9.20}$$

The system is adiabatically prepared to an eigenstate $|1_0, \psi_{\text{ref}}\rangle$, using a very large negative bias $\varepsilon_0 \ll -2J$, a large coupling $J > k_B T$ that guarantees a large separation from other states. This state will be our reference energy E_{ref}. We will then adiabatically shift the bias to a higher value $\varepsilon_0 + \delta\omega$, until $E_{\text{ref}} + \delta\omega = E_n$ and our reference state becomes resonant with $|0_0, n\rangle$, an eigenstate of the problem Hamiltonian $H_p\,|n\rangle = E_n\,|n\rangle$. At resonance, the tunneling Δ_0 becomes relevant and enables oscillations between states, $|1_0, \psi_{\text{ref}}\rangle$ and $|0_0, n\rangle$, which can be detected by measuring the probe qubit. The probability of a probe qubit flip as a function of ε_0 will show peaks centered on those resonances, with linewidths the depend on the decoherence rates of the qubits.

This method was applied by Lanting et al. (2014) to study the eigenenergies and eigenstates of a Hamiltonian H_p with two and eight qubits from a unit cell in the Chimera graph. Their spectroscopy was in perfect agreement with the predictions of an exact diagonalization model. It revealed all regimes of the multiqubit spin model, the competition between interactions J_{ij} and quantum fluctuations Δ_i, and the reduction of the gap around the phase transition point. Moreover, those studies showed that, for such a small number of qubits, the gap was large enough that an adiabatic passage could prepare the ground state with great fidelity $F \simeq 1$.

Another way to probe the quantum properties of the annealer is to study how well it simulates the Hamiltonian $\hat{H}_p$, and other equivalent models, from a purely condensed-matter physics point of view, such as studying phase transitions. In a remarkable work, Harris et al. (2018) simulated a complete $8 \times 8 \times 8$ 3D spin lattice, embedding the 3D Hamiltonian of 512 qubits into a Chimera architecture with 2 048 qubits. Engineering the right annealing passages, they could explore the three-dimensional phase space determined by the effective temperature T, the spin–spin interaction J, and the transverse magnetic field Δ. They could probe all paramagnetic, anti-ferromagnetic, and spin-glass phases, as well as find the phase boundaries predicted by quantum Monte Carlo.

The work by King et al. (2018) explores a different type of physics, looking at a topological phase transition that appears for 2D Ising models with transverse field. Once more, they worked with the embedding method to simulate frustrated spin lattices with 1 800 logical qubits on a 2D cylindrical topology and different connectivities (triangular lattice and square-octagonal) on a superconducting processor with 2 048 physical qubits. Using a very low-temperature adiabatic passage $T \simeq 8$ mK, they could explore the Berezinksii–Kosterlitz–Thouless phase transition from a paramagnetic phase – a large transverse magnetic field Δ – into the critical region, comparable Δ and J – where vortices appear. They could probe the appearance of a long-range, algebraically decaying topological order – the vortex phase – and compare with simulations based on state-of-the-art quantum Monte Carlo methods, which matched with the effective temperature $T = 8.4$ mK.

Both of these works represent a deep exploration of quantum simulation with large systems, only comparable to contemporary experiments with ultracold atoms in optical lattices. Moreover, these works demonstrate the value of large quantum processors for studying fundamental physics, and confirm that the ideas from adiabatic quantum computing are useful even outside of the computing world.

Decoherence

Despite the evidence of coherent macroscopic quantum tunneling events (Lanting et al., 2014), and simulation of quantum many-body phase transitions (Harris et al., 2018; King et al., 2018), D-Wave quantum processors up until 2019 have too large decoherence rates to be considered an adiabatic quantum processor.

The role of decoherence can be studied at the single-qubit level. Harris et al. (2010b) analyzed the tunneling between current states in a qubit at $\Delta \neq 0$, to reveal fluctuations in the qubit's flux and environmentally induced dephasing. Later, Berkley et al. (2013) used qubit tunneling spectroscopy to analyze the spectra and coherence of one qubit. Both studies suggested a large decoherence rate, with qubit linewidths $\simeq$0.7 GHz, only an order of magnitude away from the qubit–qubit interactions. This sets the decoherence time $T_2^* \sim 1$ ns, three to six orders of magnitude worse than state-of-the-art superconducting qubits (cf. Section 6.6), and much shorter than the shortest passage times $\sim$5 μs.

Harris et al. (2010b) further studied the behavior of a small eight-qubit quantum simulator when subject to a control designed for an adiabatic passage. They showed that the dynamics of the qubits could be modeled using a master equation that combines the interpolating Hamiltonian with a finite heating rate – given by the effective temperature of the chip – and a strong dephasing rate – essentially the

T_2^* of a single qubit. Even if the low temperatures were unable to bridge the gap of this relatively small system, and the experiment could prepare ground states with high fidelity, the study evidences that the superconducting processor does not implement an adiabatic quantum computation in the sense of Section 9.2. Instead, we can describe this dynamics as an implementation of *quantum annealing*.

Quantum Annealing versus AQC

The term *annealing* originated in the world of metallurgy and material science, and refers to a smooth change of the temperature of a material, starting from a hot, noisy, highly "destructive" environment capable of creating many excitations, down to a cool and quiet environment. When done properly, classical annealing can take highly excited material down to an almost perfect equilibrium state, where atoms or molecules arrange into perfect lattices with few defects.

One extension of this concept to the computational world is known as *simulated annealing*. The algorithm takes as input a function $F(\mathbf{s})$ to be optimized – e.g., a classical NP-hard spin glass (9.15) or a QUBO function. It reinterprets this function as the energy of a classical system that is in contact with a bath at temperature T. Using Markov-chain Monte Carlo or some other technique, it simulates the thermal fluctuations of the variables $\mathbf{s}$ in a process of continuously decreasing temperature. After a sufficiently long time – the *time to solution* – the classical algorithm will visit with high probability the minimum of $F(\mathbf{s})$ or a good approximation thereof.

Quantum annealing is a related idea. It was introduced by Kadowaki and Nishimori (1998) as a mean to approximate classically the ground states of quantum Hamiltonians. Instead of looking just at the temperature, quantum annealing simulates quantum fluctuations with progressively decreasing strength. The method emulates, by quantum Monte Carlo techniques, the physics of the interpolating Hamiltonian (9.18), starting with a Hamiltonian in the regime of large transverse fields Δ, and crossing over to a regime of dominating interactions, where we recover the ground state of the problem Hamiltonian $\hat{H}_p$.

Unlike adiabatic quantum computation or the idea of adiabatic passages, quantum annealing also includes the possibility of additional sources of fluctuations, such as temperature. Probably for this reason, we find the term quantum annealer already in early descriptions of quantum simulators (Harris et al., 2010b), which implement adiabatic-like passages with interpolating Hamiltonians, in nonideal conditions – interactions with the environment, dephasing, heating or cooling due to finite temperatures, etc. We will now discuss how theoretical and experimental evidence supports this analogy, and how even the idea of quantum annealing affects the computational power of the simulator.

Evidence of Quantum Annealing

Quantum annealing, adiabatic quantum computers, and simple Landau–Zener passages all exploit quantum fluctuations (quantum tunneling) to access other states and configurations in a more efficient way than classical techniques.

Consider the LZ passage for a flux qubit. When we change the bias field ε from a large negative value to a large positive one, we are changing from a potential in which the minimum energy state is the $|L\rangle$ current (cf. Figure 6.6b) to the opposite configuration, with a minimum around $|R\rangle$. Classically, the qubit would only shift from L to R when the potential is so tilted that the $|L\rangle$ state becomes unstable. However, the quantum mechanical fluctuations introduced by the capacitor enable the tunneling of the qubit between minima, and already at $\varepsilon = 0$, halfway through the qubit hyperbola, 50% of the probability has been transferred from one well to another.

One school of thought supports the idea that quantum fluctuations, when applied in the right Hamiltonians and annealing passages, can provide a computational advantage in optimization problems. The argument is that quantum fluctuations enable tunneling across local minima, avoiding incorrect solutions and reaching the true ground state faster than through classical optimization techniques. A legitimate question is whether this advantage is present in existing NISQ devices and for the type of NP-hard problems that we can simulate (9.18).

We already have evidence that quantum fluctuations are present in the adiabatic passages with flux-qubit processors. The experiments by Johnson et al. (2011) and Dickson et al. (2013) demonstrated quantum effects in spin models with 8 and 16 qubits. Boixo et al. (2013) studied also small devices with eight qubits and showed that the annealing passages were consistent with exact simulations based on master equations – that is, evolution that combines an adiabatic passage Hamiltonian (9.18) with some decoherence. The signatures of the passage were also incompatible with models based on instantaneous thermalization or classical evolution.

Boixo et al. (2014) extended this study to passages with 108 spins. Because the system was much larger, instead of master equations, they had to simulate the quantum annealing using a stochastic method known as simulated quantum annealing (SQA), which is an adaptation of quantum Monte Carlo to emulate the annealing schedule. Their work found that SQA correlated very well with the outcomes of the D-Wave annealer. Thus, instances which are hard – i.e., low success probability – for Monte Carlo SQA are also hard for the flux-qubit simulator, and vice versa. This correlation was not found for classical methods based on simulated annealing.

Nowadays, there is some evidence that both fluctuations *and* quantum correlations are relevant in the actual device. Albash et al. (2015) showed that passages with 20 qubits in the D-Wave 2 processor are consistent with correlated models based on master equations, as in Boixo et al. (2013). This work also disproved that experiments could be fully explained using classical models of spin-dynamics (SD) and spin-vector Monte Carlo[16] (SVMC), based on the statistics of the output states and excited state dynamics during the passages. Later, Boixo et al. (2016) and Denchev et al. (2016) have shown the relevance of finite-range tunneling – quantum tunneling events where more than one qubit change state simultaneously – in passages with specially crafted Hamiltonians and up to 945 qubits.

Computational Advantages

One original motivation for constructing the flux-qubit quantum simulator of the spin-glass Hamiltonian (9.18) was the possibility of solving NP-complete problems, and whether quantum effects – fluctuations, entanglement – can improve the efficiency in finding those solutions. It is for this reason that many works have analyzed the scaling of the physical quantum annealer, comparing it to other classical methods, with various results.

A first step in these analyses is figuring out the right metric. It turns out that this is the *time to solution,* which is the shortest execution time, including repetitions, that guarantees a solution with a high success probability, such as 99% (see Problem 9.7). A second step is to choose the right *classical* algorithms for the comparison. Here the focus has been on generic algorithms that work for all problems, namely spin dynamics, simulated annealing, spin vector Monte Carlo, and Monte Carlo simulation of quantum annealing passages. This ignores to some extent the fact that some problems have specific tailored methods that are very efficient.[17]

Denchev et al. (2016) is one of the first works to prove a constant speedup in the physical quantum annealing (QA) as compared to the classical method, with up to eight orders of magnitude. They found a similar scaling between SQA and QA with the problem size, but the Monte Carlo method was about 10^8 times slower in real time. Later works by Mandrá et al. (2016) and Mandrá and Katzgraber (2018) have extended the comparison to other sequential and parallelized methods, considering also some tailored algorithms. These works find also a similar scaling between SQA and QA, with a constant speedup factor in time to solution, although they also realize that this speedup could be compromised by minimum passage time and hint at the need to find the optimal time over all possible passages. Albash and Lidar (2018) performed the optimal time-to-solution analysis for the first time, comparing hard problems with a planted solution. This study confirms the speedup

16 This is a method introduced by Shin et al. (2014) that represents the quantum register as a product state of uncorrelated spins, updated in an easy-to-simulate classical method that accounts for some of the quantum fluctuations.

17 For instance, there is a very clever and very efficient algorithm that is specific for solving QUBO problems in the Chimera graph topology, due to Selby (2014).

of QA over classical simulated annealing. It also finds that QA is quantitatively faster, but it finds for the first time that SQA and SVMC have a slightly better asymptotic behavior – though in absolute terms it is larger, the optimal TTS grows slightly slower than QA with the number of qubits.

9.5 Summary and Outlook

In this chapter, we have introduced a new computational model based on the preparation of ground states from problem Hamiltonians. We have introduced the adiabatic quantum algorithm as a specific method to solve this task and presented sample Hamiltonians that encode useful and universal problems. In the second half of the chapter, we have set our focus on a superconducting circuit architecture that simulates adiabatic passages with NP-complete Hamiltonians. We have discussed the technology and limitations of this architecture and investigated how the device performs in actual experiments, and what those experiments teach us about adiabatic computations and quantum annealing.

Unfortunately, this discussion is a bit muddied by the intrinsic difficulties and challenges posed by the choice of Hamiltonians. Some key messages that we can distill from the broad literature are the following:

- NP-complete spin-glasses are stoquastic models that can be efficiently simulated using Monte Carlo methods and similar techniques. As such, even if experiments show that quantum fluctuations have a computational advantage, there is no exponential quantum advantage, because those fluctuations can be simulated by classical computers.
- D-Wave's design and implementation represent a very important technological milestone in the development of quantum simulators and quantum technologies. Those devices have still interesting fundamental applications, can be used to probe interesting physics, and may involve a de facto computational speedup with respect to other simulation techniques.
- Existing quantum annealers are based on relatively old qubit technology, with relatively short coherence as compared to modern qubits, long readout times, slow controls, and sparse, limited connectivity.

There are some obvious paths to move ahead, both in the field of quantum annealing and in quantum simulation (Hauke et al., 2020):

- Study new annealing schemes, which may involve the use of different schedules for different qubits, random perturbations, or different interaction schemes. Some of these techniques may eliminate first-order phase transitions, or allow those to happen in smaller regions that get connected and produce better solutions.

- Achieve better coherence times in qubits that are compatible with quantum annealer designs. Some examples are recent designs of capacitively shunted flux qubits, which feature slower dephasing rates (Steffen et al., 2010; Yan et al., 2016).
- Engineer new interaction schemes between superconducting qubits, to create nonstoquastic couplings of $\sigma_i^x \sigma_j^x$ simultaneous with $\sigma_i^z \sigma_j^z$. This is not a simple task, but there are evidences for such couplings with flux qubits by Ozfidan et al. (2020), and there are even designs to enable three- and four-body couplings (Chancellor et al., 2017).

However, if we have learned something from the discussion and research with D-Wave's processors, it is that sometimes we can make something very good for the wrong reasons. Superconducting quantum annealers were born out of commercial interest around computational applications, which failed at demonstrating a quantum advantage. However, the result is still a unique demonstration of the potential for scalable quantum simulations with superconducting quantum circuits. Maybe one possible future of this technology – microwave circuits and qubits – lies precisely in avenues that explore this potential, researching fundamental and practical problems at the interface between quantum optics, condensed-matter physics, and other fields with complex quantum systems. I hope you, as a reader, will be able to bring this new insight into the field, and use the tools from this book to construct those new applications.

Exercises

9.1 Let us analyze the Landau–Zener model (9.2).

(1) Show that $\hat{H}(\varepsilon) = \frac{1}{2}\Delta(\varepsilon) U \sigma^z U^\dagger$ where $U = \exp(-i\theta(\varepsilon)\sigma^y/2)$.

(2) Rewrite the wavefunction of the two-level system in a rotating frame with $|\psi(t)\rangle = U(t)\,|\xi(t)\rangle$ and show that the rotated state $|\xi(t)\rangle$ evolves with the Hamiltonian:

$$\hat{H}_{\text{eff}} = \left[\frac{1}{2}\Delta(\varepsilon)\sigma^z - \frac{\Delta_{\text{min}}\dot{\varepsilon}}{2\Delta^2}\sigma^y\right]|\xi(t)\rangle. \tag{9.21}$$

(3) Implement a change of variables $t \to \varepsilon$, using $\partial_t = \dot{\varepsilon}\partial_\varepsilon$, to recover the effective equation (9.5).

(4) Provide an upper bound on the rate of change of the prefactor in front of σ^y and compare it with the σ^z term.

(5) Approximately diagonalize the effective Hamiltonian, showing that the $|0\rangle$ eigenstate of σ^z gets a correction of order $\frac{\Delta_{\text{min}}}{\Delta^2} \times \frac{2\dot{\varepsilon}}{\Delta} \leq \frac{2\dot{\varepsilon}}{\Delta_{\text{min}}}$.

9.2 Let's assume that we have a qubit with frequency Δ, interacting with a microwave cavity of frequency ω through a coupling strength g, as described by the Jaynes–Cummings model (7.33). Initially the qubit is off-resonant but with a small frequency $\Delta \ll \omega$, and we wish to move the qubit so that it has a larger frequency $\Delta > \omega$. How fast do we have to increase Δ so that the state of the qubit and the cavity remains unchanged? What determines whether the speed is adiabatic or not?

9.3 If we implement the spin-glass model (9.15) in one dimension, we have that the only nonzero elements are of the form $J_{i,i+1}$. Show that in this case the problem is trivial. Hint: Eliminate the signs in J with a gauge transformation $\sigma_i^z \to \sigma_i^x \sigma_i^z \sigma_i^x$ on a subset of spins.

9.4 Aharonov et al. (2004) proved that an adiabatic computation with a sparse Hamiltonian can be efficiently simulated by a quantum circuit. Essentially, we can simulate the time evolution by decomposing with circuits that simulate rotations based on $\hat{H}_0$ and $\hat{H}_p$ for brief intervals:

$$U = \prod_{n=1}^{L} \exp(-i\hat{H}_0 \Delta t(1 - \varepsilon(t_n))) \exp(-i\hat{H}_p \Delta t \varepsilon(t_n)). \tag{9.22}$$

Naturally, the precision of the simulation increases when we decrease the time step $\Delta t = T/L$ and increase the total passage time T.

(1) Show that evolution with $\hat{H}_0$ decomposes into local rotations.

(2) How many quantum gates do we need to simulate evolution with the spin-glass model (9.15) if the qubits are organized in the 2D and 3D lattices studied by Barahona (1982)?

9.5 Take the QMA-complete Hamiltonian from (9.17). Show that the energy $\langle \hat{H} \rangle$ of any quantum state can be estimated by repeating M times the measurement[18] of either all σ_i^z or all σ_i^x Pauli observables on all qubits. Show that the uncertainty in these measurements scales as $1/\sqrt{M}$. Assume we want to keep the experimental precision below some bound ε. Prove that the number of measurements scales at most as $\mathcal{O}(N^4/\varepsilon)$, assuming the worst-case scenario of N^2 terms in (9.17).

9.6 Assume we want to establish an interaction J between two logical qubits, assigned to positions 1 and $N+1$, which are connected by an intermediate chain of N physical qubits, as in

$$\hat{H} = \frac{1}{2}\hbar\Delta\left(\sigma_1^z + \sigma_{N+1}^z\right) - J_{\mathrm{f}}\left(\sigma_1^x\sigma_2^x + \cdots + \sigma_{N-1}^x\sigma_N^x\right) + J\sigma_N^x\sigma_{N+1}^x. \tag{9.23}$$

Compute the minimum gap of this model in the limit of large ferromagnetic coupling $|J_{\mathrm{f}}|$. If we make an adiabatic passage with this Hamiltonian, using

[18] This is called a *two-setup experiment,* because experimentalists only have to do two kinds of experiments that only differ in the final measurement at the end.

$\hat{H}_0$ (9.12), what is the probability that the state of qubit 1 is copied to the N intermediate qubits?

9.7 The usual metric in annealing experiments is the *time to solution,* that is, the time t_{TTS} required to find the solution with a fixed probability, typically $p_{\mathrm{TTS}} = 99\%$. Assume that we perform passages with a time t_{ramp} and find the true ground state with a probability $p_{\mathrm{ramp}} < 99\%$. Compute the number of times M that we have to repeat the annealing experiment to obtain a success probability $p \geq 99.99\%$. Discuss what happens to t_{TTS} when the ramp time is lower-bounded due to the restrictions of the device.

Appendix A
Hamiltonian Diagonalizations

A.1 Tridiagonal Matrix Diagonalization

A.1.1 Periodic Boundary Conditions

We need to diagonalize the matrix

$$B = \begin{pmatrix} a & b & \dots & 0 & b \\ b & a & \dots & 0 & 0 \\ \vdots & \vdots & \ddots & & \\ 0 & 0 & \dots & a & b \\ b & 0 & \dots & b & a \end{pmatrix} \in \mathbb{R}^{N \times N}. \tag{A.1}$$

The eigenmodes are defined by the eigenvalue equations

$$b(u_{m-1} + u_{m+1}) + a u_m = \lambda u_m, \quad m = 1 \dots N, \tag{A.2}$$

where we need the identification $u_{m+N} = u_m$ associated to periodic boundary conditions. We can solve the eigenvalue equations with plane waves:

$$u_m^{(n)} = \frac{1}{\sqrt{N}} \exp(i k_n m), \quad k_n = \frac{2\pi n}{N} \in \frac{2\pi}{N} \times \{0, 1, 2, \dots N-1\}, \tag{A.3}$$

that naturally satisfy the boundary conditions, $u_{m+N} = u_m \exp(ikN) = u_m \exp(i2\pi n)$. The eigenvalues associated form a band $\lambda_n = a + 2b\cos(k_n)$.

It is quite often more convenient to define the quasimomenta k_n so that they explore the interval $(-\pi, \pi]$. For even sizes $N = 2M$,

$$k_n \in \frac{\pi}{2M} \times \{-M+1, -M+2, \dots -1, 0, 1, \dots M\}, \tag{A.4}$$

while for odd one $N = 2M+1$

$$k_n \in \frac{\pi}{2M+1} \times \{-M, -M+1, \dots -1, 0, 1, \dots M\}. \tag{A.5}$$

We can also use real wavefunctions if we instead diagonalize B with a real orthogonal transformation, $B = O^T \Lambda O$, $O \in \mathbb{R}^{N\times N}$, $O^T O = 1$. The eigenfunctions are then linear sinusoidal functions:

$$u_m^{(even,n)} = \sqrt{\frac{2}{N}} \cos(k_n m), \quad k_n \in (0,\pi) \tag{A.6}$$
$$u_m^{(odd,n)} = \sqrt{\frac{2}{N}} \sin(k_n m),$$

plus the two extra solutions:

$$u_m^{(even,0)} = \sqrt{\frac{1}{N}}, \quad k_0 = 0, \tag{A.7}$$
$$u_m^{(even,n)} = \sqrt{\frac{1}{N}}(-1)^m, \quad k_n = \pi. \tag{A.8}$$

A.1.2 Open Boundary Conditions

We need to diagonalize the tridiagonal matrix

$$B = \begin{pmatrix} a & b & \dots & 0 & 0 \\ b & a & \dots & 0 & 0 \\ \vdots & \vdots & \ddots & & \\ 0 & 0 & \dots & a & b \\ 0 & 0 & \dots & b & a \end{pmatrix} \in \mathbb{R}^{N\times N}. \tag{A.9}$$

The eigenmodes are defined by the eigenvalue equations:

$$b(u_{m-1} + u_{m+1}) + bu_m = \lambda u_m, \quad m = 2 \dots N-1 \tag{A.10}$$
$$bu_2 + au_1 = \lambda u_1,$$
$$bu_{N-1} + au_N = \lambda u_N.$$

These equations are equivalent to a periodic boundary condition problem (A.2) of size $2(N+1)$, where one imposes $u_{N+1} = u_{2(N+1)} = 0$. This solution is constructed from two degenerate plane waves with opposite momenta and the same eigenenergy:

$$u_m \propto e^{ikm} - e^{-ikm}. \tag{A.11}$$

The final solution, including normalization factors, reads

$$u_m = \sqrt{\frac{2}{N}} \sin(k_n m), \quad k_n \in \frac{\pi}{N+1} \times \{1, 2, \dots N\} \tag{A.12}$$
$$\lambda_n = a + 2b\cos(k_n).$$

Unlike the periodic case, we can no longer shift the linear momenta, which are constrained to the interval $[0,\pi)$.

A.2 Harmonic Chain Diagonalization

We are going to diagonalize a problem of coupled Harmonic oscillators, which we write in the following form:

$$H = \frac{1}{2C}\hat{\mathbf{p}}^T\hat{\mathbf{p}} + \frac{1}{2L}\hat{\mathbf{x}}^T B\hat{\mathbf{x}}. \tag{A.13}$$

Let us assume that matrix $B \in \mathbb{R}^{N\times N}$, which is real and symmetric, is diagonalized with a unitary or orthogonal transformation:

$$B = U^\dagger \Lambda U, \quad \text{with} \begin{cases} \Lambda_{ij} = \lambda_i\delta_{ij} \\ U^T U = \mathbb{1}. \end{cases} \tag{A.14}$$

We define the position and momentum operators using N Fock operators:

$$\hat{x}_m = \sum_n a_n \frac{1}{\sqrt{2}}\left(U_{mn}\hat{b}_n + U^*_{mn}\hat{b}^\dagger_n\right), \tag{A.15}$$

$$\hat{p}_m = \sum_n \frac{\hbar}{a_n}\frac{i}{\sqrt{2}}\left(U^*_{mn}\hat{b}^\dagger_n - U_{mn}\hat{b}_n\right), \tag{A.16}$$

which satisfy the usual commutation relations $[\hat{b}_n, \hat{b}^\dagger_m] = \delta_{nm}$. The constants $a_n > 0$ are "length" scales to be adjusted.

Since $U^*_{nm}U_{nr} = \delta_{mr}$, these operators satisfy the canonical commutation relations $[\hat{x}_m, \hat{x}_r] = 0$, $[\hat{p}_m, \hat{p}_r] = 0$, $[\hat{x}_m, \hat{p}_r] = i\hbar\delta_{mr}$. Thanks to this orthogonality condition, we can write

$$\hat{\mathbf{p}}^T\hat{\mathbf{p}} = \sum_m \frac{\hbar^2}{2a_n^2}\left(\hat{b}^\dagger_n\hat{b}_n + \hat{b}_n\hat{b}^\dagger_n - \hat{b}^{\dagger\,2}_n - \hat{b}^2_n\right) \tag{A.17}$$

$$\hat{\mathbf{x}}^T B\hat{\mathbf{x}} = \sum_m \frac{a_n^2\lambda_n}{2}\left(\hat{b}^\dagger_n\hat{b}_n + \hat{b}_n\hat{b}^\dagger_n + \hat{b}^{\dagger\,2}_n + \hat{b}^2_n\right). \tag{A.18}$$

We impose $\frac{\hbar^2}{2Ca_n^2} = \frac{a_n^2\lambda}{2L} = \frac{1}{2}\hbar\omega_n$, to cancel the $\hat{b}^2$ and $\hat{b}^{\dagger\,2}$ terms. This gives us the canonical frequencies and the oscillator lengths:

$$\omega_n = \sqrt{\frac{\lambda_n}{CL}}, \qquad a_n = \sqrt{\frac{\hbar}{C\omega_n}}, \tag{A.19}$$

and makes the Hamiltonian diagonal:

$$\hat{H} = \sum_n \hbar\omega_n\left(\hat{b}^\dagger_n\hat{b}_n + \frac{1}{2}\right). \tag{A.20}$$

In a problem with periodic boundary conditions, with $B_{ij} = B_{i+1j+1}$, the matrix B can be diagonalized with either sines and cosines or complex exponentials (Section A.1.1). If we choose the plane waves, we write

$$\hat{x}_m = \sum_k a_n \frac{1}{\sqrt{2N}}\left(e^{ikm}\hat{b}_k + e^{-ikm}\hat{b}^\dagger_k\right), \tag{A.21}$$

$$\hat{p}_m = \sum_k \frac{\hbar}{a_n}\frac{i}{\sqrt{2N}}(e^{-ikm}\hat{b}^\dagger_n - e^{ikm}\hat{b}_n), \tag{A.22}$$

where the sum takes place over the allowed values of the quasimomentum k. The diagonalized Hamiltonian takes the form $\hat{H} = \sum_k \hbar\omega_k \left(\hat{b}_k^\dagger \hat{b}_k + \frac{1}{2}\right)$, with the degeneracy $\omega_k = \omega_{-k}$.

A.3 Schrieffer–Wolff Perturbation Theory

In this book, and in many physics problems, we often face Hamiltonians that can be decomposed into a dominant Hamiltonian and a perturbation:

$$\hat{H} = \hat{H}_0 + g\hat{V}. \tag{A.23}$$

The Hamiltonian $\hat{H}_0$ is a simple problem – e.g., a bunch of noninteracting qubits and microwave resonators – whose eigenstates $|n\rangle$ and eigenenergies E_n we know with complete accuracy. *Perturbation theory* is a mathematical strategy to approximate the eigenstates and eigenenergies of $\hat{H}$ as "corrections" of the original states and eigenvalues.

Schrieffer–Wolff or Kato perturbation theory is a family of perturbation theory methods that have become popular as a technique designing quantum simulators and engineering quantum interactions. The assumption of this particular theory is that, for weak enough interactions $g\hat{V}$, the eigenspaces of $\hat{H}_0$ are only slightly deformed so that they can be related to those of $\hat{H}$ by a unitary transformation, to be approximated perturbatively. We will now discuss two variants of this situation, with increasing level of complexity.

A.3.1 Nondegenerate Perturbation Theory

Assume that we write a one-to-one mapping between the eigenstates:

$$\left.\begin{aligned} \hat{H}_0 &= \textstyle\sum_n E_n \, |n\rangle\langle n| \\ \hat{H} &= \textstyle\sum_n \tilde{E}_n(g) \, |n'\rangle\langle n'| \end{aligned}\right\}, \quad \text{with } |n'\rangle = \hat{U}(g)\, |n\rangle\,. \tag{A.24}$$

This can only work if the eigenenergies of each Hamiltonian are nondegenerate and they never cross – e.g., $\tilde{E}_n(g) < \tilde{E}_{n+1}(g)\ \forall n$. If g is small enough that this level crossing never happens, we can study the unitary transformation as being generated by a non-Hermitian operator $\hat{U}(g) = \exp(-g\hat{S}(g))$, expanding the function $\hat{S}(g)$ as a perturbative series in the small parameter g.

Our derivation proceeds by analyzing the effective Hamiltonian $\hat{H}_{\text{eff}} = e^{g\hat{S}}\hat{H}e^{-g\hat{S}}$ that results from mapping $\hat{H}$ back into the original unperturbed space $\hat{H}_0$:

$$\hat{H}_{\text{eff}} = \hat{H}_0 + g\hat{V} + g\big[\hat{S}, \hat{H}_0\big] + g^2\big[\hat{S}, \hat{V}\big] + \frac{1}{2}\Big[g\hat{S}, \big[g\hat{S}, \hat{H}_0\big]\Big] + \mathcal{O}(g^3). \tag{A.25}$$

We assume that $\hat{V}$ is off-diagonal on the original eigenbasis. We can always justify this assumption by incorporating the diagonal terms into $\hat{H}_0$ as an effective renormalization of the qubits and photons in that model:[1]

$$\hat{H}_0 \to \hat{H}_0 + g\sum_n |n\rangle\langle n|\hat{V}|n\rangle\langle n|, \tag{A.26}$$

[1] Alternatively, you may use more sophisticated derivations such as the one by Bravyi et al. (2011), on which this appendix is based.

$$\hat{V}_{\text{od}} := \sum_{n\neq m} |n\rangle\langle n|\hat{V}|m\rangle\langle m| \,. \tag{A.27}$$

This allows us to assume $V_{nm} := \langle n|\hat{V}_{od}|m\rangle = 0$ when $n \neq m$. We can then expand our generator $g\hat{S} = g\hat{S}_0 + \mathcal{O}(g^2)$, with an operator $\hat{S}_0$ that cancels all first-order terms in the expansion of $\hat{H}_{\text{eff}}$, producing $[g\hat{S}, \hat{H}_0] = -g\hat{V}_{\text{od}}$.

In the nondegenerate perturbation theory limit, this equation can be solved formally, projecting onto the original eigenstate basis:

$$g(E_m - E_n)\langle n|\hat{S}_0|m\rangle = -g\,\langle n|\hat{V}|m\rangle\,, \tag{A.28}$$

which is solved by

$$\hat{S}_0 = \sum_{n\neq m} \frac{V_{nm}}{E_n - E_m} \times |n\rangle\langle m| \,. \tag{A.29}$$

This gives us the second-order expansion of the effective model:

$$\begin{aligned}\hat{H}_{\text{eff}} &= \hat{H}_0 + \frac{1}{2}g^2[\hat{S}_0, \hat{V}] + \mathcal{O}(g^3) \\ &\simeq \sum_n (E_n + V_{nn})\,|n\rangle\langle n| + \frac{g^2}{2}\sum_{n,r\neq m}\left(\frac{V_{nm}V_{mr}}{E_n - E_m} + \frac{V_{nm}V_{mr}}{E_r - E_m}\right)|n\rangle\langle r| \,.\end{aligned} \tag{A.30}$$

We will use these expansions in several qubit designs, such as when justifying the truncation to a two-level subspace in the charge qubit (Problem 6.6), or when analyzing the residual interactions in two highly detuned qubits (see Section 6.5.3 and Problems 6.16 and 6.17).

A.3.2 Degenerate Perturbation Theory

Sometimes we want to consider interaction strengths that go slightly beyond the limits we have mentioned, or we want to study Hamiltonians $\hat{H}_0$ with eigenstates that are energetically very close to each other and are strongly mixed by the perturbation. To make perturbation theory work in this particular limit, we have to assume that the spectrum of $\hat{H}_0$ can be divided into eigenspaces of states that, even during the perturbation, will remain close to each other and not mingle with other subspaces. This complicates the notation somewhat, forcing us to label eigenstates by subspace α, as in $|n,\alpha\rangle$ and $|n',\alpha\rangle$, and introduce projectors onto the original and deformed models:

$$\left.\begin{aligned} P_{0,\alpha} &= \textstyle\sum_n |n,\alpha\rangle\langle n,\alpha| \\ P_\alpha(g) &= \textstyle\sum_n |n',\alpha\rangle\langle n',\alpha| \end{aligned}\right\}, \quad \text{with } P_\alpha(g) = \hat{U}(g)P_{0,\alpha}\hat{U}(g)^\dagger. \tag{A.31}$$

As discussed by Bravyi et al. (2011), we can still develop a Schrieffer–Wolff transformation building on the expansion (A.25), but now H_{eff} has to be treated as a box-diagonal Hamiltonian:

$$H_{\text{eff}} = \sum_\alpha P_{0,\alpha} H_{\text{eff}} P_{0,\alpha} = \sum_n H_{\text{eff}}^\alpha. \tag{A.32}$$

This structure changes the equation we have to solve for $\hat{S}_0$ and $\hat{V}$. We need to split $\hat{V} = \hat{V}_\mathrm{d} + \hat{V}_\mathrm{od}$ into contributions that "live" within the same energy subspace and that will be incorporated into $\hat{H}_0$ and contributions $\hat{V}_\mathrm{od}$ that connect different subspaces:

$$P_{0,\beta}[\hat{S}, \hat{H}_0] P_{0,\alpha} = P_{0,\beta} \hat{V} P_{0,\alpha}, \quad \alpha \neq \beta. \tag{A.33}$$

These equations can be solved, producing the anti-Hermitian operator:

$$\hat{S} = \sum_{\alpha\neq\beta}\sum_{n,m} \frac{1}{E_{n,\alpha} - E_{m,\beta}} |n,\alpha\rangle\langle n,\alpha|\hat{V}|n,\beta\rangle\langle n,\beta| . \tag{A.34}$$

We can continue further with the perturbation theory argument, substituting this operator into the series for $\hat{H}_\mathrm{eff}$ and inspecting the box within one particular subspace. The result is a generalization of the nondegenerate perturbation series:

$$\hat{H}^\alpha_\mathrm{eff} = P_{0\alpha}\big(\hat{H}_0 + g\hat{V}_\mathrm{d}\big)P_{0\alpha} + \frac{1}{2}g^2 P_{0\alpha}\big[\hat{S}, \hat{V}_\mathrm{od}\big]P_{0\alpha} + \mathcal{O}(g^3). \tag{A.35}$$

Explicitely in terms of eigenstates,

$$\begin{aligned}\hat{H}^\alpha_\mathrm{eff} \simeq & \sum_n E_{n,\alpha}\, |n,\alpha\rangle\langle n,\alpha| + \sum_{n,m} \langle n,\alpha|\hat{V}|m,\alpha\rangle\, |n,\alpha\rangle\langle n,\alpha| \\ & + \frac{g^2}{2}\sum_{\beta\neq\alpha}\sum_{n,r\neq m}\left(\frac{1}{E_{n,\alpha} - E_{m,\beta}} + \frac{1}{E_{r,\alpha} - E_{m,\beta}}\right) \\ & \qquad \times |n,\alpha\rangle\langle n,\alpha|\hat{V}|m,\beta\rangle\langle m,\beta|\hat{V}|r,\alpha\rangle\langle r,\alpha| .\end{aligned} \tag{A.36}$$

This series combines the matrix elements of the perturbation within a given subspace (see the first line), with additional terms that result from virtual transitions to other subspaces (generically labeled $\beta \neq \alpha$ in the second and third lines). We use these expansions when studying the dispersive coupling between cavities and resonators in Section 7.3.7.

A.3.3 Considerations

Equations (A.30) and (A.36) are mathematical recipes for an effective Hamiltonian that is expressed in the basis of eigenstates of the unperturbed model $\hat{H}_0$. From this point of view, the appeal of this method to the design of quantum simulators and quantum engineering is clear: This theory allows us to design effective interactions, simply by studying how various perturbations to a system or circuit $\hat{V}$ transform into qubit–qubit or qubit–photon couplings.

In essence, this is what we do in Sections 6.5.1 and 6.5.2 when introducing the dipolar magnetic couplings. These interactions were treated as *diagonal* perturbations $\hat{V}_\mathrm{d}$ that took place in the qubit space. However, this method allows us to consider the effect of the interaction via the connection of the qubit eigenspace to higher excited states, or to take the already developed interaction and understand how it can be approximated by other, much simpler models – as we did in Section 6.5.3 and (6.67). Finally, perturbation theory also provides us with effective interactions in limits, such as the dispersive qubit–cavity coupling from Section 7.3.7, where the interactions are only mediated by excited states and $\hat{V}_\mathrm{d} = 0$.

A word of caution is needed when applying the Schrieffer–Wolff transformation to develop effective models. This relates to the fact that the eigenstates of the perturbed problem result from deformations of the original qubit states, and that the Hamiltonian $\hat{H}_{\text{eff}}$ is a unitary transformation of the Hamiltonian that actually rules the dynamics in the lab.

For instance, consider we have some state $|\Psi\rangle$ that may be either an eigenstate of the effective model

$$\hat{H}^{\alpha}_{\text{eff}} |\Psi\rangle = E_r(g) |\Psi\rangle , \tag{A.37}$$

or the solution of a Schrödinger equation at a given time:

$$i\partial_t |\Psi(t)\rangle = \hat{H}_{\text{eff}} |\Psi(t)\rangle . \tag{A.38}$$

In both cases, we will have those states expressed in the basis of non-interacting states:

$$|\Psi\rangle = \sum \Psi_n |n, \alpha\rangle . \tag{A.39}$$

However, this is a "fake" frame of reference. If at some point we need to make predictions about the observables and properties of the state $|\Psi\rangle$, we will need to undo the Schrieffer–Wolff transformation:

$$|\Psi_{\text{lab}}\rangle = e^{-g\hat{S}} |\Psi\rangle \simeq \sum \Psi_n (1 - g\hat{S}_0) |n, \alpha\rangle + \mathcal{O}(g^2). \tag{A.40}$$

In doing so, we see that the application or $\hat{S}$ or $\hat{S}_0$ deforms the basis in which we expanded our state, from a superposition of the original bare states of the qubit or cavity, to contain some *hybridization* with excitations $|m, \beta\rangle$ injected by the $\hat{S}_0$ operator. The actual probability of those excitations grows as $\mathcal{O}(g^2)$, but it may be something to consider, especially when doing high-precision experiments, such as quantum gates and quantum computations.

Appendix B

Open Quantum Systems

B.1 Nonunitary Evolution

The theory of open quantum systems studies the evolution of quantum mechanical systems that are in contact with other larger systems, which we call environments. This theory assumes the postulates of quantum mechanics, but understands that when we look at a piece of a larger object, it is not described by a wavefunction but a reduced density matrix, and does not satisfy a Schrödinger equation but some other type of linear differential equation that describes its incoherent evolution.

As with any other treatment, the need to develop a theory of open systems is contingent to the experimental timescales. We can argue that no quantum object is actually isolated – for instance, all quantum objects are in contact with the electromagnetic field that permeates our universe. However, when we perform an experiment with an atom, for example, the time for that atom to get entangled with the universe – or even the laboratory, or its trap – may be larger than our observation time. Under such circumstances, the formalism of quantum mechanics and Schrödinger equations can provide an effective and correct description of the experiment.

However, this book is full of experimental setups where we connect quantum circuits to waveguides and measurement devices, which inject and extract energy and information from qubits and cavities alike. Moreover, those circuits are themselves imperfect objects that decohere due to interaction with the chip's substrate, with trapped charges, and with the surrounding electromagnetic fields. Both timescales, measurement and decoherence, will be long, but not so long that they can be neglected.

When we cannot ignore the interaction between our experiment and the environment, we have to describe our system as the reduced density matrix of a larger and more complicated wavefunction, as explained in Section 2.3. The evolution of these density matrices is no longer unitary, but described with a *trace-preserving completely positive map*. These are linear transformations operating in the space of bounded linear operators that include density matrices. They are linear $\varepsilon(\alpha\rho + \beta\rho') = \alpha\varepsilon(\rho) + \beta\varepsilon(\rho')$. They are positive[1] because

[1] The term *completely positive* is a technicality meaning that when we compose our Hilbert space with another one, the extended operator remains positive. In other words, $\varepsilon \otimes \mathbb{1}_B$ is positive for any dimension of the identity $\mathbb{1}_B$.

they transform nonnegative operators $\rho \geq 0$ into nonnegative operators $\varepsilon(\rho) \geq 0$. And they are trace preserving because they maintain the normalization of density matrices,[2] $\mathrm{tr}(\varepsilon(\rho)) = \mathrm{tr}(\rho)$.

Most physical operations are trace-preserving completely positive maps that mutate density matrices into other density matrices. These *superoperators* – transformations of other operators, in this case density matrices – admit a *Kraus decomposition*, a compact expression of the map as a sum of operators multiplying the density matrix:

$$\varepsilon(\rho) = \sum_i^n \hat{K}_i^\dagger \rho \hat{K}_i, \quad \text{with} \sum_i \hat{K}_i^\dagger \hat{K}_i = \mathbb{1}. \tag{B.1}$$

The matrices $\{\hat{K}_i, \hat{K}_i^\dagger\}$ are called *Kraus operators.* The number of Kraus operators in a decomposition can grow quadratically with the Hilbert space dimension. However, many transformations require very few terms. These are then efficient representations of noise, errors, and prototypical decohererence models – such as the ones in Problem 8.6 – that can be experimentally calibrated and used to engineer error-correcting methods.

In the study of superconducting circuits, we are not so much concerned with specific instances of a positive map. Our interest is more focused on the non-unitary evolution of density matrices, given by differentiable maps $\rho(t) = \varepsilon(\rho(0), t)$. Here the Kraus decomposition is not so useful, as having a differential equation for $\rho(t)$. Lindblad (1976) studied such *master equations* and found that they are linear ordinary differential equations with a common structure:

$$\partial_t \rho = \mathcal{L}\rho = -\frac{i}{\hbar}[\hat{H}, \rho] + \sum_{i=1}^{d^2-1} \frac{1}{2} \sum_i \left(\hat{l}_i \rho \hat{l}_i^\dagger - \hat{l}_i^\dagger \hat{l}_i \rho - \rho \hat{l}_i^\dagger \hat{l}_i \right). \tag{B.2}$$

The generator of the evolution $\mathcal{L}$ is the Lindblad superoperator. It separates into a Hamiltonian term $-i[\hat{H}, \rho]/\hbar$, and up to $d^2 - 1$ decoherence terms, where d is the dimension of our system. Both the Hermitian operator $\hat{H}$ and the dissipators $\{\hat{l}_i, \hat{l}_i^\dagger\}$ can depend on time, but we will focus on the *Markovian dynamics* with constant dissipators. This is a relevant situation that appears when the environment is very large and "forgets" very quickly the history of the system with which it interacts. We will see in Section B.2 how in such cases we can derive master equations, with very few Lindblad terms, that are both simple and intuitively appealing. Examples include Pauli operators $\hat{l}_i \sim \sqrt{\gamma}\sigma^-$ modeling the spontaneous relaxation of a qubit (see Section 6.1.6), Fock operators $\hat{l}_i \sim \sqrt{\kappa}\hat{a}$ that model the cooling of a microwave resonator in a zero-temperature environment (see Section 5.5.4), etc.

B.2 Master Equations

Our goal in this section is to derive a master equation for the reduced density matrix of an open system. This equation arises when we trace out the environment in the Schrödinger equation for the combined system-environment wavefunction. For that, we assume a weak coupling between the system and the environment coupling, treating the Schrödinger dynamics up to second order in perturbation theory. The result is an integro

[2] The only circumstance where we would accept *loss of trace* would be when we want to model experiments with postselection, where some fraction of outcomes are discarded. Then we could accept a decrease in the norm, interpreting the trace of the density matrix as the total success probability of postselection.

differential equation that becomes local in time under a Markovian approximation – that the environment forgets the history of our subsystem. With a slight massaging, we will end up with a simple Lindblad equation that applies to microwave resonators and superconducting qubits. But we will also offer prescriptions to study more complicated problems, such as composite, driven, or highly nonlinear quantum systems.

B.2.1 Lindblad Equation

Assume a system–bath Hamiltonian that separates into a solvable term $\hat{H}_0$ and a small perturbation $\hat{H}_1$. The unitary operator for the combined system-bath wavefunction $U(t)$ satisfies a Schrödinger equation,

$$i\hbar\frac{\mathrm{d}}{\mathrm{d}t}U(t) = (\hat{H}_0 + \hat{H}_1)U(t), \tag{B.3}$$

with initial condition $U(0) = 1$. We can split the evolution operator into two factors $U(t) = \exp(-i\hat{H}_0 t/\hbar)W(t)$. The first factor is the exponential of the problem that we know how to solve $\hat{H}_0$. It becomes the exact solution when $\hat{H}_1$ vanishes. The second factor W is a correction generated by the perturbation $\hat{H}_1$. It satisfies the *interaction picture* equation:

$$i\hbar\dot{W}(t) = e^{i\hat{H}_0 t/\hbar}\hat{H}_1 e^{-i\hat{H}_0 t/\hbar}W(t) =: \hat{H}_1(t)W(t). \tag{B.4}$$

We can apply this interaction picture also to density matrices, splitting $\rho(t) = \exp(-i\hat{H}_0 t/\hbar)\rho_I(t)\exp(i\hat{H}_0 t/\hbar)$. The interaction picture $\tilde{\rho}_I$ satisfies a master equation generated by the perturbations $\frac{\mathrm{d}}{\mathrm{d}t}\rho_I(t) = -i[\hat{H}_1(t)/\hbar, \rho_I(t)]$. Usually, the dynamics generated by $\hat{H}_1(t)$ is much slower than the unitary evolution induced by $\exp(-i\hat{H}_0 t/\hbar)$. This justifies a Dyson series for the solution of the master equation, which we truncate up to second order:

$$\begin{aligned}\rho_I(t+\delta t) \simeq \rho(t) &- \frac{i}{\hbar}\int_t^{t+\delta t}\left[\hat{H}_1(t'), \rho(t)\right]\mathrm{d}t' \\ &- \frac{1}{\hbar^2}\int_t^{t+\delta t}\int_t^{t'}\left[\hat{H}_1(t'), \left[\hat{H}_1(t''), \rho(t)\right]\right]\mathrm{d}t''\mathrm{d}t'.\end{aligned} \tag{B.5}$$

The next approximation is to neglect the feedback of the system onto the environment. We impose that expectation values over the bath and system are uncorrelated, and that the former are approximately constant.[3] For an off-diagonal coupling[4] $\mathrm{Tr}_B\{\hat{H}_1(t')\tilde{\rho}(t)\} = 0$,

$$\tilde{\rho}_S(t+\delta t) \simeq \tilde{\rho}_S(t) - \frac{1}{\hbar^2}\int_t^{t+\delta t}\int_t^{t'}\mathrm{Tr}_B\left[\hat{H}_1(t'), \left[\hat{H}_1(t''), \tilde{\rho}_S(t)\otimes\tilde{\rho}_B\right]\right]\mathrm{d}t''\mathrm{d}t'.$$

To continue further, we need to impose some structure in $\hat{H}_1$. The following section introduces a coupling that is typical of linear systems.

[3] Traditionally, it has been said that the system and the environment adopt a product state structure where the bath does not change, $\tilde{\rho}(t) = \tilde{\rho}_S(t)\otimes\tilde{\rho}_B$, but this is not strictly needed.

[4] We will see that this is not really a restriction, as typically $\hat{H}_1 \sim O_S \otimes O_B$, where the system and bath operators O_S and O_B have zero diagonal elements.

B.2.2 Linear System–Bath Coupling

Many superconducting circuits consist of a small system, such as a qubit or a cavity, which is linearly coupled to a large number of bosonic operators, embodying the electromagnetic enviroment, a transmission line, an antenna, or any reservoir the circuits thermalize to. Such system–bath couplings take the form of products between system and bath operators $\hat{A}$ and $\hat{b}_k$:

$$\hat{H}_1 = \hat{A}^\dagger \sum_k g_k \hat{b}_k + \hat{A} \sum_k g_k^* \hat{b}_k. \tag{B.6}$$

The Hamiltonian $\hat{H}_0$ typically has a simple effect on these operators:

$$\left.\begin{array}{l} e^{i\hat{H}_0 t/\hbar}\hat{A}e^{-i\hat{H}_0 t/\hbar} = e^{i\Delta t}\hat{A} \\ e^{i\hat{H}_0 t/\hbar}\hat{b}_k e^{-i\hat{H}_0 t/\hbar} = e^{i\omega_k t}\hat{b}_k \end{array}\right\} \Rightarrow \hat{H}_1(t) = \hat{A}^\dagger e^{i\Delta t}\sum_k g_k \hat{b}_k e^{-i\omega_k t} + \text{H.c.} \tag{B.7}$$

When we insert this expression into the formula (B.6), we obtain four combinations of $\{\hat{A}, \hat{A}^\dagger\}$ and $\{\hat{b}_k, \hat{b}_k^\dagger\}$ acting on $\tilde{\rho}_S$:

$$\begin{aligned} \Delta\tilde{\rho}_S = \int_t^{t+\delta t} \mathrm{d}t \left\{ I_{B^\dagger B^\dagger}\mathcal{L}_{\hat{A}\hat{A}} + I_{B^\dagger B}\mathcal{L}_{\hat{A}\hat{A}^\dagger} + I_{BB^\dagger}\mathcal{L}_{\hat{A}^\dagger\hat{A}} + I_{BB}\mathcal{L}_{\hat{A}^\dagger\hat{A}^\dagger}\right\}\tilde{\rho}_S(t), \\ \text{with } \mathcal{L}_{\hat{X}\hat{Y}}\tilde{\rho} = 2\hat{X}\tilde{\rho}\hat{Y} - \hat{Y}\hat{X}\tilde{\rho} - \tilde{\rho}\hat{Y}\hat{X}. \end{aligned} \tag{B.8}$$

The prefactors are four integrals of expectation values over the bath:

$$\begin{aligned} I_{BB} &= \sum_{pq} g_p g_p \int_t^{t+\delta t} \langle \hat{b}_p \hat{b}_q \rangle e^{-i(\omega_p-\Delta)t - i(\omega_q - \Delta)t'} = I^*_{B^\dagger B^\dagger} \\ I_{BB^\dagger} &= \sum_{pq} g_p g_p^* \int_t^{t+\delta t} \langle \hat{b}_p \hat{b}_q^\dagger \rangle e^{-i(\omega_p-\Delta)t + i(\omega_q - \Delta)t'}, \\ I_{B^\dagger B} &= \sum_{pq} g_p^* g_p \int_t^{t+\delta t} \langle \hat{b}_p^\dagger \hat{b}_q \rangle e^{+i(\omega_p-\Delta)t - i(\omega_q - \Delta)t'}. \end{aligned} \tag{B.9}$$

For a large enough bath, these integrals and the superoperators are approximately constant. The equation can be summarized as the action of one superoperator $\tilde{\rho}_S(t+\delta t) = \tilde{\rho}_S(t) + \delta t \times \mathcal{L}\tilde{\rho}_S(t)$, which in the limit $\delta t \to 0$ generates a *Lindblad equation*:

$$\frac{\mathrm{d}}{\mathrm{d}t}\tilde{\rho}_S = \mathcal{L}\tilde{\rho}_S. \tag{B.10}$$

B.2.3 System in a Thermal Bath: Cooling and Heating

If we focus on thermal baths, only $I_{BB^\dagger}$ and $I_{B^\dagger B}$ survive, because

$$\langle \hat{b}_q \hat{b}_p \rangle = \langle \hat{b}_q^\dagger \hat{b}_p^\dagger \rangle = 0, \qquad \langle \hat{b}_q^\dagger \hat{b}_p \rangle = \bar{n}(\omega_p)\delta_{q,p}. \tag{B.11}$$

All integrals can be expressed in terms of the spectral function (5.63), e.g.,

$$I_{B^\dagger B} = \int_t^{t+\delta t} \mathrm{d}t' \int \frac{\mathrm{d}\omega}{2\pi} J(\omega)\bar{n}(\omega)e^{i(\omega-\Delta)(t-t')} = \int_t^{t+\delta t} \mathrm{d}t' F(t-t'). \tag{B.12}$$

Following Walls and Milburn (1994), the *Markov approximation* enters by assuming that J and $\bar{n}$ are smooth functions, almost constant over a broad range of frequencies. As in Section B.3.1, the function $F(t-t')$ vanishes quickly beyond a short memory time $\tau \ll \delta t$. We can thus take the upper integration limits to $\pm\infty$, obtaining a distribution

$$\int_0^\infty \mathrm{d}\tau e^{\pm i\epsilon\tau} = \pi\delta(\epsilon) \pm i\mathcal{PV}\left(\frac{1}{\epsilon}\right), \tag{B.13}$$

where $\tau = t - t'$ and $\epsilon = \omega - \Delta$. Using this distribution, we conclude

$$I_{B^\dagger B} = \frac{\gamma}{2}\bar{n}(\Delta) + i\delta', \quad \text{and} \quad I_{BB^\dagger} = \frac{\gamma}{2}[\bar{n}(\Delta)+1] + i\delta, \tag{B.14}$$

with the decay rate $\gamma = J^{\mathrm{QO}}(\Delta)$ and two different Lamb shifts δ' and δ that are usually neglected. Inserting these values in (B.8) gives the usual quantum optics textbook master equation:

$$\begin{aligned}\frac{\mathrm{d}}{\mathrm{d}t}\tilde{\rho}_S = {}& [\bar{n}(\Delta)+1]\frac{\gamma}{2}\left(2\hat{A}\rho\hat{A}^\dagger - \hat{A}^\dagger\hat{A}\rho - \rho\hat{A}^\dagger\hat{A}\right) \\ &+ \bar{n}(\Delta)\frac{\gamma}{2}\left(2\hat{A}^\dagger\rho\hat{A} - \hat{A}\hat{A}^\dagger\rho - \rho\hat{A}\hat{A}^\dagger\right),\end{aligned} \tag{B.15}$$

with the cooling and heating Lindblad operators. At this point, it is interesting to remark that this equation can be brought back to the Schrödinger image, to compute the reduced density matrix of the system in the laboratory frame, $\rho_S(t) = \mathrm{tr}_B\,\rho(t) = e^{-i\hat{H}_0 t/\hbar}\tilde{\rho}_S(t)e^{i\hat{H}_0 t/\hbar}$. In a compact form,

$$\frac{\mathrm{d}}{\mathrm{d}t}\rho_S = -\frac{i}{\hbar}[\hat{H}_0, \rho_S] + [\bar{n}(\Delta)+1]\frac{\gamma}{2}\mathcal{L}_{\hat{A}\hat{A}^\dagger}\rho_S + \bar{n}(\Delta)\frac{\gamma}{2}\mathcal{L}_{\hat{A}^\dagger\hat{A}}\rho_S. \tag{B.16}$$

B.2.4 Perturbations and Generalizations

Our derivation assumed that the coupling operator $\hat{A}$ has a simple interaction picture (B.7). This happens for resonators, where $\hat{A}(t) = e^{i\Delta t}\hat{a}$ is a bosonic operator (see Section 5.5.2), and two-level systems, where $\hat{A}(t) = e^{i\Delta t}\sigma^-$ (Section 6.1.6). However, there are three important problems that do not conform to this scheme: a quantum system under a time-dependent control, a weakly nonlinear system, and a composite quantum system. The canonical superconducting circuits for each case are the resonator with a coherent microwave drive (5.8), the transmon qubit (6.34), and the Jaynes–Cummings model (7.33) for a qubit in a microwave cavity. In these setups, it is too difficult to compute $\hat{A}$ as we originally defined it. Instead, we use the part of $\hat{H}_0$ that is easy to build the interaction picture. The corrections – driving $\Omega(t)(\hat{a}+\hat{a}^\dagger)$, nonlinearity $\alpha\hat{a}^{\dagger 2}\hat{a}^2$, or coupling $g(\sigma^+\hat{a}+\sigma^-\hat{a}^\dagger)$ – are small terms that we incorporate into $\hat{H}_1$ and our perturbative expansion.

We illustrate this with the qubit-resonator model (7.33). The cavity and the qubit are coupled to two independent baths – the transmission line for the cavity and some source of intrinsic losses for the qubit. We develop the interaction picture using $\hat{H}_0 = \hbar\omega\hat{A}^\dagger\hat{A} + \frac{1}{2}\hbar\Delta\sigma^z + \sum_k \hbar\omega_k\hat{b}_k^\dagger\hat{b}_k$ and group all other terms into the "coupling" part:

$$\hat{H}_1 = g\sigma^-\hat{a}^\dagger + \hat{a}^\dagger\sum_k g_k\hat{b}_k + \sigma^-\sum_k f_k\hat{c}_k + \mathrm{H.c.} \tag{B.17}$$

This perturbation contains three uncorrelated terms $\hat{H}_1 = \hat{H}_{1a} + \hat{H}_{1b} + \hat{H}_{1c}$, such that $\mathrm{tr}(H_{1x}H_{1y}\rho_B) = 0$ if $x \neq y$. The Dyson series for $\Delta\tilde{\rho}_S$ has three independent contributions that originate three Lindblad operators in the master equation $\frac{\mathrm{d}}{\mathrm{d}t}\tilde{\rho}_S = (\mathcal{L}_a + \mathcal{L}_b + \mathcal{L}_c)\tilde{\rho}_S$, one of which, $\mathcal{L}_a$, will be a Hamiltonian contribution. A long and tedious but technically simple process confirms this idea. It results in a model (7.34) with the original JC Hamiltonian, accompanied by dissipation terms for the cavity and the qubit. Note that the decay rates $\kappa = J_b^{\mathrm{QO}}(\omega)$ and $\gamma = J_c^{\mathrm{QO}}(\Delta)$ are evaluated in the spectral functions of the respective *independent* environments. If the cavity and the qubit *share* the environment, the derivation of the master equation will reveal interesting cross terms, $\mathcal{L}_{\hat{a}\sigma^+}, \mathcal{L}_{\hat{a}^\dagger\sigma^-}$, but these are out of the scope of this book and have not been measured in experiments.

B.2.5 Strong Nonlinearity and Multilevel systems

When the system is strongly nonlinear or it is made of strongly coupled components, we must use an alternative derivation of the master equation using the eigenstates of the system Hamiltonian $\hat{H}_S = \sum_n E_n \left|E_n\right\rangle\left\langle E_n\right|$. We begin expanding the coupling operator in this basis $\hat{A} = \sum_{m<n} A_{mn} \left|E_m\right\rangle\left\langle E_n\right|$, with decay operators that connect states $E_m < E_n$. The system–bath coupling will enable all these transitions. In the interaction picture,

$$\hat{H}_1(t) = \sum_{m\neq n}\sum_k A^*_{mn} g_k e^{i\Delta_{nm}t - i\omega_k t} \left|E_n\right\rangle\left\langle E_m\right| \hat{b}_k^\dagger + \text{H.c.}, \tag{B.18}$$

where $\hbar\Delta_{nm} = E_n - E_m > 0$ are the positive frequency gaps. The analysis of $\hat{H}_1$ leads to a small explosion of terms in (B.8), but they all support the same approximations. The result is a generalized master equation:

$$\frac{\mathrm{d}}{\mathrm{d}t}\rho_S = -\frac{i}{\hbar}[\hat{H}_0, \rho_S] + \sum_{m,n} \frac{\Gamma^{\downarrow}_{mn}}{2}\mathcal{L}_{nm}\rho_S + \frac{\Gamma^{\uparrow}_{mn}}{2}\mathcal{L}_{mn}\rho_S. \tag{B.19}$$

The Lindblad operators model decay and excitation:

$$\mathcal{L}_{mn}\rho = 2\left|E_n\right\rangle\left\langle E_m\right| \rho \left|E_m\right\rangle\left\langle E_n\right| - \left|E_m\right\rangle\left\langle E_m\right| \rho - \rho\left|E_n\right\rangle\left\langle E_n\right|. \tag{B.20}$$

The rates satisfy detailed balance and are proportional to the original coupling strengths, that is,

$$\begin{aligned}\Gamma^{\downarrow}_{mn} &= [\bar{n}(\Delta_{nm} + 1)]|A_{mn}|^2 J^{\mathrm{QO}}(\Delta_{nm}), \\ \Gamma^{\uparrow}_{nm} &= \bar{n}(\Delta_{nm})|A_{mn}|^2 J^{\mathrm{QO}}(\Delta_{nm}).\end{aligned} \tag{B.21}$$

Note how (B.19) includes the simpler model from (B.16). However, the new equation describes the ultrastrong coupling limit of a qubit in a microwave resonator (see Section 7.4.2), something that the trivially corrected master equation from Section B.2.4 fails to do.

B.3 Input–Output Theory

We derive master equations by focusing on the system and ignoring its effect on the environment. This is often justified: When a superconducting qubit emits a microwave photon, the contribution of this photon to the universe's observables is tiny. This said, we may be interested in that photon, in capturing it, measuring it, and extracting the information it carries about the system that created it.

In this section, we develop a formalism known as *input–output theory* that is complementary to the master equation approach. In this formalism, we study the observables of a system and of its environment. Using the Heisenberg picture, we develop a stochastic equation for the system (5.64), where the environment acts as a source of noise. We also obtain an input–output relation (5.65) between the field of the environment before and after it interacts with the system. The first equation makes prediction about the systems's dynamics, while the second equation tells us how to verify these predictions, measuring the excitations that are emitted by the system when it decoheres. This is a powerful framework that describes spectroscopy, amplifiers, and measurement devices, and which applies to many superconducting circuit experiments.

B.3.1 Memory Function

Our study of input–output formalism starts with a resonator that interacts with a bosonic environment. We use the RWA Hamiltonian (5.60) to derive Heisenberg equations for the cavity and bath operators:

$$\frac{\mathrm{d}}{\mathrm{d}t}\hat{a} = -i\omega\hat{a} - i\sum_k g_k^*\hat{b}_k, \quad \text{and} \quad \frac{\mathrm{d}}{\mathrm{d}t}\hat{b}_k = -i\omega_k\hat{b}_k - ig_k\hat{a}. \tag{B.22}$$

The bath equations are solved implicitely, using initial conditions in the faraway past $t_0 \to -\infty$ or in the faraway future $t_f \to +\infty$. The first type of solutions reads

$$\hat{b}_k(t) = e^{-i\omega_k(t-t_0)}\hat{b}_k(t_0) - ig_k\int_{t_0}^{t} e^{-i\omega_k(t-\tau)}\hat{a}(\tau)\mathrm{d}\tau. \tag{B.23}$$

It contains the field leaking from the cavity $\hat{a}(t)$, together with the field that was injected long ago into the line $\{b_k(t_0)\}$ and that is now reaching the cavity. Using this solution, we obtain an integrodifferential equation for $\hat{a}(t)$ (5.61) that depends on (i) the bare properties of the cavity, ω; (ii) the noise operator for the incoming field,

$$\hat{\xi}(t) = \sum_k g_k^* e^{-i\omega_k(t-t_0)}\hat{b}_k(t_0); \tag{B.24}$$

and (iii) the memory function of the bath $K(t-\tau)$. The memory function is the Fourier transform of the spectral function J^{QO} (5.62). This function is broad and changes slowly around the frequencies of interest. This produces a Fourier transform $K(t)$ that concentrates around $t = 0$ and becomes zero past the *memory time* of the bath.

Let us illustrate this argument using a superconducting transmission line. The line has a spectral function that is approximately Ohmic (i.e., linear) around the cavity resonance $J^{\mathrm{QO}}(\bar{\omega}) \propto \bar{\omega}$. In order to avoid ultraviolet divergencies, we regularize this function with an exponential high-energy cutoff ω_c constraining the range of frequencies that pass through the guide. Apart from mathematical convenience, this is motivated by the filters on the line and ultimately by the superconducting gap. Using this cutoff, we compute a memory function:

$$J^{\mathrm{QO}}(\bar{\omega}) = \pi\alpha\bar{\omega}e^{-\bar{\omega}/\omega_c} \Rightarrow K(t) = \frac{\alpha\omega_c^2}{(it\omega_c+1)^2}, \tag{B.25}$$

which decays on a time of order $t_{\mathrm{memory}} \simeq 1/\omega_c$. The memory time is the shortest timescale in the problem – shorter than $2\pi/\omega$ or than the time for the cavity to emit all its energy.

We talk about the *Markovian regime* as a limit in which the bath immediately loses all memory of its interaction with the system. In this limit, the *Markovian approximation* is to take the system operators as approximately constant during t_{memory}, replacing the memory function with a Dirac delta $K(t) \sim \delta(t)$.

B.3.2 Markovian Approximation: Decay Rate and Lamb Shift

The procedure to simplify (5.61) receives the name of *separation of time scales*. It rewrites the cavity operator as a product of a slow modulation $\hat{a}_{\text{slow}}$ and a fast oscillation $\exp(-i\omega' t)$ with unknown frequency ω'. Introducing the variable $u = t - \tau$,

$$\begin{aligned}\int_{t_0}^{t} K(t-\tau)\hat{a}(\tau)\mathrm{d}\tau &\simeq \hat{a}_{\text{slow}}(t)\int_{t_0}^{t} K(t-\tau)e^{-i\omega'\tau}\mathrm{d}\tau \\ &= \hat{a}(t)\int_{0}^{+\infty} K(u)e^{i\omega' u}\mathrm{d}u = \left(i\delta\omega + \frac{1}{2}\kappa\right)\hat{a}(t),\end{aligned} \tag{B.26}$$

we simplify the whole integral down to a linear term that is local in time and depends on the Lamb shift $\delta\omega = \omega' - \omega$ and the decay rate κ. We can further relate these constants to the spectral function:

$$i\delta\omega + \frac{\gamma}{2} = \int_0^\infty K(u)e^{-i\omega' u}\mathrm{d}u \tag{B.27}$$

$$= \frac{1}{2\pi}\int_0^\infty \int J^{\text{QO}}(\omega)e^{i(\omega'-\omega)u}\mathrm{d}\omega\mathrm{d}u \tag{B.28}$$

$$= \frac{1}{2\pi}\int \left[\pi\delta(\omega'-\omega) + i\mathcal{PV}\left(\frac{1}{\omega'-\omega}\right)\right] J^{\text{QO}}(\omega)\mathrm{d}\omega. \tag{B.29}$$

The real and imaginary parts of the last integral give us the decay rate of the cavity as the spectral function evaluated at the dressed cavity resonance $\kappa = J^{\text{QO}}(\omega')$, and the Lamb shift as the principal value part:

$$\omega' - \omega = -\frac{1}{2\pi}\lim_{\epsilon\to 0}\left[\int_{-\infty}^{-\epsilon} + \int_{+\epsilon}^{+\infty}\right]\frac{J^{\text{QO}}(\omega'+x)}{x}\mathrm{d}x, \tag{B.30}$$

$$\kappa = J^{\text{QO}}(\omega'). \tag{B.31}$$

These are implicit relations that have to be solved for ω' and κ. However, in usual experiments we only have access to ω', which is obtained after fitting spectra, and not to ω. Hence, many quantum opticians ignore ω and define κ in terms of the "real frequency" of the resonator.

B.3.3 Input–Output Relations

We still have not discussed the noise operator $\hat{\xi}(t)$. To give some meaning to this term, we are going to first solve the Langevin equation (5.64) in the asymptotic limit $\exp(\kappa t_0) \to 0$:

$$\hat{a}(t) = e^{(-i\omega+\kappa/2)t}\hat{a}(0) - i\sum_k \frac{e^{-i\omega_k(t-t_0)}}{i(\omega'-\omega_k)+\kappa/2}g_k^*\hat{b}_k(t_0). \tag{B.32}$$

Note how the oscillator feeds only from a narrow band κ around the resonance ω'. This band is so small that we can assume that the coupling strength g_k, the spectral function $J^{\text{QO}}(\omega)$, and the density of states are more or less constant. Applying this idea to the study of the input operators gives

$$[\hat{\xi}_{\text{in}}(t), \hat{\xi}_{\text{in}}^\dagger(t')] = \sum_k |g_k|^2 e^{-i\omega_k(t-t')} = \int d\omega J^{\text{QO}}(\omega) e^{-i\omega(t-t')} \tag{B.33}$$

$$\simeq J^{\text{QO}}(\omega') \int d\omega e^{-i\omega(t-t')} = \kappa\delta(t-t'). \tag{B.34}$$

We replace the input operators with new ones $\hat{b}^{\text{in}}(t) = \hat{\xi}_{\text{in}}/\sqrt{\kappa}$ that satisfy causal commutation relations, producing the Langevin equation (5.64). Note that the cavity operator $\hat{a}(t)$ is dimensionless, but $\kappa\, \hat{b}^{\text{in}\,\dagger}\hat{b}^{\text{in}}$ has dimensions of photon flux $[\text{time}]^{-1}$. Hence $[\hat{b}^{\text{in}}] = [\text{time}]^{-1/2}$.

We can repeat the whole preceding study, but using initial conditions that are far away in the future, $t_f \to +\infty$. The equation is slightly different:

$$\frac{d}{dt}\hat{a} = (-i\omega' + \kappa/2)\hat{a} - i\sqrt{\kappa}\,\hat{b}^{\text{out}}, \tag{B.35}$$

and now involves the field reflected by the cavity $\hat{b}^{\text{out}}$. The problem with this equation is that we do not really know the state of the environment in the future, $\hat{b}_k(+\infty)$, so that we cannot compute $\hat{b}^{\text{out}}$. Fortunately, we can guess $\hat{b}^{\text{out}}$ indirectly, realizing that both boundary conditions, $\hat{b}^{\text{out}}$ and $\hat{b}^{\text{in}}$, must give the same value for the field experienced by the cavity (B.23):

$$\sum_k g_k^* b_k(t) = \sqrt{\kappa}\hat{b}^{\text{in}}(t) - i\int_{-\infty}^{t} K(t-\tau)\hat{a}(\tau)d\tau \tag{B.36}$$

$$= \sqrt{\kappa}\hat{b}^{\text{out}}(t) + i\int_{0}^{+\infty} K(t-\tau)\hat{a}(\tau)d\tau. \tag{B.37}$$

Combining both equations and using $\int_{\mathbb{R}} K(t-\tau)\hat{a}(\tau)d\tau \simeq \kappa\hat{a}(t)$, we arrive at the *input–output relation*:

$$\hat{b}^{\text{out}}(t) = \hat{b}^{\text{in}}(t) - i\sqrt{\kappa}a(t). \tag{B.38}$$

Let us close this section with some remarks. The first one is to point out that the input–output relations vary with how we write the coupling Hamiltonian. If instead of (5.60) we use $\hbar i(g_k a b_k^\dagger - g_k^* b_k a^\dagger)$, this amounts to a phase shift $\hat{a} \to i\hat{a}$ that suppresses some factors:

$$\frac{d}{dt}\hat{a}(t) = (-i\omega - \kappa/2)\hat{a}(t) - \sqrt{\kappa}\,\hat{b}^{\text{in}}(t), \tag{B.39}$$

$$\hat{b}^{\text{out}}(t) = \hat{b}^{\text{in}}(t) + \sqrt{\kappa}\hat{a}(t). \tag{B.40}$$

Another important variation is when we want to consider multiple input and output channels. This happens in transmission experiments, when the LC resonator is inserted in the middle of a line, or when it is connected to multiple lines. In that case, the input–output equations read

$$\frac{d}{dt}\hat{a}(t) = \left(-i\omega - \sum_r \kappa^r/2\right)\hat{a}(t) - i\sum_n \sqrt{\kappa^r}\,\hat{b}^{\text{in}}(t), \tag{B.41}$$

$$\hat{b}_r^{\text{out}}(t) = \hat{b}_r^{\text{in}}(t) - i\sqrt{\kappa^r}\hat{a}(t). \tag{B.42}$$

The index $r = 1, 2, \ldots R$ runs over all input–output channels, which may have very different decay rates κ^r.

B.3.4 Spectroscopy

Throughout this book, we offer many examples of spectroscopy: experiments where we illuminate a quantum system with microwaves and study the radiation that it absorbs, emits, and scatters. Such experiments fit naturally in the input–output framework that we just developed, provided it is extended to treat more general setups, such as qubits, cavity-QED, and other composite and nonlinear systems.

The generalization of input–output theory parallels Section B.2.5. We consider an arbitrary quantum system with discrete eigenstates, coupled to R input–output channels:

$$\hat{H}_1 = \sum_{r=1}^{R} \sum_{m\neq n} \sum_{k} A^r_{mn} g_{r,k} \, |E_n\rangle\langle E_m| \, \hat{b}^\dagger_{r,k} + \text{H.c.} \tag{B.43}$$

Applying the same treatment and approximations as before, we arrive at a series of input–output relations:

$$\hat{b}^{\text{out}}_r(t) = \hat{b}^{\text{in}}_r(t) - i\sqrt{\gamma^r_{mn}}\hat{\sigma}_{mn}, \tag{B.44}$$

with system ladder operators $\hat{\sigma}_{mn}(t = 0) = |E_m\rangle\langle E_n|$ that feed the different lines at different rates $\gamma^r_{mn} \propto |A^r_{mn}|^2 J^{\text{QO}}(\Delta_{nm})$.

Studying the evolution of the system operators $\hat{\sigma}_{mn}$ is rather complicated, and it is rarely done in the Heisenberg picture. Most often, we are just interested in the expectation values and correlations of the output fields, $\langle \hat{b}^{\text{out}}_r \rangle$. In this case, we can get away with any method to estimate quantities such as $\langle \hat{\sigma}_{mn}(t) \rangle$. The most popular method is to calculate the coherences using a master equation such as (B.19), adapted to reproduce the conditions of a spectroscopy protocol.

The simulations presume that the waveguides are close to zero temperature and host some input fields $\langle \hat{b}^{\text{in}}_r(t) \rangle \sim \Omega_r(t)$. The Hamiltonian part of the master equation accommodates the coherent driving with which we illuminate our system. The losses are extended to add up all spectroscopy channels and the intrinsic decoherence of our sample. We typically neglect heating. In other words,

$$\hat{H}_0 = \sum_m E_m \, |E_m\rangle\langle E_m| + \sum_{r,m,n} \sqrt{\gamma^r_{mn}} \Omega_r(t)^* \, |E_m\rangle\langle E_n| + \text{H.c.}, \tag{B.45}$$

$$\Gamma^{\downarrow}_{mn} = \gamma^{int}_{mn} + \sum_r \gamma^r_{mn} \text{ and } \Gamma^{\uparrow}_{nm} \simeq 0.$$

In the common case in wich we drive monochromatically through one channel $\Omega_0(t) \sim \Omega_0 \exp(-i\omega t)$, this model predicts a Lorentzians response in all channels:

$$\langle \hat{b}^{\text{out}}_r(t) \rangle \sim \Omega_0 e^{-i\omega t} \left[\delta_{r0} - \frac{\sqrt{\gamma^r_{mn}\gamma^0_{mn}}}{i(\omega - \Delta_{nm}) + \Gamma_{mn}/2} \right]. \tag{B.46}$$

The Lorentzians are centered on the different transition frequencies Δ_{mn}. The weight and the linewidth of these resonances give information on the decoherence rates and on the transition strengths, $\Gamma_{mn} \propto |A_{mn}|^2$. This is a lot of information that can be used to confirm and fit our microscopic models, such as the strong coupling and ultrastrong coupling Hamiltonians from our circuit-QED models (Section 7.4.1).

References

Abdo, B., Suchoi, O., Segev, E., et al. 2009. Intermodulation and parametric amplification in a superconducting stripline resonator integrated with a dc-SQUID. *EPL (Europhysics Letters)*, **85**(6), 68001.

Adesso, Gerardo, Ragy, Sammy, and Lee, Antony R. 2014. Continuous variable quantum information: Gaussian states and beyond. *Open Systems & Information Dynamics*, **21**(01–02), 1440001.

Aharonov, Dorit, van Dam, Wim, Kempe, Julia, Landau, Zeph, Lloyd, Seth, and Regev, Oded. 2004. Adiabatic quantum computation is equivalent to standard quantum computation. *SIAM Review*, **50**(4), 755–787.

Albash, Tameem, and Lidar, Daniel A. 2018. Demonstration of a scaling advantage for a quantum annealer over simulated annealing. *Physical Review* X 8, 031016 (2018).

Albash, Tameem, Vinci, Walter, Mishra, Anurag, Warburton, Paul A., and Lidar, Daniel A. 2015. Consistency tests of classical and quantum models for a quantum annealer. *Physical Review* A 91, 042314 (2015).

Anderson, P. W., and Rowell, J. M. 1963. Probable observation of the Josephson superconducting tunneling effect. *Physical Review Letters*, **10**(6), 230–232.

Arute, Frank, Arya, Kunal, Babbush, Ryan, et al. 2019. Quantum supremacy using a programmable superconducting processor. *Nature*, **574**(7779), 505–510.

Arute, Frank, Arya, Kunal, Babbush, Ryan, et al. 2020. Hartree–Fock on a superconducting qubit quantum computer. *Science*, **369**(6507), 1084–1089.

Astafiev, O., Zagoskin, A. M., Abdumalikov, A. A., et al. 2010. Resonance fluorescence of a single artificial atom. *Science (New York, N.Y.)*, **327**(5967), 840–843.

Ballentine, L. E. 1970. The statistical interpretation of quantum mechanics. *Reviews of Modern Physics*, **42**(4), 358–381.

Ballentine, Leslie E. 1998. *Quantum Mechanics*. World Scientific.

Bapst, V., Foini, L., Krzakala, F., Semerjian, G., and Zamponi, F. 2013. The quantum adiabatic algorithm applied to random optimization problems: the quantum spin glass perspective. *Physics Reports*, **523**(3), 127–205.

Barahona, F. 1982. On the computational complexity of Ising spin glass models. *Journal of Physics A: Mathematical and General*, **15**(10), 3241–3253.

Bardeen, J., Cooper, L. N., and Schrieffer, J. R. 1957a. Microscopic theory of superconductivity. *Physical Review*, **106**(1), 162–164.

Bardeen, J., Cooper, L. N., and Schrieffer, J. R. 1957b. Theory of superconductivity. *Physical Review*, **108**(5), 1175–1204.

Barends, R., Kelly, J., Megrant, A., et al. 2014. Superconducting quantum circuits at the surface code threshold for fault tolerance. *Nature*, **508**(7497), 500–503.

Baumgratz, T., Cramer, M., and Plenio, M. B. 2014. Quantifying coherence. *Physical Review Letters*, **113**(14), 140401.

Benioff, Paul. 1980. The computer as a physical system: a microscopic quantum mechanical Hamiltonian model of computers as represented by Turing machines. *Journal of Statistical Physics*, **22**(5), 563–591.

Bennett, Douglas A., Longobardi, Luigi, Patel, Vijay, Chen, Wei, and Lukens, James E. 2007. rf-SQUID qubit readout using a fast flux pulse. *Superconductor Science and Technology*, **20**(11), S445–S449.

Berkley, A. J., Johnson, M. W., Bunyk, P., et al. 2010. A scalable readout system for a superconducting adiabatic quantum optimization system. *Superconductor Science and Technology*, **23**(10), 105014.

Berkley, A. J., Przybysz, A. J., Lanting, T., et al. 2013. Tunneling spectroscopy using a probe qubit. *Physical Review B*, **87**(2), 020502.

Bialczak, R. C., Ansmann, M., Hofheinz, M., et al. 2010. Quantum process tomography of a universal entangling gate implemented with Josephson phase qubits. *Nature Physics*, **6**(6), 409–413.

Biamonte, Jacob D., and Love, Peter J. 2008. Realizable Hamiltonians for universal adiabatic quantum computers. *Physical Review A*, **78**(1), 012352.

Bishop, Lev S., Bravyi, Sergey, Gambetta, Jay M., and Smolin, John. 2017. *Quantum Volume. Technical Report.*

Blais, Alexandre, Huang, Ren-Shou, Wallraff, Andreas, Girvin, S., and Schoelkopf, R. 2004. Cavity quantum electrodynamics for superconducting electrical circuits: an architecture for quantum computation. *Physical Review A*, **69**(6), 062320.

Boixo, Sergio, Albash, Tameem, Spedalieri, Federico M., Chancellor, Nicholas, and Lidar, Daniel A. 2013. Experimental signature of programmable quantum annealing. *Nature Communications*, **4**(1), 2067.

Boixo, Sergio, Rønnow, Troels F., Isa kov, Sergei V., et al. 2014. Evidence for quantum annealing with more than one hundred qubits. *Nature Physics*, **10**(3), 218–224.

Boixo, Sergio, Smelyanskiy, Vadim N., Shabani, Alireza, et al. 2016. Computational multiqubit tunnelling in programmable quantum annealers. *Nature Communications*, **7**(Jan), 10327.

Bombin, H., and Martin-Delgado, M. A. 2006. Topological Quantum Distillation. *Physical Review Letters*, **97**(Oct), 180501.

Bouchiat, V., Vion, D., Joyez, P., Esteve, D., and Devoret, M. H. 1998. Quantum coherence with a single Cooper pair. *Physica Scripta*, **T76**(1), 165–170.

Bourassa, J., Gambetta, J., Abdumalikov, A., Astafiev, O., Nakamura, Y., and Blais, A. 2009. Ultrastrong coupling regime of cavity QED with phase-biased flux qubits. *Physical Review A*, **80**(3), 032109.

Bravyi, S. B., and Kitaev, A. Yu. 1998. Quantum codes on a lattice with boundary. *e-print* arXiv:quant-ph-9811052.

Bravyi, Sergey, DiVincenzo, David P., and Loss, Daniel. 2011. Schrieffer–Wolff transformation for quantum many-body systems. *Annals of Physics*, **326**(10), 2793–2826.

Brecht, T., Chu, Y., Axline, C., et al. 2017. Micromachined integrated quantum circuit containing a superconducting qubit. *Physics Review Applied*, **7**(Apr), 044018.

Brecht, Teresa, Pfaff, Wolfgang, Wang, et al. 2016. Multilayer microwave integrated quantum circuits for scalable quantum computing. *npj Quantum Information*, **2**(1), 16002.

Brink, Markus, Corcoles-Gonzalez, Antonio, Gambetta, Jay M., Rosenblatt, Sami, and Solgun, Firat. 2019 (May 28). *Low loss architecture for superconducting qubit circuits*. US Patent 10,305,015.

Brown, K. R., Wilson, A. C., Colombe, Y., et al. 2011. Single-qubit-gate error below $\mathbf{10^{-4}}$ in a trapped ion. *Physics Review A*, **84**(Sep), 030303.

Cai, Jun, Macready, William G., and Roy, Aidan. 2014. A practical heuristic for finding graph minors. *e-print* arXiv:1406.2741.

Calderbank, A. R., and Shor, Peter W. 1996. Good quantum error-correcting codes exist. *Physical Review A*, **54**(Aug), 1098–1105.

Castellanos-Beltran, M. A., Irwin, K. D., Vale, L. R., Hilton, G. C., and Lehnert, K. W. 2009. Bandwidth and dynamic range of a widely tunable Josephson parametric amplifier. *IEEE Transactions on Applied Superconductivity*, **19**(3), 944–947.

Chancellor, N., Zohren, S., and Warburton, P. A. 2017. Circuit design for multi-body interactions in superconducting quantum annealing systems with applications to a scalable architecture. *npj Quantum Information*, **3**(1), 21.

Chen, Yu, Neill, C., Roushan, P., et al. 2014. Qubit architecture with high coherence and fast tunable coupling. *Physical Review Letters*, **113**(Nov), 220502.

Chen, Zijun, Satzinger, Kevin J., Atalaya, Juan, et al. 2021. Exponential suppression of bit or phase errors with cyclic error correction. *Nature*, **595**(7867), 383–387.

Chiorescu, I., Bertet, P., Semba, K., Nakamura, Y., Harmans, C. J. P. M., and Mooij, J. E. 2004. Coherent dynamics of a flux qubit coupled to a harmonic oscillator. *Nature*, **431**(7005), 159–162.

Chow, J. M., Gambetta, J. M., Tornberg, L., et al. 2009. Randomized benchmarking and process tomography for gate errors in a solid-state qubit. *Physical Review Letters*, **102**(9), 090502.

Chow, Jerry M., Córcoles, A. D., Gambetta, Jay M., et al. 2011. Simple all-microwave entangling gate for fixed-frequency superconducting qubits. *Physical Review Letters*, **107**(8), 080502.

Chow, Jerry M., Gambetta, Jay M., Córcoles, A. D., et al. 2012. Universal quantum gate set approaching fault-tolerant thresholds with superconducting qubits. *Physical Review Letters*, **109**(6), 060501.

Cirac, J. I., and Zoller, P. 1995. Quantum computations with cold trapped ions. *Physical Review Letters*, **74**(20), 4091–4094.

Cirac, J. I., Ekert, A. K., Huelga, S. F., and Macchiavello, C. 1999. Distributed quantum computation over noisy channels. *Physical Review A*, **59**(6), 4249–4254.

Clarke, John, and Braginski, Alex I. (eds). 2004. *The SQUID Handbook: Fundamentals and Technology of SQUIDs and SQUID Systems, Volume I*. Wiley-VCH.

Clarke, John, and Wilhelm, Frank K. 2008. Superconducting quantum bits. *Nature*, **453**(7198), 1031–1042.

Cohen-Tannoudji, Claude, Diu, Bernard, and Laloë, Franck. 1977. *Quantum Mechanics*. Wiley.

Consani, Gioele, and Warburton, Paul A. 2020. Effective Hamiltonians for interacting superconducting qubits: local basis reduction and the Schrieffer–Wolff transformation. *New Journal of Physics*, **22**(5), 053040.

Cooper, Leon N. 1956. Bound electron pairs in a degenerate Fermi gas. *Physical Review*, **104**(4), 1189–1190.

Cory, D. G., Fahmy, A. F., and Havel, T. F. 1997. Ensemble quantum computing by NMR spectroscopy. *Proceedings of the National Academy of Sciences of the United States of America*, **94**(5), 1634–1639.

Cross, Andrew W., Bishop, Lev S., Sheldon, Sarah, Nation, Paul D., and Gambetta, Jay M. 2019. Validating quantum computers using randomized model circuits. *Physical Review A*, **100**(Sep), 032328.

Cullen, A. L. 1960. Theory of the travelling-wave parametric amplifier. *Proceedings of the IEE Part B: Electronic and Communication Engineering*, **107**(32), 101.

Deaver, Bascom S., and Fairbank, William M. 1961. Experimental evidence for quantized flux in superconducting cylinders. *Physical Review Letters*, **7**(2), 43–46.

Denchev, Vasil S., Boixo, Sergio, Isakov, Sergei V., et al. 2016. What is the computational value of finite-range tunneling? *Physical Review X*, **6**(3), 031015.

Deutsch, David Elieser. 1985. Quantum theory, the Church–Turing principle and the universal quantum computer. *Proceedings of the Royal Society A: Mathematical, Physical and Engineering Sciences*, **400**(1818), 97–117.

Deutsch, David Elieser. 1989. Quantum computational networks. *Proceedings of the Royal Society of London. A. Mathematical and Physical Sciences*, **425**(1868), 73–90.

Devoret, M. H. 1995. Quantum fluctuations in electrical circuits. *Les Houches, Session LXIII*, 351–386.

DiCarlo, L., Chow, J. M., Gambetta, J. M., et al. 2009. Demonstration of two-qubit algorithms with a superconducting quantum processor. *Nature*, **460**(7252), 240–244.

DiCarlo, L., Reed, M. D., Sun, L., et al. 2010. Preparation and measurement of three-qubit entanglement in a superconducting circuit. *Nature*, **467**(7315), 574–578.

Dickson, N. G., Johnson, M. W., Amin, M. H., et al. 2013. Thermally assisted quantum annealing of a 16-qubit problem. *Nature Communications*, **4**(1), 1903.

Dieks, D. 1982. Communication by EPR devices. *Physics Letters A*, **92**(6), 271–272.

DiVincenzo, D. P. 1995. Quantum computation. *Science*, **270**(5234), 255–261.

DiVincenzo, David P. 2000. The physical implementation of quantum computation. *Fortschritte der Physik*, **48**(9-11), 771–783.

Dowling, Jonathan P., and Milburn, Gerard J. 2003. Quantum technology: the second quantum revolution. *Philosophical Transactions of the Royal Society of London. Series A: Mathematical, Physical and Engineering Sciences*, **361**(1809), 1655–1674.

Egger, D. J., Werninghaus, M., Ganzhorn, M., et al. 2018. Pulsed reset protocol for fixed-frequency superconducting qubits. *Physical Review Applied*, **10**(Oct), 044030.

Eichler, C., Bozyigit, D., and Wallraff, A. 2012. Characterizing quantum microwave radiation and its entanglement with superconducting qubits using linear detectors. *Physical Review A*, **86**(3), 032106.

Eichler, C., Bozyigit, D., Lang, C., Steffen, L., Fink, J., and Wallraff, A. 2011. Experimental state tomography of itinerant single microwave photons. *Physical Review Letters*, **106**(22), 220503.

Farhi, Edward, Goldstone, Jeffrey, Gutmann, Sam, and Sipser, Michael. 2000. Quantum computation by adiabatic evolution. *arXiv*, **quant-ph:0**(jan).

Feynman, Richard P. 1982. Simulating physics with computers. *International Journal of Theoretical Physics*, **21**(6-7), 467–488.

Forn-Díaz, P., Lisenfeld, J., Marcos, D., et al. 2010. Observation of the Bloch–Siegert shift in a qubit-oscillator system in the ultrastrong coupling regime. *Physical Review Letters*, **105**(23), 237001.

Forn-Díaz, P., García-Ripoll, J. J., Peropadre, B., et al. 2016. Ultrastrong coupling of a single artificial atom to an electromagnetic continuum in the nonperturbative regime. *Nature Physics*, **13**(1), 39–43.

Forn-Díaz, P., Warren, C. W., Chang, C. W. S., Vadiraj, A. M., and Wilson, C. M. 2017. On-demand microwave generator of shaped single photons. *Physical Review Applied*, **8**(5), 054015.

Fowler, Austin G., Mariantoni, Matteo, Martinis, John M., and Cleland, Andrew N. 2012. Surface codes: towards practical large-scale quantum computation. *Physical Review A*, **86**(Sep), 032324.

Fowler, Austin G., Stephens, Ashley M., and Groszkowski, Peter. 2009. High-threshold universal quantum computation on the surface code. *Physical Review A*, **80**(Nov), 052312.

Gaebler, J. P., Meier, A. M., Tan, T. R., et al. 2012. Randomized benchmarking of multiqubit gates. *Physical Review Letters*, **108**(26), 260503.

Galindo, Alberto, and Pascual, Pedro. 1991. Time-dependent perturbation theory. Pages 161–199 of: *Quantum Mechanics II*. Springer Berlin Heidelberg.

García-Ripoll, J. J., Peropadre, B., and De Liberato, S. 2015. Light–matter decoupling and A2 term detection in superconducting circuits. *Scientific Reports*, **5**(1), 16055.

García-Ripoll, J. J., Ruiz-Chamorro, A., and Torrontegui, E. 2020. Quantum control of frequency-tunable transmon superconducting qubits. *Physical Review Applied*, **14**(4), 044035.

Gardiner, C. W., and Zoller, P. 2004. *Quantum Noise*. 3rd ed. Berlin, Heidelberg: Springer-Verlag Berlin Heidelberg.

Gershenfeld, Neil A., and Chuang, Isaac L. 1997. Bulk spin-resonance quantum computation. *Science (New York, N.Y.)*, **275**(5298), 350–356.

Gilchrist, Alexei, Langford, Nathan K., and Nielsen, Michael A. 2005. Distance measures to compare real and ideal quantum processes. *Physical Review A*, **71**(6), 062310.

Gor'kov, Lev P. 1959. Microscopic derivation of the Ginzburg–Landau equations in the theory of superconductivity. *JETP* **36**(9), 1364.

Gottesman, Daniel. 1998. Theory of fault-tolerant quantum computation. *Physical Review A*, **57**(Jan), 127–137.

Haroche, Serge. 2013. Nobel lecture: controlling photons in a box and exploring the quantum to classical boundary. *Reviews of Modern Physics*, **85**(3), 1083–1102.

Harris, R., Berkley, A. J., Johnson, M. W., et al. 2007. Sign- and magnitude-tunable coupler for superconducting flux qubits. *Physical Review Letters*, **98**(17), 177001.

Harris, R., Johansson, J., Berkley, A. J., et al. 2010a. Experimental demonstration of a robust and scalable flux qubit. *Physical Review B*, **81**(13), 134510.

Harris, R., Johnson, M. W., Lanting, T., et al. 2010b. Experimental investigation of an eight-qubit unit cell in a superconducting optimization processor. *Physical Review B*, **82**(2), 024511.

Harris, R., Lanting, T., Berkley, A. J., et al. 2009. Compound Josephson-junction coupler for flux qubits with minimal crosstalk. *Physical Review B*, **80**(5), 052506.

Harris, R., Sato, Y., Berkley, A. J., et al. 2018. Phase transitions in a programmable quantum spin glass simulator. *Science*, **361**(6398), 162–165.

Hauke, Philipp, Katzgraber, Helmut G, Lechner, Wolfgang, Nishimori, Hidetoshi, and Oliver, William D. 2020. Perspectives of quantum annealing: methods and implementations. *Reports on Progress in Physics*, **83**(5), 054401.

Heisenberg, W. 1925. Über quantentheoretische Umdeutung kinematischer und mechanischer Beziehungen. *Zeitschrift für Physik*, **33**(1), 879–893.

Hime, T., Reichardt, P. A., Plourde, B. L. T., et al. 2006. Solid-state qubits with current–controlled coupling. *Science (New York, N.Y.)*, **314**(5804), 1427–1429.

Hita-Pérez, María, Jaumá, Gabriel, Pino, Manuel, and García-Ripoll, Juan José. 2022. Ultrastrong capacitive coupling of flux qubits. *Physical Review Applied* **17**, 014028.

Hofheinz, Max, Wang, H., Ansmann, M., et al. 2009. Synthesizing arbitrary quantum states in a superconducting resonator. *Nature*, **459**(7246), 546–549.

Hoi, Io-Chun, Kockum, Anton F., Palomaki, Tauno, et al. 2013. Giant Cross–Kerr effect for propagating microwaves induced by an artificial atom. *Physical Review Letters*, **111**(5), 053601.

Hoi, Io-Chun, Palomaki, Tauno, Lindkvist, Joel, Johansson, Göran, Delsing, Per, and Wilson, C. M. 2012. Generation of nonclassical microwave states using an artificial atom in 1d open space. *Physical Review Letters*, **108**(26), 263601.

Hoi, Io-Chun, Wilson, C. M., Johansson, Göran, Palomaki, Tauno, Peropadre, Borja, and Delsing, Per. 2011. Demonstration of a single-photon router in the microwave regime. *Physical Review Letters*, **107**(7), 073601.

Jansen, Sabine, Ruskai, Mary-Beth, and Seiler, Ruedi. 2007. Bounds for the adiabatic approximation with applications to quantum computation. *Journal of Mathematical Physics*, **48**(10), 102111.

Jaynes, E. T., and Cummings, F. W. 1963. Comparison of quantum and semiclassical radiation theories with application to the beam maser. *Proceedings of the IEEE*, **51**(1), 89–109.

Jeffrey, Evan, Sank, Daniel, Mutus, J. Y., et al. 2014. Fast accurate state measurement with superconducting qubits. *Physical Review Letters*, **112**(19), 190504.

Johnson, M. W., Amin, M. H. S., Gildert, S., et al. 2011. Quantum annealing with manufactured spins. *Nature*, **473**(7346), 194–198.

Johnson, M. W., Bunyk, P., Maibaum, F., et al. 2010. A scalable control system for a superconducting adiabatic quantum optimization processor. *Superconductor Science and Technology*, **23**(6), 065004.

Josephson, B. D. 1962. Possible new effects in superconductive tunnelling. *Physics Letters*, **1**(7), 251–253.

Kadowaki, Tadashi, and Nishimori, Hidetoshi. 1998. Quantum annealing in the transverse Ising model. *Physical Review E*, **58**(5), 5355–5363.

Kandala, Abhinav, Temme, Kristan, Córcoles, Antonio D., Mezzacapo, Antonio, Chow, Jerry M., and Gambetta, Jay M. 2019. Error mitigation extends the computational reach of a noisy quantum processor. *Nature*, **567**(7749), 491–495.

Kato, Tosio. 1950. On the adiabatic theorem of quantum mechanics. *Journal of the Physical Society of Japan*, **5**(6), 435–439.

Kelly, J., Barends, R., Fowler, A. G., et al. 2015. State preservation by repetitive error detection in a superconducting quantum circuit. *Nature*, **519**(7541), 66–69.

Kempe, Julia, Kitaev, Alexei, and Regev, Oded. 2006. The complexity of the local Hamiltonian problem. *SIAM Journal on Computing*, **35**(5), 1070–1097.

King, Andrew D., Carrasquilla, Juan, Raymond, Jack, et al. 2018. Observation of topological phenomena in a programmable lattice of 1,800 qubits. *Nature*, **560**(7719), 456–460.

Kitaev, A., Shen, A., and Vyalyi, M. 2002. *Classical and Quantum Computation*. Graduate Studies in Mathematics. American Mathematical Society.

Knill, E., Leibfried, D., Reichle, R., et al. 2008. Randomized benchmarking of quantum gates. *Physical Review A*, **77**(1), 012307.

Koch, Jens, Yu, Terri, Gambetta, Jay, et al. 2007. Charge-insensitive qubit design derived from the Cooper pair box. *Physical Review A*, **76**(4), 042319.

Krinner, S., Lazar, S., Remm, A., et al. 2020. Benchmarking coherent errors in controlled-phase gates due to spectator qubits. *Physical Review Applied*, **14**(2), 024042.

Kurcz, Andreas, Bermudez, Alejandro, and García-Ripoll, Juan José. 2014. Hybrid quantum magnetism in circuit QED: from spin-photon waves to many-body spectroscopy. *Physical Review Letters*, **112**(18), 180405.

Landau, L. 1932. Zur theorie der energieubertragung II. *Physik. Z. Sowjet*, **2**, 46–50.

Lanting, T., Przybysz, A. J., Smirnov, A. Yu., et al. 2014. Entanglement in a quantum annealing processor. *Physical Review X*, **4**(2), 021041.

Laurat, Julien, Keller, Gaëlle, Oliveira-Huguenin, José Augusto, et al. 2005. Entanglement of two-mode Gaussian states: characterization and experimental production and manipulation. *Journal of Optics B: Quantum and Semiclassical Optics*, **7**(12), S577–S587.

Law, C. K., and Eberly, J. H. 1996. Arbitrary control of a quantum electromagnetic field. *Physical Review Letters*, **76**(7), 1055–1058.

Leggett, A., Chakravarty, S., Dorsey, A., Fisher, Matthew, Garg, Anupam, and Zwerger, W. 1987. Dynamics of the dissipative two-state system. *Reviews of Modern Physics*, **59**(1), 1–85.

Leibfried, D., Blatt, R., Monroe, C., and Wineland, D. 2003. Quantum dynamics of single trapped ions. *Reviews of Modern Physics*, **75**(1), 281–324.

Lidar, Daniel A., Rezakhani, Ali T., and Hamma, Alioscia. 2009. Adiabatic approximation with exponential accuracy for many-body systems and quantum computation. *Journal of Mathematical Physics*, **50**(10), 102106.

Lindblad, G. 1976. On the generators of quantum dynamical semigroups. *Communications in Mathematical Physics*, **48**(2), 119–130.

London, F., London, H., and Lindemann, Frederick Alexander. 1935. The electromagnetic equations of the supraconductor. *Proceedings of the Royal Society of London. Series A - Mathematical and Physical Sciences*, **149**(866), 71–88.

Loss, Daniel, and DiVincenzo, David P. 1998. Quantum computation with quantum dots. *Physical Review A*, **57**(1), 120–126.

Macklin, C., O'Brien, K., Hover, D., et al. 2015. A near-quantum-limited Josephson traveling-wave parametric amplifier. *Science (New York, N.Y.)*, **350**(6258), 307–310.

Magesan, Easwar, Gambetta, J. M., and Emerson, Joseph. 2011. Scalable and robust randomized benchmarking of quantum processes. *Physical Review Letters*, **106**(18), 180504.

Magesan, Easwar, Gambetta, Jay M., Johnson, B. R., et al. 2012. Efficient measurement of quantum gate error by interleaved randomized benchmarking. *Physical Review Letters*, **109**(8), 080505.

Magnard, P., Kurpiers, P., Royer, B., et al. 2018. Fast and unconditional all-microwave reset of a superconducting qubit. *Physical Review Letters*, **121**(Aug), 060502.

Magnard, P., Storz, S., Kurpiers, P., et al. 2020. Microwave quantum link between superconducting circuits housed in spatially separated cryogenic systems. *Physical Review Letters*, **125**(26), 260502.

Majer, J., Chow, J. M., Gambetta, J. M., et al. 2007. Coupling superconducting qubits via a cavity bus. *Nature*, **449**(7161), 443–447.

Makhlin, Yuriy, Schön, Gerd, and Shnirman, Alexander. 2001. Quantum-state engineering with Josephson-junction devices. *Reviews of Modern Physics*, **73**(2), 357–400.

Mallet, F., Castellanos-Beltran, M. A., Ku, H. S., et al. 2011. Quantum state tomography of an itinerant squeezed microwave field. *Physical Review Letters*, **106**(22), 220502.

Mandrá, Salvatore, and Katzgraber, Helmut G. 2018. A deceptive step towards quantum speedup detection. *Quantum Science and Technology*, **3**(4), 04LT01.

Mandrá, Salvatore, Zhu, Zheng, Wang, Wenlong, Perdomo-Ortiz, Alejandro, and Katzgraber, Helmut G. 2016. Strengths and weaknesses of weak-strong cluster problems: a detailed overview of state-of-the-art classical heuristics versus quantum approaches. *Physical Review A*, **94**(2), 022337.

McEwen, M., Kafri, D., Chen, Z., et al. 2021. Removing leakage-induced correlated errors in superconducting quantum error correction. *Nature Communications*, **12**(1), 1761.

Menzel, E. P., Deppe, F., Mariantoni, M., et al. 2010. Dual-path state reconstruction scheme for propagating quantum microwaves and detector noise tomography. *Physical Review Letters*, **105**(10), 100401.

Menzel, E. P., Di Candia, R., Deppe, F., et al. 2012. Path entanglement of continuous-variable quantum microwaves. *Physical Review Letters*, **109**(25), 250502.

Merkel, Seth T., Gambetta, Jay M., Smolin, John A., et al. 2013. Self-consistent quantum process tomography. *Physical Review A*, **87**(Jun), 062119.

Mi, Xiao, Ippoliti, Matteo, Quintana, Chris, Greene, et al. 2022. Observation of time-crystalline eigenstate order on a quantum processor. *Nature* **601**, 531.

Mirrahimi, Mazyar, Leghtas, Zaki, Albert, et al. 2014. Dynamically protected cat-qubits: a new paradigm for universal quantum computation. *New Journal of Physics*, **16**(4), 045014.

Mlynek, J. A., Abdumalikov, A. A., Eichler, C., and Wallraff, A. 2014. Observation of Dicke superradiance for two artificial atoms in a cavity with high decay rate. *Nature Communications*, **5**(1), 5186.

Mooij, J. E., Orlando, T. P., Levitov, L., Tian, L., van der Wal, C. H., and Lloyd, S. 1999. Josephson persistent-current qubit. *Science*, **285**(5430), 1036–1039.

Nakamura, Y., Chen, C. D., and Tsai, J. S. 1997. Spectroscopy of energy-level splitting between two macroscopic quantum states of charge coherently superposed by Josephson coupling. *Physical Review Letters*, **79**(12), 2328–2331.

Nakamura, Y., Pashkin, Yu. A., and Tsai, J. S. 1999. Coherent control of macroscopic quantum states in a single-Cooper-pair box. *Nature*, **398**(6730), 786–788.

Nataf, Pierre, and Ciuti, Cristiano. 2010. No-go theorem for superradiant quantum phase transitions in cavity QED and counter-example in circuit QED. *Nature Communications*, **1**(6), 1–6.

Neeley, Matthew, Ansmann, M., Bialczak, Radoslaw C., et al. 2008. Process tomography of quantum memory in a Josephson-phase qubit coupled to a two-level state. *Nature Physics*, **4**(7), 523–526.

Neill, C., McCourt, T., Mi, X., et al. 2021. Accurately computing the electronic properties of a quantum ring. *Nature*, **594**(7864), 508–512.

Nielsen, M. A, and Chuang, I. L. 2011. *Quantum Computation and Quantum Information*. Cambridge University Press.

Nielsen, Michael A. 2002. A simple formula for the average gate fidelity of a quantum dynamical operation. *Physics Letters A*, **303**(4), 249–252.

Niemczyk, T., Deppe, F., Huebl, H., et al. 2010. Circuit quantum electrodynamics in the ultrastrong-coupling regime. *Nature Physics*, **6**(10), 772–776.

Nigg, Simon E., Paik, Hanhee, Vlastakis, et al. 2012. Black-box superconducting circuit quantization. *Physical Review Letters*, **108**(24), 240502.

Niskanen, A O, Harrabi, K, Yoshihara, F, Nakamura, Y, Lloyd, S, and Tsai, J S. 2007. Quantum coherent tunable coupling of superconducting qubits. *Science (New York, N.Y.)*, **316**(5825), 723–726.

O'Brien, J. L., Pryde, G. J., Gilchrist, A., et al. 2004. Quantum process tomography of a controlled-NOT gate. *Physical Review Letters*, **93**(Aug), 080502.

O'Brien, William, Vahidpour, Mehrnoosh, Whyland, Jon Tyler, et al. 2017. Superconducting caps for quantum integrated circuits. *e-print* arXiv:1708.02219.

Ofek, Nissim, Petrenko, Andrei, Heeres, Reinier, et al. 2016. Extending the lifetime of a quantum bit with error correction in superconducting circuits. *Nature*, **536**(7617), 441–445.

Olivares, S. 2012. Quantum optics in the phase space. *European Physical Journal Special Topics*, **203**(1), 3–24.

Olver, F. W. J., Lozier, D. W., Boisvert, R. F., and Clark, C. W. (eds). 2010. *NIST Handbook of Mathematical Functions*. Cambridge University Press.

O'Malley, P. J. J., Kelly, J., Barends, R., et al. 2015. Qubit metrology of ultralow phase noise using randomized benchmarking. *Physical Review Applied*, **3**(Apr), 044009.

Orlando, Terry P. 1991. *Foundations of Applied Superconductivity*. Addison-Wesley Publishing Company.

Ortuño, M., Somoza, A. M., Vinokur, V. M., and Baturina, T. I. 2015. Electronic transport in two-dimensional high dielectric constant nanosystems. *Scientific Reports*, **5**(1), 9667.

Otterbach, J. S., Manenti, R., Alidoust, N., et al. 2017. Unsupervised machine learning on a hybrid quantum computer. *e-print* arXiv:1712.05771.

Ozfidan, I., Deng, C., Smirnov, A.Y., et al. 2020. Demonstration of a nonstoquastic Hamiltonian in coupled superconducting flux qubits. *Physical Review Applied*, **13**(3), 034037.

Paauw, F., Fedorov, A., Harmans, C., and Mooij, J. 2009. Tuning the gap of a superconducting flux qubit. *Physical Review Letters*, **102**(9), 090501.

Paik, Hanhee, Schuster, D. I., Bishop, Lev S., et al. 2011. Observation of high coherence in Josephson junction qubits measured in a three-dimensional circuit QED architecture. *Physical Review Letters*, **107**(24), 240501.

Pashkin, Yu. A., Yamamoto, T., Astafiev, O., Nakamura, Y., Averin, D. V., and Tsai, J. S. 2003. Quantum oscillations in two coupled charge qubits. *Nature*, **421**(6925), 823–826.

Pechal, M., Huthmacher, L., Eichler, C., et al. 2014. Microwave-controlled generation of shaped single photons in circuit quantum electrodynamics. *Physical Review X*, **4**(4), 041010.

Pegg, D., and Barnett, S. 1989. Phase properties of the quantized single-mode electromagnetic field. *Physical Review A*, **39**(4), 1665–1675.

Peropadre, B., Forn-Díaz, P., Solano, E., and García-Ripoll, J. 2010. Switchable ultrastrong coupling in circuit QED. *Physical Review Letters*, **105**(2), 023601.

Peropadre, B., Zueco, D., Porras, D., and García-Ripoll, J. 2013. Nonequilibrium and nonperturbative dynamics of ultrastrong coupling in open lines. *Physical Review Letters*, **111**(24), 243602.

Pino, Manuel, and García-Ripoll, Juan José. 2018. Quantum annealing in spin-boson model: from a perturbative to an ultrastrong mediated coupling. *New Journal of Physics*, **20**(11), 113027.

Pitaevskii, Lev, and Stringari, Sandro. 2016. *Bose-Einstein Condensation and Superfluidity*. Oxford University Press.

Place, Alexander P. M., Rodgers, Lila V. H., Mundada, Pranav, et al. 2021. New material platform for superconducting transmon qubits with coherence times exceeding 0.3 milliseconds. *Nature Communications*, **12**(1), 1779.

Preskill, John. 2018. Quantum computing in the NISQ era and beyond. *Quantum*, **2**(Aug.), 79.

Raimond, J. M., Brune, M., and Haroche, S. 2001. Manipulating quantum entanglement with atoms and photons in a cavity. *Reviews of Modern Physics*, **73**(3), 565–582.

Ramos, Tomás, and García-Ripoll, Juan José. 2018. Correlated dephasing noise in single-photon scattering. *New Journal of Physics*, **20**(10), 105007.

Reed, M. D., DiCarlo, L., Johnson, B. R., et al. 2010. High-fidelity readout in circuit quantum electrodynamics using the Jaynes–cummings nonlinearity, 173601. *Physical Review Letters*, **105**(17).

Rigetti, Chad, and Devoret, Michel. 2010. Fully microwave-tunable universal gates in superconducting qubits with linear couplings and fixed transition frequencies. *Physical Review B*, **81**(13), 134507.

Rigetti, Chad, Gambetta, Jay M., Poletto, Stefano, et al. 2012. Superconducting qubit in a waveguide cavity with a coherence time approaching 0.1 ms. *Physical Review B*, **86**(10), 100506.

Rol, M. A., Battistel, F., Malinowski, F. K., et al. 2019. Fast, high-fidelity conditional-phase gate exploiting leakage interference in weakly anharmonic superconducting qubits. *Physical Review Letters*, **123**(Sep), 120502.

Romero, G., García-Ripoll, J. J., and Solano, E. 2009a. Microwave photon detector in circuit QED. *Physical Review Letters*, **102**(17), 173602.

Romero, Guillermo, García-Ripoll, Juan José, and Solano, Enrique. 2009b. Photodetection of propagating quantum microwaves in circuit QED. *Physica Scripta*, **T137**(T137), 014004.

Rosenblum, S., Reinhold, P., Mirrahimi, M., Jiang, Liang, Frunzio, L., and Schoelkopf, R. J. 2018. Fault-tolerant detection of a quantum error. *Science*, **361**(6399), 266–270.

Roy, Ananda, and Devoret, Michel. 2016. Introduction to parametric amplification of quantum signals with Josephson circuits. *Comptes Rendus Physique*, **17**(7), 740–755.

Sánchez-Burillo, E., Martín-Moreno, L., García-Ripoll, J. J., and Zueco, D. 2016. Full two-photon down-conversion of a single photon. *Physical Review A*, **94**(5), 053814.

Sandberg, M., Wilson, C. M., Persson, F., et al. 2008. Tuning the field in a microwave resonator faster than the photon lifetime. *Applied Physics Letters*, **92**(20), 203501.

Schneider, Christian Markus Florian. 2014. *On-Chip Superconducting Microwave Beam Splitter*. Ph.D. thesis, TU Munich.

Schrödinger, E. 1926. Quantisierung als Eigenwertproblem. *Annalen der Physik*, **384**(4), 361–376.

Schuster, D. I., Houck, A. A., Schreier, J. A., et al. 2007. Resolving photon number states in a superconducting circuit. *Nature*, **445**(7127), 515–518.

Schuster, D. I., Wallraff, A., Blais, A., et al. 2005. ac stark shift and dephasing of a superconducting qubit strongly coupled to a cavity field. *Physical Review Letters*, **94**(12), 123602.

Selby, Alex. 2014. Efficient subgraph-based sampling of Ising-type models with frustration. *e-print*, sep, arXiv:1409.3934.

Sharafiev, Aleksei, Juan, Mathieu L., Gargiulo, Oscar, et al. 2021. Visualizing the emission of a single photon with frequency and time resolved spectroscopy. *Quantum*, **5**(Jun), 474.

Shi, Tao, Chang, Yue, and García-Ripoll, Juan José. 2018. Ultrastrong coupling few-photon scattering theory. *Physical Review Letters*, **120**(15), 153602.

Shin, Seung Woo, Smith, Graeme, Smolin, John A., and Vazirani, Umesh. 2014. How "quantum" is the D-Wave machine? *e-print* arXiv:1712.05771.

Shnirman, Alexander, Schön, Gerd, and Hermon, Ziv. 1997. Quantum manipulations of small Josephson junctions. *Physical Review Letters*, **79**(12), 2371–2374.

Shor, Peter W. 1995. Scheme for reducing decoherence in quantum computer memory. *Physical Review A*, **52**(Oct), R2493–R2496.

Silver, A., and Zimmerman, J. 1967. Quantum states and transitions in weakly connected superconducting rings. *Physical Review*, **157**(2), 317–341.

Steane, Andrew. 1996. Multiple-particle interference and quantum error correction. *Proceedings of the Royal Society of London. Series A: Mathematical, Physical and Engineering Sciences*, **452**(1954), 2551–2577.

Steffen, Matthias, Ansmann, M., Bialczak, Radoslaw C., et al. 2006. Measurement of the entanglement of two superconducting qubits via state tomography. *Science*, **313**(5792), 1423–1425.

Steffen, Matthias, Kumar, Shwetank, DiVincenzo, David P., et al. 2010. High-coherence hybrid superconducting qubit. *Physical Review Letters*, **105**(Sep), 100502.

Sundaresan, Neereja M., Liu, Yanbing, Sadri, Darius, et al. 2015. Beyond strong coupling in a multimode cavity. *Physical Review X*, **5**(2), 021035.

Susskind, L., and Glogower, J. 1964. Quantum mechanical phase and time operator. *Physics*, **1**, 49–61.

Teufel, Stefan. 2003. *Adiabatic Perturbation Theory in Quantum Dynamics*. Lecture Notes in Mathematics, vol. 1821. Springer Berlin Heidelberg.

Townsend, Christopher, Ketterle, Wolfgang, and Stringari, Sandro. 1997. Bose–Einstein condensation. *Physics World*, **10**(3), 29–36.

van den Brink, Alec Maassen, Berkley, A. J., and Yalowsky, M. 2005. Mediated tunable coupling of flux qubits. *New Journal of Physics*, **7**(1), 230–230.

van der Ploeg, S. H. W., Izmalkov, A., van den Brink, Alec Maassen, et al. 2007. Controllable Coupling of Superconducting Flux Qubits. *Physical Review Letters*, **98**(5), 057004.

van der Wal, C. H., ter Haar, A. C., Wilhelm, F. K., et al. 2000. Quantum superposition of macroscopic persistent-current states. *Science*, **290**(5492), 773–777.

Venturelli, Davide, Mandrá, Salvatore, Knysh, Sergey, O'Gorman, Bryan, Biswas, Rupak, and Smelyanskiy, Vadim. 2015. Quantum optimization of fully connected spin glasses. *Physical Review X*, **5**(3), 031040.

Vlastakis, Brian, Kirchmair, Gerhard, Leghtas, Zaki, et al. 2013. Deterministically encoding quantum information using 100-photon Schrödinger cat states. *Science (New York, N.Y.)*, **342**(6158), 607–610.

Wallraff, A., Schuster, D. I., Blais, A., et al. 2004. Strong coupling of a single photon to a superconducting qubit using circuit quantum electrodynamics. *Nature*, **431**(7005), 162–167.

Wallraff, A., Schuster, D. I., Blais, A., et al. 2005. Approaching unit visibility for control of a superconducting qubit with dispersive readout. *Physical Review Letters*, **95**(6), 0650501.

Walls, D. F., and Milburn, G. J. 1994. *Quantum Optics*. Springer Berlin Heidelberg.

Walter, T., Kurpiers, P., Gasparinetti, S., et al. 2017. Rapid high-fidelity single-shot dispersive readout of superconducting qubits. *Physical Review Applied*, **7**(5), 054020.

Wang, C., Gao, Y. Y., Reinhold, P., et al. 2016. A Schrodinger cat living in two boxes. *Science*, **352**(6289), 1087–1091.

Williams, Colin P. 2011. *Quantum Gates*. Springer London. Pages 51–122.

Wilson, C. M., Johansson, G., Pourkabirian, A., et al. 2011. Observation of the dynamical Casimir effect in a superconducting circuit. *Nature*, **479**(7373), 376–379.

Wootters, W. K., and Zurek, W. H. 1982. A single quantum cannot be cloned. *Nature*, **299**(5886), 802–803.

Wu, Yulin, Bao, Wan-Su, Cao, Sirui, Chen, et al. 2021. Strong quantum computational advantage using a superconducting quantum processor. *Physical Review Letters* **127**, 180501.

Yamamoto, T., Neeley, M., Lucero, E., et al. 2010. Quantum process tomography of two-qubit controlled-Z and controlled-NOT gates using superconducting phase qubits. *Physical Review B*, **82**(18), 184515.

Yamamoto, T., Pashkin, Yu. A., Astafiev, O., Nakamura, Y., and Tsai, J. S. 2003. Demonstration of conditional gate operation using superconducting charge qubits. *Nature*, **425**(6961), 941–944.

Yan, F., Gustavsson, S., Kamal, A., et al. 2015. The flux qubit revisited to enhance coherence and reproducibility. *e-print*, arXiv:1508.06299.

Yan, Fei, Gustavsson, Simon, Kamal, Archana, Birenbaum, et al. 2016. The flux qubit revisited to enhance coherence and reproducibility. *Nature Communications*, **7**(1), 12964.

Yoshihara, Fumiki, Fuse, Tomoko, Ashhab, Sahel, Kakuyanagi, Kosuke, Saito, Shiro, and Semba, Kouichi. 2016. Superconducting qubit–oscillator circuit beyond the ultrastrong-coupling regime. *Nature Physics*, **13**(1), 44–47.

Yurke, Bernard, and Denker, John. 1984. Quantum network theory. *Physical Review A*, **29**(3), 1419–1437.

Zener, C. 1932. Non-adiabatic crossing of energy levels. *Proceedings of the Royal Society A: Mathematical, Physical and Engineering Sciences*, **137**(833), 696–702.

Zueco, David, and García-Ripoll, Juanjo. 2019. Ultrastrongly dissipative quantum Rabi model. *Physical Review A*, **99**(Jan), 013807.

Index